AF522151

ENCYCLOPEDIA OF BIOPROCESS TECHNOLOGY

ENCYCLOPEDIA OF BIOPROCESS TECHNOLOGY

Vol. 1

ADVANCE BIOPROCESS TECHNOLOGY

By

Dr. Arvind N. Shukla

School of Studies of Zoology & Biotechnology

Vikram University

Ujjain

(India)

DISCOVERY PUBLISHING HOUSE PVT. LTD.

NEW DELHI-110 002

Published by:
Tilak Wasan
DISCOVERY PUBLISHING HOUSE PVT. LTD.
4383/4A, Ansari Road, Darya Ganj
New Delhi-110 002 (India)
Phone: +91-11-23279245, 43596064-65
Fax : +91-11-23253475
E-mail: parul.wasan@gmail.com
discoverypublishinghouse@gmail.com
web : www.discoverypublishinggroup.com

***First Edition:* 2012**

ISBN: 978-93-5056-026-6 (Set)
978-93-5056-027-3 (Vol. 1)
978-93-5056-028-0 (Vol. 2)
978-93-5056-029-7 (Vol. 3)
978-93-5056-030-3 (Vol. 4)
978-93-5056-031-0 (Vol. 5)

ncyclopedia of Bioprocess Technology

Art Press

Preface

The present title **"Encyclopedia of Bioprocess Technology"** has been written for undergraduate, post-graduate students and those engaged in pharmaceutical, environmental biotechnology, industrial and clinical research. Actually the explosion of new technologies with their vast potential has brought with it the need for emerging scientists to equip themselves and their laboratories with a whole array of new expertise. With the high discriminating power of the DNA systems has come high potential in evidentiary terms, high profile status for many investigations, and not least, a high degree of professional scruting of evidence produced by such technology.

This book is addressed to all who are curious and who want to understand fundamental biological processes that ensure our survival. The person who might be interested is probably someone just starting research who wants to learn about aspects of regulation, outside his chosen field. It could become a supplementary text for graduate students and post-graduate in biochemistry and molecular biology.

To make the work more comprehensive and informative, the author has consulted many authoritative books, research journals, abstracts, monographs etc., so there can be no claim to originality except in the manner of treatment.

The author express his thanks to his friends and colleagues whose continue inspirations have initiated him to bring out this book.

The author expresses her gratitude to Mr. Wasan and staff of M/s Discovery Publishing House Pvt. Ltd. for their whole hearted co-operation in the publication of this book.

The author will remain sincerely responsible for any shortcomings of the book and be grateful to the readers for their suggestions and constructive criticism for the continuous betterment of the book. He takes this opportunity to appeal to the readers to send their suggestions straightaway to this Publisher.

Arvind N. Shukla

CONTENTS

SAMPLING TECHNIQUES

To optimize fermentation processes, that is, to ensure maximum growth and/or productivity, the fermentation conditions should be controlled and maintained at their optimum values. However, the media for fermentations and other bioprocesses are very complex in nature. These are often multiphase systems. The liquid phase contains soluble nutrients (e.g., sugar, proteins, vitamins, salts, and products of metabolism such as amino acids, alcohols, and organic acids). Furthermore, there may be a solid phase of suspended material (insoluble substrate/product, carrier materials, etc.). The gas phase is in equilibrium with dissolved gases in the medium. Analysis of any of these compounds by traditional techniques is tedious and time-consuming and cannot provide on-line information of the process. Recently, the increasing demand of high-value products has focused interest on well-controlled, optimized processes. This has initiated the development of new on-line measurement techniques. The ideal situation would be to have specific probes that can operate in situ during the fermentation. However, at present there is only a limited number of such probes available for measuring pH, dissolved oxygen, and carbon dioxide. There is still no suitable technology for operating biosensors containing a biological macromolecule. In the last few years research on biosensors for bioprocess control has been growing. These sensors offer the ability to measure concentrations of various biomolecules, as well as biological parameters, very specifically using immobilized enzymes or cells or other biochemical/biological entities. Unfortunately, these sensors are inactivated by steam sterilization. Because sensor technology is not yet sufficiently advanced to permit this type of monitoring, the majority of analysis is still carried out outside of the bioreactor, a process requiring sampling and sample-handling strategies.

CRITERIA FOR SAMPLING

For analysis of any bioprocess, off-line or on-line sampling is a prerequisite and is the basis for the final analytical result. The following strategies/criteria are important when designing a sampling device.

1. The sterility of the process must be maintained during sampling.
2. The delay between sampling and analysis may result in changes in the composition of the sample. Precautions must be taken to prevent or compensate for that.
3. The analysis must be accurate and reproducible.
4. When constructing the sampling and sample treatment unit, the selection of material must be governed by compatibility criteria such that the process being monitored is not negatively influenced.
5. Protein adsorption and cell colonization on the sampling and sample-handling device may lead to an imparted mass transport and formation of unwanted immobilized catalytic activities that may severely influence the composition of the sample before analysis can be carried out.

Considering these parameters, a sampling system for on-line monitoring of fermentation processes must meet the following criteria.

1. The system can take representative samples from fermenter but should not disturb the fermentation process.
2. The system should be sterilizable and provide a sterile barrier to avoid contamination.
3. Separation of cells, solids, or macromolecules should be carried out with as short a delay as possible within the sampling devices.

There are many different types of sampling units available to achieve these goals. This review highlights a detailed description of some of the sampling devices used for on-line monitoring of bio-processes, especially when biosensors are used in the analytical step. It should, however, be stressed that the sampling and sample handling as described here will be applicable when setting up any kind of analysis to monitor a bioprocess.

DIFFERENT KINDS OF ANALYTES

The character of the reaction mixture to be sampled determines the choice of sampling method. Different kinds of analytes require different sampling protocols.

Analysis of Volatile Compounds

Volatile compounds may be present in the head space of a fermentation or be dissolved in the liquid medium. At equilibrium, the concentration in the head space reflects that in the liquid. The composition of the gas phase is usually less complex than that of the aqueous phase; therefore, when the metabolites are volatile, it is probably simpler to monitor the concentration in the gas phase. The risks of infection of the fermentation are lower when sampling from the gas phase. This has made it attractive to use analysis of the gas phase when monitoring industrial fermentations. There are two main modes described in the literature for the sampling of gases and volatile compounds:

1. Analysis of the outlet gases or of the headspace gases.
2. Analysis of the volatile gaseous compounds dissolved in the culture medium.

A sampling method is applied that is based on a technology for supplying oxygen to bioprocesses that are disturbed by bubbling oxygen. Oxygen is supplied in a tubing made of hydrophobic material. The oxygen penetrates the walls of the tubing and is released to the cultivation broth. Similar phenomena can take place for gaseous compounds with an opposite concentration gradient. Tubings made from materials that are permeable to organic solvent molecules are immersed in the fermentation broth, and an inert gas that flows through the lumen of the tubings transports the solvent molecules to an analyzer (e.g., a gas-sensitive semiconductor sensor placed downstream of the tubing). Recently a new approach for monitoring carbon dioxide during fermentation was reported based on a piezoelectric crystal. A change in the oscillation frequency of the crystal occurs due to selective sorption of a gaseous compound to the surface, and the change in frequency is directly proportional to the mass deposited. Fast response, simple use, and low cost may lead to a more extensive use of this technique.

Analysis of Nonvolatile Compounds

Sampling systems

Automatic sampling, either continuously or by a frequently repeated procedure, is a prerequisite for on-line or "at-line" monitoring. The sampling system normally serves also as a sterile barrier between the bioreactor and the analytical system. Different analytical systems can be used. Flow injection analysis (FIA) is a common and convenient way to carry out the analysis. The samples are heterogeneous, with compounds dissolved in the medium, some forming complexes with other constituents, some not fully soluble, some particulate, and so forth. Manual sampling and off-line analysis are still in practice; however, they result in delays in analysis, risks of infection, and less reproducible sample handling. Sampling can be done using sampling units mounted either inside or outside the bioreactor. Several systems of both kinds have been reported recently.

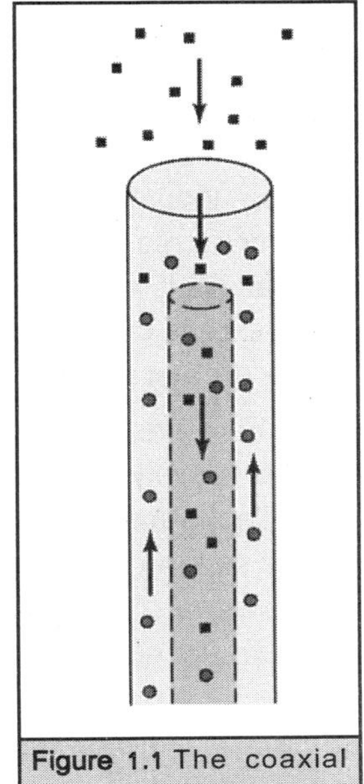

Figure 1.1 The coaxial catheter.

Direct sampling

Direct sampling from a bioreactor is possible through a catheter with a small inner diameter. It is possible to connect this sample stream directly to a FIA system or any other analytical device. However, for precise measurements, the metabolic activity in the sample must be stopped immediately after sampling, otherwise unwanted reactions may take place during the period after sampling but before analysis. If a deactivation is carried out with addition of an inhibitory agent, one has to examine the possibility of interference from the reagent on the analysis. One device that addresses these challenges is the coaxial catheter unit, which is connected to the bioreactor for direct sample withdrawal. It consists of two concentrically placed tubings, an inner and an outer; the outer protrudes beyond the inner a few millimeters. The sampling is done via the inner tubing. An inhibitor is fed into the space between the tubings, and the broth plus the inhibitor are mixed in the small mixing zone at the tip of the catheter. Inhibition thus takes place at the moment of sampling. A major advantage of the direct sampling system is that a complete sample is obtained, and it is therefore possible to measure the biomass concentration on-line and even perform analysis on intracellular compounds. The coaxial catheter has not yet been used for sample withdrawal from cultures of filamentous fungi and highly viscous fluids. This method is so far limited to studies of unicellular organisms.

Aseptic barrier

The sampling system acts as an aseptic barrier between the bioreactor and the analytical equipment. The basic idea behind this type of sampling device is to utilize a membrane for separation during the sampling process. In most cases separations based on differences in size are carried out. The membrane-based sampling techniques can also be applied to cultures of filamentous fungi and in other viscous media. The major disadvantage of the membrane module is clogging of the membrane, mainly due to deposition of proteins on the membrane surface. When using a membrane sampling module, there are three types of separation forces that can be utilized: pressure-driven difference, concentration difference, and electrical potential difference. Applying any of these forces, transport over the membrane will occur and may serve several purposes:

1. Cells are removed from the sample, thus stopping the metabolic activity.
2. Compounds that may interfere with the analysis are retained.
3. Membrane sampling acts as a barrier between the analytical system and the bioreactor.

Membrane modules can be categorized into two groups: those placed in a recycle loop connected to the bioreactor (filtration outside the reactor), and those placed in situ (as with dialysis filtration). When the membrane unit is placed outside the reactor, fermentation broth is usually pumped through the membrane unit and then recycled to the fermentor. What is special with the external loop membrane unit is that upon passage of the fermentation broth through the loop, special conditions may prevail. Thus, oxygen limitations may be severe, and, damaging shear forces may appear. This type of device is thus less suited to dealing with high-cell-density fermentations of aerobic cells or with shear-stress-sensitive cells. The great advantage to using an external membrane is the greater flexibility. Thus, a membrane may be easily replaced when needed.

There are two basic modes of operation: the permeate may be collected and used for analysis, or alternatively, a receiving buffer is pumped through the unit on the collection side of the membrane to collect substances passing the membrane. It is important that analyte collection is carried out under sterile conditions. This is usually secured by intermittent sterilization. The membranes used for external filtration are made of a variety of materials (e.g., nylon, polypropylene, acrylnitril copolymer, cellulose acetate, cellulose nitrate, polyvinylpyrroliden fluoride) and are available in different pore sizes.

Normally, membranes with pore sizes of 0.2 μm are preferred because they guarantee the separation of medium from the biomass and other suspended material. Sampling units with external filters are available commercially. The Bio-Pem consists of a chamber (vol. 175 mL) containing a magnetically stirred rod. The stirrer ensures a tangential flow across the membrane, which is placed in the bottom of the chamber. This unit was successfully used for sample withdrawal during a cultivation process involving unicellular microorganisms but was found unsuitable for sampling cultures of fungi. A major draw back of the Bio-Pem is the large dead volume in the membrane module, which may cause substrate depletion in the loop as well as high response times. An alternative method involves a cross-flow filtration module in which the dead volume is less, resulting in a much shorter response time (<10 s). This module is applied efficiently in fed-batch penicillin fermentations on complex media and for wastewater treatment. There is still another unit based on a planar membrane. The membrane is clamped between two metal blocks, which are then screwed together.

During filtration, the incoming broth (feed) is pumped into grooves in the upper block, and the filtrate is collected in the lower chamber. The broth is separated into two streams: the particular-free filtrate (permeate), which passes the membrane, and the particulate-enriched retentate, which is returned to the fermentor. The filtrate is then led to an injection loop for later injection into the analytical system. After passing the loop, two possibilities arise: the filtrate is either returned to the bioreactor or sent to the waste collector. The latter would quickly empty a small fermentor. To avoid this, the filtrate can be recycled to the fermentor, but this increases the risk of contamination. Placing a sterile barrier (e.g., a microfilter) between the injection loop and the bioreactor reduces the risk. This may be acceptable in microbial fermentations, but in certain types of processes, such as human cell line production, such a strategy cannot be applied because such cells are very slowly growing and must be cultivated during long time periods and at the same time have a high value. An alternative approach is called discontinuous filtration. The filtrate line is closed between injections and opened shortly before the injection to flush the tubings and the loop. There are several units with planar membranes available on the market. These units have been used for different applications.

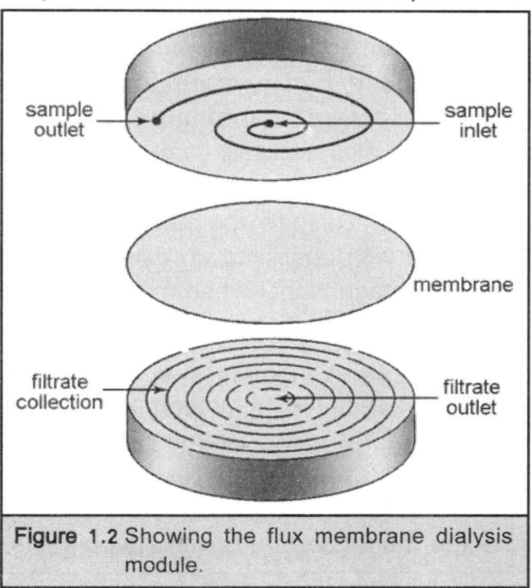

Figure 1.2 Showing the flux membrane dialysis module.

The A-SEP TM module has been used for monitoring acetate and phosphate during *Escherichia coli* fermentations, utilizing a standard 47-mm-wide 0.45-μm MF membrane. The Minitan TM system, using two membranes, has been employed for the determination of glucose in *Candida rugosa* cultivations, followed by several others for the determination of glucose, diols, and ethanol. Membrane filters may also be available in the form of hollow fibers of different diameter and porosity. Depending on the intended application, the cartridge contains different numbers of hollow fibers. In accordance with the cross-flow filtration principle, the medium is pumped at high velocity (1 m s^{-1}) through the fiber. A pressure is applied, and the filtrate is collected on the outer side of the fibers. It is also possible to pump the broth on the outer side of the fiber(s) and collect the filtrate inside. However, depending on the membrane geometry, this latter may be less attractive. When dealing with anisotropic membranes it is advisable to apply the complex medium to the smooth side of the membrane.

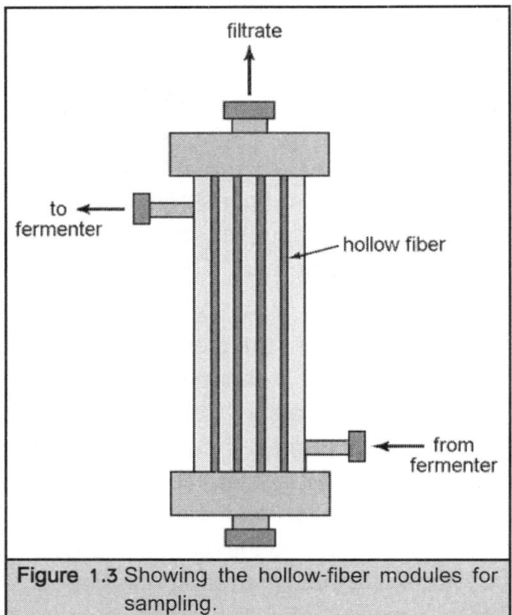

Figure 1.3 Showing the hollow-fiber modules for sampling.

A range of different applications are described in the literature. A stainless steel module with only one fiber inside has been reported. The fiber had an inner diameter of 5.5 mm and a length of about 200 mm. The dead volume of the module was 2.5 cm^3; and it was successfully used for sampling in conjunction with determination of ammonium, glucose, and phosphate. Reports on the use of several in-house built hollow-fiber modules can be found in the literature, covering vastly different applications in mammalian cell culture, lignocellulose hydrolysate fermentations, and biodegradation of naphthalene sulphonic acid. Systems are also available commercially from Bio-FIO and used for monitoring different types of fermentation components. Despite the potential complications associated with operating an external loop filtration system, it remains a popular approach. The great advantage is that when membrane clogging occurs, it is possible to intervene without ruining the fermentation.

In dialysis, a membrane is used to separate a complex solution from a more well defined solution. Pores in the membrane allow exchange of small molecules between the two liquid volumes. The driving force is concentration difference. The different modes of dialysis originate from whether the donor and acceptor phases are stagnant or moving. In equilibrium dialysis both are stagnant and, at equilibrium, the concentration in both phases will be the same of those molecules able to pass the membrane. In continuous dialysis, compounds passing the membrane are continuously removed. This results in a larger concentration gradient, and a higher flux is maintained. The membranes used for dialysis are classified according to their molecular weight cutoff. Most dialysis membranes are made from cellulose materials, but because many organisms are able to hydrolyze cellulosic materials, this limits their applicability. There are several sampling units that incorporate dialysis membranes. They are usually placed in the fermentor, and a receiving dialysis carrier buffer is circulated on the permeate side of the membrane. The carrier buffer is thus pumped into the bioreactor and, after having received the dialysate

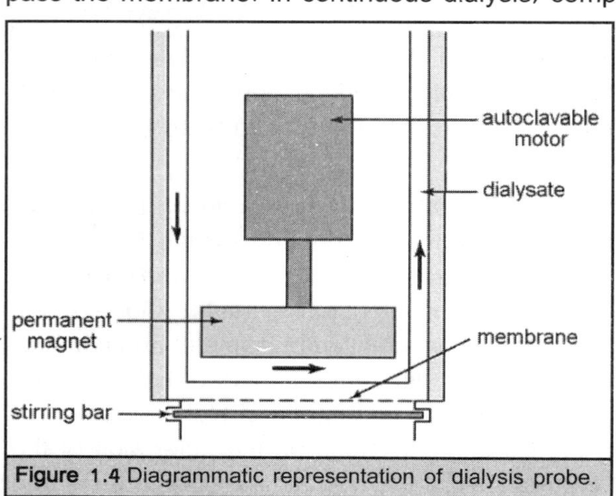

Figure 1.4 Diagrammatic representation of dialysis probe.

molecules, removed again. The major problem associated with this approach is the deposition of cells and fouling of the membrane on the fermentor side.

The accumulated layer of cells may reduce the mass transfer from the fermentor to the accepting solution and also cause decreases in concentration of metabolites passing through the membrane because they run the risk of being metabolized by the cell layer before entering the dialysis membrane. To help reduce fouling it is possible to place the dialysis probe into the zone of the fermentor where the impeller creates the greatest turbulence. Another means of obtaining similar results, but with fewer design restrictions, is to construct a dialysis probe that besides containing the membrane unit also has a stirring function, so that a high tangential liquid flow is generated over the membrane. This approach was used by Mandenius et al. in constructing a system for use in dense particulate suspensions. The membrane is placed in the front of the probe, and membrane fouling is avoided by means of the rotating magnet driven by an electrical motor inside the probe.

Zabriskie and Humprey developed a dialysis probe situated within one baffle of a stirred fermentor. They were able to take samples for glucose analysis during cultivation of yeast cells on a synthetic medium. The disadvantage of this dialyzer is that it can be used only in this type of bioreactor. Another sampling device was presented by Glaxo Laboratories, Ltd. This unit consist of an elongated support with a milled helical groove. A tubular membrane made of cellulosic materials or polytetrafluoroethylene was fixed to the support using ring seals.

An in situ dialysis membrane module was developed at the University of Hannover and is marketed by Advanced Biotechnology Corporation (ABC). The module consists of a unit supporting the membrane, and one or more extension units. The membrane module is usually inserted through a standard 19- or 25-mm port in the top of the bioreactor or, alternatively, through a 25-mm port below the liquid surface. When inserted from the top of the bioreactor, one or more extension tubes are needed to place the membrane below the liquid surface. Liquid passing the membrane is collected in 16 grooves, and these 16 permeate streams are joined in four openings placed evenly along the membrane holding unit. Through these four openings the permeate enters the central channel (10 mm i.d.), and a continuous sample stream to the analyzer leaves the unit from the top of the probe. Gibson and Woodward have also described an internal dialysis module, based on the same principle but without a stirrer. Ethanol and glucose could be sampled with this unit, and the results from the subsequent analyses agreed well with the on-line data.

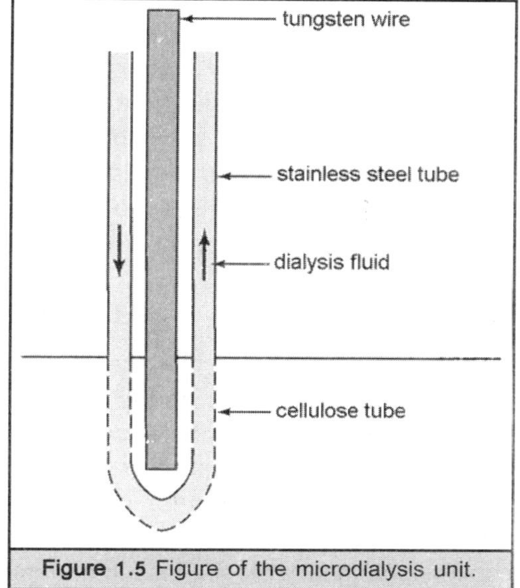

Figure 1.5 Figure of the microdialysis unit.

The microdialysis sampling unit is a probe, and it is a miniaturized form of a dialysis unit. It is manufactured in several different configurations. Another modified technique is the microdialysis used for continuous in vivo sampling. The microdialysis sampling unit is a probe, and it is a miniaturized form of a dialysis unit. It is manufactured in several different configurations. Typical flow rates of the perfusion liquid are 0.5–25 μL/min, and to avoid pulses in the flow, a peristaltic pump or preferably a syringe pump should be used.

Due to its ease of operation and the general approach, the number of applications increases rapidly. It can easily be coupled on-line to a number of separation and detection techniques. This technique was used in fermentation processes for the production of ethanol, penicillin, and so on. However, the slow flux of liquid and thus the long response time from an analysis based on this sampling technique, compared with the conventional sensing devices, should be considered. The major problem in dialysis units is the fouling and clogging by macromolecules on the membrane surface. This can be overcome by adding molecules that compete for the binding site of the protein, adding a displacer, changing the pH, or pumping the solution tangentially over the membrane surface.

REMOVAL OF OBSTACLES

As already noted a major concern in sample handling is clogging. Quite often packed-bed systems are used, and then particulate matter constitutes a major obstacle. In these cases, particulate matter has to be removed prior to analysis. However, with the introduction of expanded-bed chromatography, it is possible to handle large-scale chromatographic processes without separating out the cells. This made it tempting to investigate the possibility of using the same approach in an analytical system. Enzyme-based analysis was studied using glucose oxidase immobilized to Streamline particles. These particles are characterized by being of heterogeneous size and density distribution, such that when used in an expanded bed, a stable expanded bed can be obtained. Glucose was monitored using this technique.

A more challenging issue is whether the same system would allow binding reactions to take place. If so, one should be able to run immunoassays in whole broth and maybe even in cell homogenates. Assays of human serum albumin in the presence of increasing amounts of heat-killed yeast cells was used as a model system. Antibodies against albumin were present on the Streamline. The assays seemed successful, and therefore the model was also studied to see if the intracellular enzyme β-galactosidase could be quantified in a cell suspension or a cell homogenate. The results were so encouraging that it is safe to say that this technique allows binding assays without the preceding separation of particulate matter from the sample.

METHODS OF PRETREATMENT

Another vital part of the analysis is the pretreatment of the samples. This includes dilution, centrifugation, extraction, and filtration, depending on the nature of the samples. Certain samples can be analyzed without pretreatment; however, for complex samples extensive clean-up is required before introduction into the flow system. The reproducibility and rapidity of flow injection analysis promotes its applicability for on-line pretreatment of samples with less manual intervention. Additionally pretreatment steps by FIA are rapid. The pretreatment and analysis, including chemical conversion, usually require less than a minute. Furthermore, when samples are toxic or pathogenic,

FIA offers the advantage of removing, or at least minimizing, human exposure. An excellent review on sample pretreatment by FIA was published by Clark et al. and includes a description of the theoretical background of sample pretreatment.

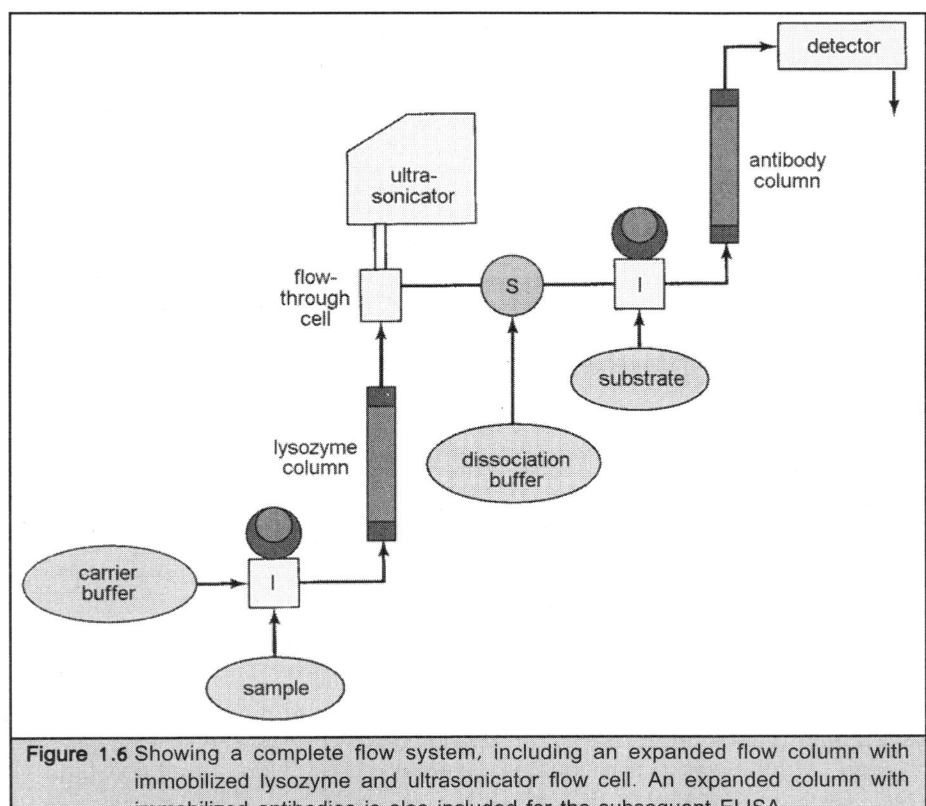

Figure 1.6 Showing a complete flow system, including an expanded flow column with immobilized lysozyme and ultrasonicator flow cell. An expanded column with immobilized antibodies is also included for the subsequent ELISA.

System of Disintegration

Monitoring levels of intracellular compounds has so far been possible only via off-line analysis. However, after the recent development of an on-line disintegration system that can be used together with expanded-bed adsorption in a flow system, a new possibility seems to have emerged. The important issues are that the whole broth is taken and treated and after that the whole homogenate can be processed in, for example, a binding assay. The whole broth is introduced into an FIA system where it first passes an expanded bed containing Streamline with immobilized lysozyme. On passage of *Micrococcus lysodeiticus* cells it has been proven that a high degree of disintegration is achieved, as judged from the amount of protein released. When adding a subsequent step involving sonication using a flow cell, the disintegration is even more efficient. A culture of *E. coli* was studied and its level of β-galactosidase production was monitored. The detection step consisted of an expanded bed containing Streamline with immobilized antibodies to β-galactosidase. Upon introduction of the whole broth, homogenization takes place by treatment in the two units; and afterward liberated enzyme is trapped by the immunosorbent. The enzyme is quantified by reading the amount of *o*-nitrophenol liberated after passing a pulse of *o*-nitrophenyl-*β*-galactopyranoside over the sorbent. After the assay the immunocomplex is broken and the system is ready for another assay.

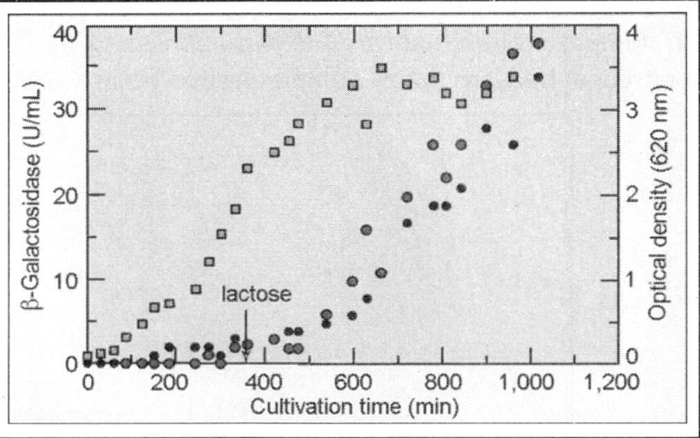

Figure 1.7 Graphical presentation of monitoring of a shake-flask experiment. On-line (●) and off-line (○) β-galactosidase activity and o.d. (□) measurements.

Mechanical Separation

Centrifugation is the mechanical separation, and is commonly used for separating cells and particulate materials from the fermentation broth. It is possible to include a small-scale centrifuge in an FIA system, with recycling of the sample to the reaction vessel.

Liquid–Liquid Extraction

Liquid–liquid extraction is based on the partitioning of the analytes between an aqueous and an organic phase and is widely used for gas chromatographic analysis. Integrating this technique into FIA is difficult because it is difficult to automate, laborious, and expensive. Solid-phase extraction (SPE) is based on sorption of analytes in solution to a sorbent packed in a cartridge or reactor, typically the size of a precolumn (10–20 x 2–4.6 mm in length and i.d., respectively). While the analyte is trapped into the packing material, compounds with little or no affinity for the sorbent pass through the column. After washing the column to further remove the interfering substances, the analytes are then desorbed and eluted by a small volume of an appropriate solvent. Recently, Vreuls et al. demonstrated the coupling of on-line SPE to gas chromatography. A precolumn packed with mixed anion/cation exchange materials was used on-line in a flow system for removal of ionic substances in aliquots of a filtrate from fermentation.

Concentration of Analyte

If the analyte concentration is less than the concentration range of the detection method, the sample must be concentrated. Most of the manual pretreatment methods are time-consuming and expensive. Microcolumns with ion exchange materials can be used to significantly reduce the volume of the analyte by adsorbing the sample into a solid support in the column and by a further elution.

Dilution of Samples

If the analyte concentration is more than that of the detectable range of the analytical system, the sample must be diluted. The same goes for a sample in which inhibiting or otherwise disturbing substances are present in high concentrations. Manual dilution is used for off-line analysis. On-line dilution is now widely applied for fermentation processes and includes the use of gradient dilution, zone sampling, and cascade dilution. In the gradient dilution, an originally homogeneous sample zone disperses during its movement through the FIA conduit. The zone undergoes a change from an initial plug shape with a concentration C_0, to an exponentially modified gaussian shape with a maximum concentration in the zone of C_{max}. If the peak maximum does not lie in the calibration range of the measurement, the zone can be made at a time on the profile that yields a decreased response. By selection of measurement times longer than the time of peak maximum (*T*), an element with a decreased concentration is measured, yielding a proportionally decreased absorbance. This measurement is reproducible from sample to sample and can be calibrated. In zone sampling a small portion of the sample zone injected into a second flowing stream results in dilution. In cascade dilution the sample volume is reduced through stream splitting using differential pumping coupled with a true dilution by merging with a subsequent carrier stream.

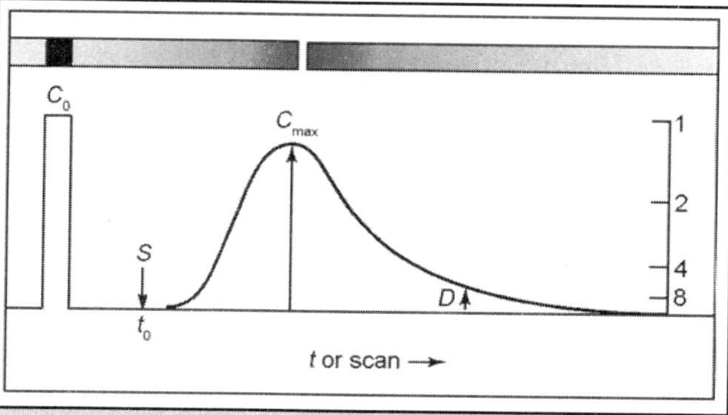

Figure 1.8 In addition to peak height a dispersed sample zone yields on infinite number of vertical readouts.

Further Handling the Samples

Sample handling and pretreatment can be achieved by using many techniques in FIA. The method of injection is important. The purpose of sample injection in FIA is to deliver a well-defined sample bolus reproducibly into a continuously flowing carrier stream. Ruzicka and Hansen described that changing the injected sample volume is a powerful way to change the dispersion. An increase in peak height and sensitivity of measurement is best achieved by increasing the volume of the injected sample solution. Conversely, dilution of overly concentrated sample material is achieved by reducing the sample volume. There are two methods of injection: time-based injection (a sample loaded for a precise time at a given flow rate) and volume-based injection (the sample fills a geometrically defined volumetric cavity). This well-defined sample bolus is then inserted into the carrier stream. Volume-based injections are performed using rotary valves, commutators, and hydrodynamic injections.

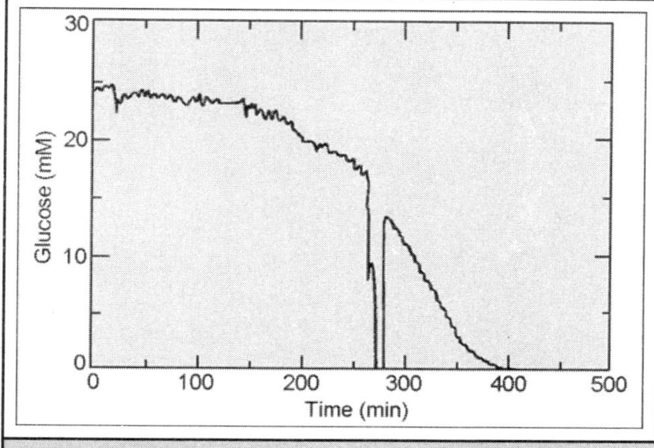

Figure 1.9 Showing the monitoring of glucose in batch cultivation of *Escherichia coli.*

Measurements

For any kind of analysis, whether it is on-line or off-line, the system needs to be calibrated to ensure that the compound of interest is being measured reliably. It is easy to calibrate an off-line system. However, the continuous systems described so far commonly have no possibility for calibration and recalibration on-line without disrupting the monitoring procedure. Thus, any changes occurring in the sample during the recalibration, are not registered. Another important limitation is the fact that the calibration solution and the sample solution are usually not treated in the same manner. In-line calibrating systems are used in medical monitoring. Attempts were made for in-line calibration during fermentation with *E. coli* using a double-lumen catheter. The calibration was done by closing the sample inlet from the fermentor and substituting it with a sample of known analyte concentration. A further development of this technology was the use of the coaxial catheter for sampling. In on-line monitoring of glucose and lactate in blood, the calibration solution was introduced as a pulse in the heparin stream fed via the outer space in the coaxial catheter. This led to an on-line monitoring of the process without having to interrupt the reading. Furthermore, by introducing the calibration solution at the tip of the sampling catheter, the whole system is calibrated. Thus a more true calibration value is obtained. Sampling and sample treatment are extremely important when applying good analytical technique to the monitoring and control of biotechnological processes. Too often much effort is spent on developing a sensitive assay, but very little or no concern has been given to sampling and sample handling. The examples given in this article clearly demonstrate that many good techniques can be applied. Which method to use will be governed by the system studied, the equipment available, and also the earlier experience of the researcher.

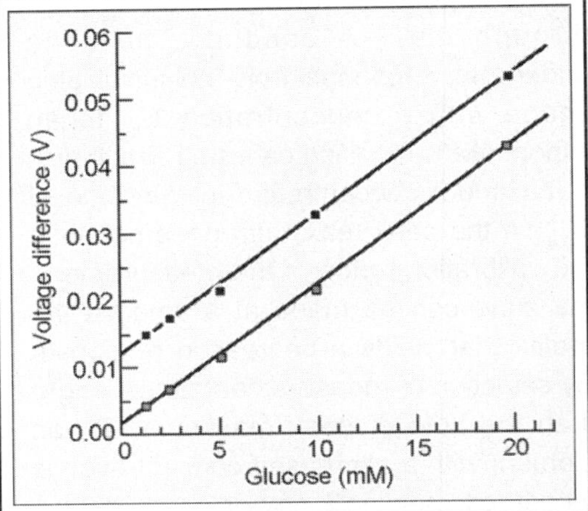

Figure 1.10 On-line outlet signal (V) as a function of glucose concentration (mM); (■) glucose standard solution with added internal standard (5 mM), $y = 1.14 \times 10^{-2} + 2.05 \times 10^{-3}$, $r^3 = 1.000$; (□) glucose standard solution, $y = 1.09 \times 10^{-3} + 2.04 \times 10^{-3}\, x$, $r^2 = 1.000$.

2

NEWTONIAN AND NON-NEWTONIAN METHODS

Many biotechnology products are manufactured using microorganisms and mammalian cells that behave in a Newtonian manner. In other cases, the product of the culture or the cellular material itself may cause the culture to be highly non-Newtonian. It is the processing of non-Newtonian materials that can cause major design and operational problems. For example, in most filamentous fermentations in submerged cultures, at the start of fermentation the broth rheology is Newtonian, not too different from water. However, often the microbiological requirements of the fermentation are such that the prevailing operating conditions lead to the development of diverse and complex non-Newtonian broth rheology, and extreme changes in the apparent viscosity of the broth can occur during fermentation. These changes can cause practical difficulties in maintaining an adequate power input per unit volume in the bioreactor. Subsequently, significant heterogeneity in dissolved oxygen tension and shear rate distribution can occur in the vessel contents. Consequently, transport processes such as heat and mass transfer and metabolic activity of the cells can suffer, cell damage can occur, and a fall in productivity can result.

The extent of non-Newtonian behavior of the broth is closely related to the growth of the biomass and the development of a three-dimensional network of highly branched hyphae in the case of mycelia microorganisms. Considerable work has been done in recent years to understand these complex structures by measuring the mycelia morphology in dispersed (filamentous) and pelleted (aggregated) forms. However, there are as yet no general predictive constitutive equations that can be used confidently for design and scale-up, and even for the same strain there is no consensus of opinion among researchers on the precise effects of mycelia morphology on broth rheology or how the rheology of such systems should be described. The uncertainties associated with the use of predictive rheological equations are to a large extent understandable considering the wide range of parameters that are thought to influence the rheology of the cultures. These include the mode of fermentation (e.g., batch, fed- batch, or continuous), growth rate, level of the dissolved oxygen concentration and pH levels of the culture, level of mixing and shear intensity in the fermentor, and presence of solid debris, impurities, surfactants, and additives. These parameters critically affect the properties of the individual hyphae, for example, their length and length distribution, diameter, and mechanical integrity (strength, rigidity, and flexibility), as well as the physicochemical interactions between the hyphae (compactness, roughness) and between the hyphae and the continuous liquid phase.

The analysis of rheological data is further complicated by the degree of hyphae aggregation (i.e., formation of clumps) and by the flocculation of clumps. The superimposed effects of all these parameters on broth rheology can be difficult to simulate. Few publications are available on the rheology of non-filamentous cultures. In cases where data are published (e.g., for yeast suspensions), there is considerable disagreement between the results of different researchers. Fermentation cultures and biological solutions that do not exhibit mycelia growth but can show non-Newtonian behavior include high concentrations of single-cell suspensions (e.g., yeasts), high molecular weight fermentation products (e.g., polysaccharides), solutions containing globular proteins, supercoiled plasmid DNA and macromolecules (e.g., membrane protein complexes), chromosomal DNA, and animal cells in solutions in high concentrations). This review article focuses largely on the rheology of the microbial suspensions in order to examine some of the key factors that are perceived to affect their flow properties. The intention is to highlight the results of some of the recent research on the interactions between mycelia morphology and culture rheology without emphasizing the effect of rheology on, for example, transport processes such as heat and mass transfer.

BIOMASS CONCENTRATION

In addition to biomass concentration, the extent of non-Newtonian behavior of fermentation broths is determined largely by the primary properties of the hyphae, which include hyphal morphology and dimensions, hyphal flexibility, and ability to form clumps. The primary properties of the hyphae in turn are critically dependent on the detailed fermentation conditions, making it difficult to develop a satisfactory theoretical simulation of the non-Newtonian behavior of broths. An alternative engineering approach to the study of flow properties of a fermentation culture is to experimentally develop the flow curve, or shear diagram, which is a plot of shear stress, τ, against shear rate, γ, in the laminar region. This information is usually obtained in a conventional laboratory viscometer such as the cup-and-bob (coaxial cylinder), capillary tube, or a cone-and-plate viscometer. However, most cultures are highly heterogeneous and contain large amounts of insoluble debris and particulate material, which can be fragile and can separate and sediment during rheological measurements. This renders many of the conventional rheometers unsuitable for characterization of biological suspensions. In such cases useful rheological data can be generated indirectly in laminar flow by using a calibrated, scaled-down, "engineering" viscometer, for example, an instrumented pipe flow viscometer or impeller viscometer. The former is widely used in the biochemical industry for flow characterization and is essentially a cup-and-bob viscometer except that the rotating inner cylinder is replaced by a mixer, (e.g., a turbine, paddle, disc, or helical ribbon).

From the rheological viewpoint, the basic difficulty in using engineering viscometers is that these instruments do not provide a well-defined and uniform shear field, and therefore, individual elements of fluid are likely to undergo periods of exposure to different levels of shear during measurements. In the case of the impeller viscometer, the evaluation of an average shear rate in the cup is based on an empirical technique originally developed for mechanically agitated vessels by Metzner and Otto and later modified for relatively simple rotating units (e.g., a disc spindle) by Mitschka. The estimation of the average shear rate, γ_{av}, in the cup of an impeller viscometer is based on the assumption that the average shear rate, γ_{av}, is directly proportional to the rotational speed of the impeller, N; the constant of proportionality, k_s, which is generally known as the impeller shear rate constant, is uniquely dependent on the geometry of the impeller viscometer but is independent of the fluid rheology. Thus, $\gamma_{av} = k_s N$. The key steps in the determination of k_s are as follows.

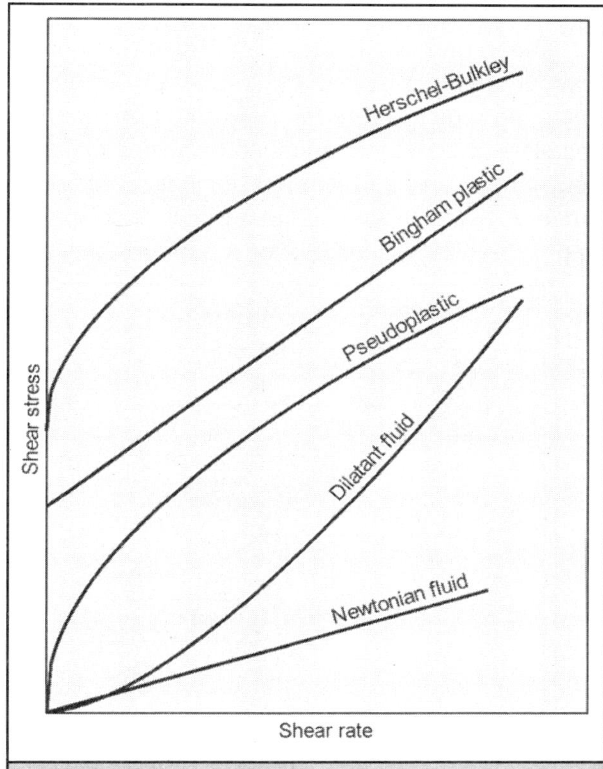

Figure 2.1 Graphical presentation of rheological models.

1. The laminar Newtonian power curve is obtained for the impeller-cup-and-bob viscometer from a series of measurements of torque and speed using one or more Newtonian liquids.
2. The flow curve is developed for a (model) non-Newtonian liquid, for example, carboxymethycellulose (CMC) solution using a conventional cup-and-bob viscometer.
3. The cup of the impeller viscometer is filled with the (model) non-Newtonian liquid.
4. The impeller speed is set at a predetermined value, N_1, and the corresponding torque, T_1, is measured. These values are used to calculate the corresponding dimensionless power number, $(P_o)_1$, $[P_o = P/(pN^3D^5)]$.
5. From the Newtonian power curve, the Reynolds number, $(Re)_1$ corresponding to $(Po)_1$ is obtained.
6. Using the definition of the general Reynolds number, $(Re = \rho ND^2/\mu_a)$ and the value of the Re_1 from step 5, the apparent viscosity, $(\mu_a)_1$, is calculated.
7. From the non-Newtonian flow curve the value of the shear rate, $(\gamma_a)_1$, is obtained corresponding to $(\mu_a)_1$.
8. Steps 4 through 7 are repeated for a range of speeds (N_2, N_3, etc.), and these values are plotted against the corresponding values of shear rates $((\gamma_a)_2$, $(\gamma_a)_3$, etc.).

The plot obtained in step 8 is often a straight line through the origin and has a slope equal to the shear rate constant, k_s. It should be noted that the above procedure is valid only if the measurements of speed and torque (power) fall in the laminar flow region of the power curve, this being the region where the power number is a function of the Reynolds number. For most turbine impellers, the Metzner and Otto technique gives a value of the shear rate constant, k_s, in the range between 10 and 13, whereas for helical ribbon and large paddle impellers its value falls in the range of 26 to 30. An important practical consideration is the selection of an appropriate rheometer (e.g., a rotating viscometer or a capillary flow rheometer).

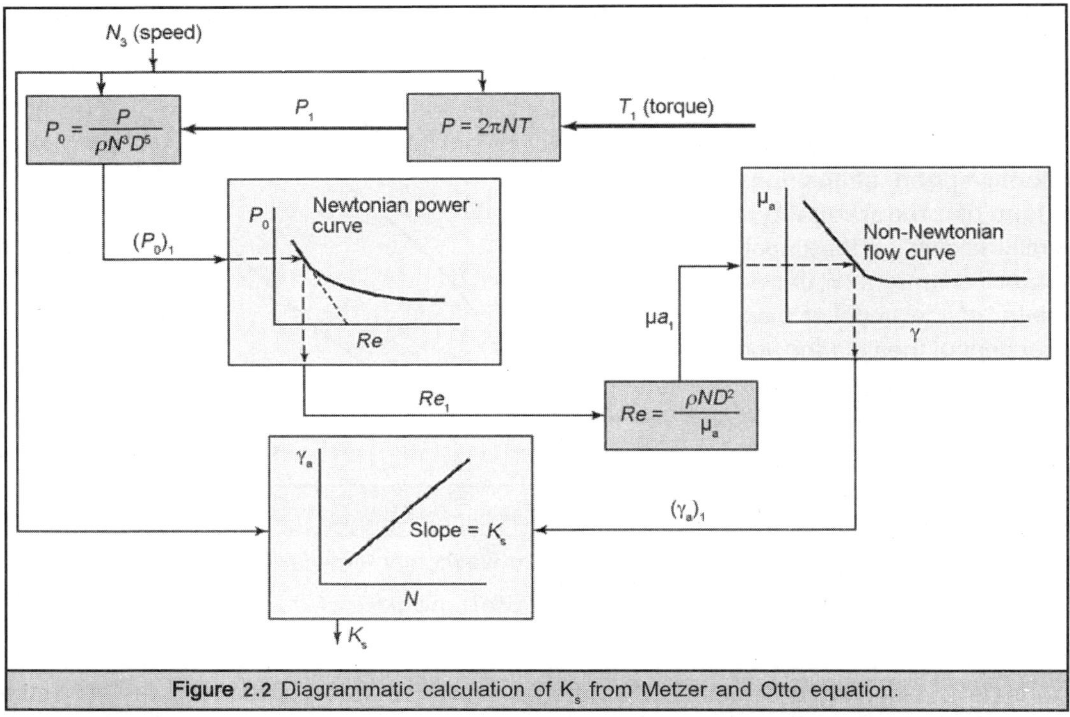

Figure 2.2 Diagrammatic calculation of K_s from Metzer and Otto equation.

Different methods of measuring the rheological parameters of a non-Newtonian liquid do not necessarily yield the same result. This is not to say that one method is better than another but merely that it reflects the fact that the liquid flow field is different in the different devices, and the response of a non-Newtonian liquid depends on the flow field to which it is exposed. Therefore, in specifying a value for the apparent viscosity, it is important to quote the means by which the parameter has been measured. Moreover, the method of measurement should be closely related to the application of the final product.

For example, for the design of a mechanically agitated bioreactor, an appropriate viscometer is the cup-and-bob or the turbine impeller viscometer, whereas the capillary tube or the instrumented pipe-flow viscometer can be used effectively to gather rheological data for pipeline transportation of non-Newtonian biological suspensions. The following discussion is based on rheological data obtained in the former group of viscometers.

Industrial Viscometers

Bohlin, Haake, Brookfield, Rheomat, Carrimed, and Rheomatrics are just a few of the more commonly used devices used for the rheological characterization of liquids. The first four viscometers are available in both laboratory and portable (battery-operated) form, and some, such as the Brookfield, can also be used on-line. The state of technology has vastly improved over the past two decades, and current instruments such as the Bohlin can be fitted with different heads (e.g., coaxial cylinders, parallel plates, or cone-and-plate geometry) and can perform a variety of measurements including controlled stress and controlled strain measurements. Additionally, the Bohlin rheometer can accurately determine the response of the liquid to more complex types of motion imposed on the sample, including oscillatory motion from which rheological information can be obtained on the viscoelastic properties of the liquid. These rheometers can be used to develop the non-Newtonian flow curve directly over a wide range of shear rates (10^{-6}–10^{7} s^{-1}) of interest to most process engineering operations. The low shear rates represent processes involving, for example, sedimentation of fine particles (10^{-6}–10^{-4} s^{-1}), leveling due to surface tension (10^{-2}–10^{-1}), and gravity drainage (10^{-1}–10^{1}). The medium range of shear rates covers processes such as extrusion, chewing and swallowing (10^{0}–10^{2} s^{-1}), mixing in mechanically agitated vessels, and pipe flow (10^{1}–10^{3} s^{-1}). Shear rates of 10^{3} s^{-1} to 10^{7} s^{-1} occur in operations such as spraying, centrifugation, and high-pressure homogenization.

EXPERIMENTAL MEASUREMENTS

For design purposes it is often desirable to describe mathematically the information contained in a flow curve by fitting empirical equations to experimental measurements. Numerous flow equations have been proposed to describe the shape of the flow curves, including the Bingham plastic, Herschel–Bulkley, Casson, and power law equations. There is, however, no theoretical justification in recommending any of these equations a priori, and the selection of an appropriate flow equation should be based on practical considerations; the model equation should give an accurate fit of the experimental flow curve, it should have a minimum number of independent constants that are readily evaluated, and preferably the constants should have some physical basis. A few researchers have reported time-dependent rheology and viscoelastic behavior for filamentous cultures. However, in the majority of cases the rheology of the cultures appears to exhibit time-independent shear thinning behavior, which can be described adequately by a two-parameter power law equation.

This is characterized by a flow curve that, over a reasonable range of shear rate, is a straight line on a logarithmic plot of shear stress against shear rate. Thus the relation between shear stress, τ, and shear rate, γ, can be written as $\tau = K\gamma n$, in which K is the consistency index and n is known as the flow behavior index and has a value less than unity for most cultures. The value of the flow behavior index gives a measure of the extent of the non-Newtonian behavior of the broth. For $n = 1.0$, the culture has Newtonian flow properties, with $K = \mu$. As n deviates from 1.0, the broth becomes increasingly non-Newtonian. Flow equations have also been recommended for fermentation cultures based on the Bingham plastic, Casson, and Herschel–Bulkley models, all of which contain a yield stress parameter that needs to be evaluated from experimental measurements.

The flow curves of many biological suspensions rise rapidly and then flatten out at relatively low shear rate. Under these conditions it is difficult to differentiate between the power law model and a model containing a yield stress, and very accurate data at quite low shear rates would be required to prove or disprove the existence of "true" yield stress. In the absence of such information, it seems prudent to use the power law model to describe the rheological data of biological materials over a specified range of shear rate.

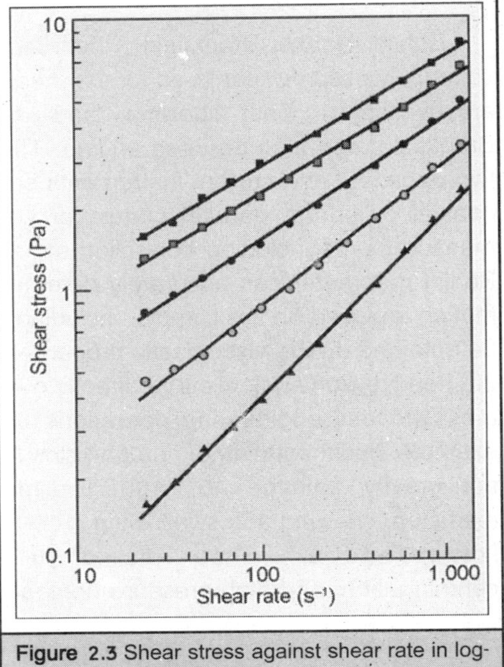

Figure 2.3 Shear stress against shear rate in log-log scale.

FILAMENTOUS CULTURES

Filamentous cultures exhibit two extreme types of morphology in submerged culture, highly dense pelleted and free mycelia, with loosely packed mycelium forming a transition between the two types. The freely dispersed mycelia and its entanglement to form clumps can cause the fermentation to become highly viscous and non-Newtonian. Past studies on morphology have attempted to distinguish between the different types by using various morphological parameters including compactness, roughness, and fractal dimension. These studies show that the compactness of *Penicillium chrysogenum* varies between about 0.3 for highly filamentous structure to close to unity for pellets. The fractal of *P. chrysogenum, Streptomyces griseus,* and *Streptomyces tendae* have been measured by Patankar et al. and in our laboratory and seem to vary between 1.9 and 2.7 in filamentous form and between 1.45 and 2.00 in pelleted form.

The pelleted type of growth consists of compact discrete masses of hyphae that generally lead to Newtonian behavior of the broth. However, in many cases as the size of the pellets increases, the diffusion of nutrients, particularly O_2, into the center of the pellets becomes limited, resulting in low productivity. Moreover, there are reported cases of cultures consisting of pelleted mycelium that exhibit clear non-Newtonian behavior. Great caution must therefore be exercised in generalizing the flow behavior of microbial cultures. The flow curves presented later are intended to serve as typical rheological examples of real biological microorganisms extensively studied in the past. The discussion examines the effects of some of the important parameters that are perceived to influence broth rheology. For these reasons, the details of fermentation conditions are not included, but where necessary, the reader is referred to other publications and references where further information on the relevant aspects of the work can be found. Additionally, with the exception of the information, a conscious decision has been made to avoid recommending specific constitutive equations because the absolute values of the constitutive parameters K and n are highly system specific and ideally should always be based on experimental flow curves obtained under the actual fermentation conditions for which they are required.

Culture of Molds

Numerous publications are available on the rheology of mold fermentation cultures; mostly the mycelium is in the filamentous form because these cultures show strong non-Newtonian behavior. The organisms most widely studied are *P. chrysogenum* and *Aspergillus niger*. The non-Newtonian behavior of most of the cultures investigated has been modeled either in terms of a flow behavior index, *n*, and consistency coefficient, *K*, of a power law shear thinning equation, or in terms of a yield stress, τ_0, and a plastic viscosity, μ_p, of the Bingham plastic model. In a few cases the Casson model is used, in which case the yield stress, τ_0, and the shear rate coefficient, K_c, are given as the two model parameters. In these studies, plastic viscosity, μ_p, and yield stress, τ_0, in the Bingham plastic model and the consistency coefficient, *K*, in the power law equation were found to increase and the flow behaviour index, *n*, decrease with increasing biomass concentration. The increase was found to be nonlinear, and its extent was highly strain dependent. An examination shows that equations are also available in which the rheological parameters have been related to the morphological properties of the individual hyphae. However, recent measurements have shown that in many cases as much as 90% of the mycelium exists in flocculated (clump) form and not in a freely dispersed form, suggesting that rheological parameters are more likely to be related to clump properties and not to the detailed characteristics of the individual hyphae. Recent measurements carried out by Tucker and Thomas and Olsvik et al. appear to support this conclusion and suggest that, for example, the compactness and the roughness of hyphae flocs can be successfully related to the rheological parameter.

Culture of Actinomycetes

In filamentous form, the sizes of the individual hyphae in the mycelium are generally less than 100 μm, and the broths exhibit considerable non-Newtonian flow behavior. However, recent experimental evidence obtained in our laboratory using three industrial strains of Actinomyces ranging in size from 15 μm to 25 μm indicated that the properties of the individual hyphae have little effect on broth rheology. The results of these experiments, discussed in more detail later, also revealed that there was little change in the size of the mean main length of the hyphae, and furthermore very few clumps were observed during hyphal characterization. As stated already, the variations of *n* and *K* of *Saccharopolyspora erythraea* (strain CA 340) mycelia cultures as a function of time of fermentation.

The data were obtained in a standard coaxial cylinder viscometer using fresh samples of the culture that were removed during fermentation in a 7-L stirred bioreactor. The plots are effectively snap shots of a developing rheology, showing the dramatic changes in the non-Newtonian behavior of the fermentation broth during the growth period. The data for *S. rimosus* were obtained from an industrial-scale mechanically agitated bioreactor; in the case of *S. erythraea* and *A. roseorufa*, samples were taken during fermentation in a small, laboratory-scale mechanically stirred unit. In all three cases, the broths contained a significant amount of particulate matter, which made the use of a standard cup-and-bob viscometer inappropriate. The rheological parameters were therefore obtained by using a Brookfield 2HA-TDVII Synchro-Lectric rheometer fitted with a disc-spindle impeller, for which the average shear rate at each rotational speed was obtained by using Mitschka's technique, which is described fully elsewhere. The experimental data shown in the figure, the mean main hyphal length as a function of fermentation time. The width of the plotted lines indicates the 95% confidence limits of the data.

The mean main hyphal length fluctuates throughout the fermentation time, but no observable trend in the data can be detected for any of the microorganisms. Additionally these plots indicate no correlation between the mean main hyphal length and broth rheology. The mean main hyphal length of the three microorganisms are significantly smaller than those reported previously for *Streptomyces clavuligerus* (45–120 mm), and *P. chrysogenum* (e.g., 130–420 μm and 60–120 μm). An examination of the information shows that, compared with the fungal cultures, the rheological parameters, K, τ_o, μ_p, and K_c, for example, are less sensitive to changes in the biomass concentration. The similarity of the two profiles is evident from these plots, and a series of experiments were conducted to examine the factors responsible for the observed dependency of the broth consistency coefficient, K, on DCW. Two fermentation batches were prepared and run in the same way except that only one batch was inoculated with *S. erythraea*. Several samples were taken periodically from each batch for rheological measurements and DCW evaluation. Some of the samples were also centrifuged prior to analysis to remove *S. erythraea* cells and/or medium particulate and insoluble debris.

However, once the *S. erythraea* biomass was removed by centrifugation (curve B) the broth rheology changed dramatically, variation in K disappeared, and the flow behavior index increased from its value of about 0.2 to a value close to unity, indicating a change from a highly shear thinning to an almost Newtonian behavior. Samples from the uninoculated batch showed very little change in their rheology, which remained Newtonian practically throughout. Similar results were observed for the fermentation of *A. roseorufa,* confirming that the high apparent viscosity of the filamentous cultures depends on the presence and concentration of the organisms, and that other constitutive components (e.g., insoluble particulate matter) appear to have little effect on broth rheology, at least for the cases presented here.

The effect of pH of the culture on broth rheology was also investigated, and it was found to have a strong effect on the extent of non-Newtonian behavior of the culture. This effect was somewhat unexpected and was further investigated for one of the microorganisms, *S. rimosus.* A large sample of the broth was removed from the bioreactor and divided into several specimens. Using concentrated hydrochloric acid and concentrated sodium hydroxide (60% by weight) the pH of each sample was altered to cover the range of pH between 3 and 10, in increments of 0.5 pH. The rheological properties of each sample were then measured using the Brookfield viscometer. Measurements of the hyphal morphology (not shown) revealed no effect of pH on the mean main hyphal length of the individual hyphae. However, is that alteration in charges on and/or close to the surface of the microorganisms, arising from variations in pH, could play a role in determining the observed changes in broth rheology.

The zeta potential characterizes the nature and the extent of surface charges, and the zeta potential–pH profiles of the microorganisms should reflect these effects. No electrophoretic measurements are available in the literature concerning the zeta potential–pH profiles of microorganisms. These profiles should be developed experimentally and studied to examine how they are affected by broth conditions. The results could provide valuable information on the interrelationship between broth rheology, hyphal morphology, and biomass concentration. Considerable experimental work has been published on the rheology of biological suspensions since the 1960s. Advances made in image analysis techniques in the early 1990s resulted in further work on relating the rheological properties to the morphological characteristics of filamentous microorganisms. No consistent opinion has yet emerged on the interaction between the two. Recent developments on the measurement of fractal properties of mycelium may provide a useful basis for further elucidation of these complex interactions.

3

ROLLER BOTTLE BIOREACTOR

The settling behavior of biological cells with densities different than the fluid is a key factor affecting the performance of roller bottle bioreactors. The fully three-dimensional velocity field in a roller bottle bioreactor was simulated using a finite-element method, and the settling behavior of cells was simulated for both steady and time-dependent bottle rotation rates using particle dynamics algorithms. These simulations were validated experimentally using fluorescent particle tracers. Under steady flow conditions the flow is divided into two regions, one in which the particles settle to the wall and one in which the particles remain suspended indefinitely, but unsteady flow conditions (characterized by repeated reversal of the flow direction) lead to complete cell settling. These observations showed qualitative agreement with culture infection kinetic data in which changes in the roller bottle rotation rate resulted in enhanced infection. The main mixing limitation observed experimentally in roller bottles is very slow mixing in the axial direction, caused by flow segregation at the central plane of symmetry of the bottle. This limitation is readily overcome by imposing a small- amplitude vertical rocking motion to the bottle, forcing fluid to flow through the segregation plane and leading to a completely homogeneous mixture in just a few revolutions of the bottle.

HISTORY

Manual and automated roller bottle systems have been used for more than 40 years in the pharmaceutical, biochemical, and medical fields for processes such as cell growth and infection, heterologous glycoprotein production, vaccine preparation, and high-density plant cell cultivation. Despite efforts by numerous investigators to develop unit operations such as microcarrier culture for the production of anchorage-dependent cells or cell products, roller bottle systems still persist in research and industry. Additionally, for industrial-scale cell culture production (e.g., vaccines), cells are frequently expanded in roller bottles before transfer to microcarrier cultures for the final growth phase even when unit operation-based systems are utilized. Widespread use of the roller bottle has several reasons. Most notably, the process relies on very simple technology: a horizontal cylindrical vessel is filled approximately one- fourth full and rotated axially; increased production is accomplished simply by increasing the number of bottles. Thus, scale-up development is straightforward, resulting in reduced developmental time lines for industry and faster introduction to market for new products.

The system allows constant fluid–gas contact and easy addition of nutrients without interruption of the process. In addition, the process is capable of maintaining sterile conditions for prolonged times, contamination of one or more roller bottles does not necessarily result in the failure of an entire lot, precise control of nutrient and waste product levels is possible, and the direct monitoring of cell growth and metabolism is relatively simple. On the other hand, roller bottles are limited in surface area available for growth and in the volume of harvest fluid obtained. Manpower and facility space requirements are higher than for unit operation systems such as microcarriers because hundreds or thousands of roller bottles can be operated for a single production run, although new automation systems are addressing the manpower issue.

In addition, cell growth and infection are believed to be suboptimal in certain vaccine products because of flow and mixing dynamics, perhaps by preventing infected cells from attaching to host cells attached to the bottle walls. Although these issues point toward an obvious need for flow analysis and process design criteria, there have been no published results to date on either of these topics. The following sections detail computational and experimental characterizations of the fluid flow profiles within rotating roller bottles, including particle trajectories, fluid mixing patterns, and unsteady-state flow strategies.

The results point to ways in which cell proliferation and infection can be optimized by simple modifications to roller bottle rotation. This article is organized as follows. First, the numerical method for obtaining the three-dimensional flow solution is described. The flow solution is then validated by comparison with experimental measurements obtained via particle imaging velocimetry. Computationally predicted mixing patterns are compared with mixing patterns obtained by visualization of fluorescent dye streak- lines. The model is then used to examine cell motion in the flow field. Computationally predicted cell motion patterns are compared with experimentally observed particle motion patterns obtained via long-time exposure photography of fluorescent particles. Then, particle settling results for the roller bottle under steady flow conditions are discussed and related to cell growth results obtained for various bottle rotation speeds. Finally, attention turns to improving the performance of roller bottles. The use of unsteady motions on the cylinder to enhance cell settling is explored, and the use of rocking flow perturbations to enhance axial mixing is demonstrated. The final portion of this chapter is devoted to conclusions and recommendations for future research.

COMPUTATIONAL MODEL OF FLUID AND CELL MOTION

Three Dimensional Flow Field

The three-dimensional flow field for a standard 850-cm^2 roller bottle geometry was obtained using a commercial CFD software package, FLUENT. This and similar software packages have been used by several researchers in the past to solve fluid dynamics problems. The equations of motion for the flow are the well-known Navier–Stokes equations, which, for a two-dimensional flow, include the momentum equation in two components and the continuity equation:

$$u/t + u\nabla u = -\nabla p + (1/\mathrm{Re})\nabla^2 u + (1/\mathrm{Fr}) \quad ..(1)$$

$$\nabla u = 0 \quad ...(2)$$

where u is the dimensionless velocity and p is the dimensionless pressure. Re in Equation 1 is the Reynolds number, $\mathrm{Re} = \rho vL/\mu$, where ρ and μ are the fluid density and viscosity, and v and L are the bottle's linear rotational velocity and the liquid height, respectively; Fr is the Froude number, $\mathrm{Fr} = gL/v^2$, where g is gravity. The rationale for selecting liquid height as the natural length scale for the flow is that the liquid height is the diffusional length scale in the system. Viscous forces can be viewed as diffusive transport of momentum, with the kinematic viscosity (μ/ρ) playing the role of a momentum diffusivity. Hence, the liquid depth is the natural length scale for the action of viscous forces.

To obtain the velocity field, it was assumed that (i) the flow is steady, (ii) the flow is in the Stokes regime (creeping flow conditions), (iii) the free boundary is both planar and horizontal (effects of surface tension and viscous drag on the boundary are neglected). For the conditions used in the biotechnology industry, Re = 10, confirming the validity of the creeping flow assumption. The physical domain was discretized using a curvilinear mesh; although the bottle has a total length to diameter ratio of 2:1, because of its symmetry only one-half of the system needed to be simulated, giving a length to diameter ratio of 1:1. The liquid height in the bottle was one-fourth the total diameter, which is typical of industrial practice. A 80 x 20 x 94 structured computation grid was used. An acceptable node density for the computational mesh was found by creating several meshes with different node densities. The number of nodes in the mesh was successively increased by 25% until an average velocity difference of less than 3% was achieved for two successive meshes. The lower node density was then used to generate the computational mesh. Velocity and pressure fields were obtained via iteration using a Sun SPARC 20 workstation. Approximately 140 Mb of RAM and 60 h of CPU time were required for solution convergence. To determine convergence, a criterion of 10^{-6} was used for each normalized residual velocity and pressure component. Residuals for each iteration were normalized versus the residual values obtained after the second iteration of the solver.

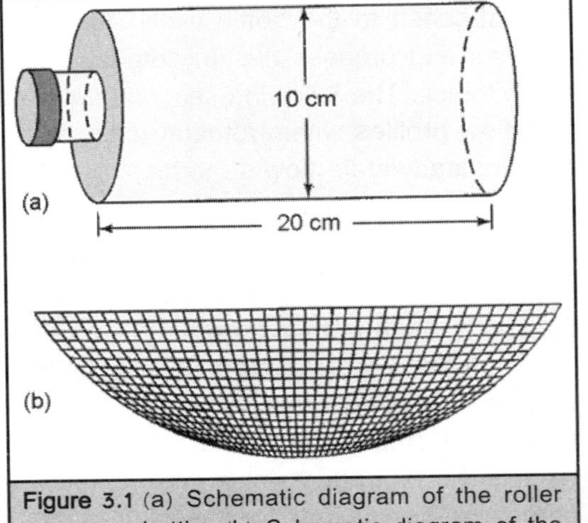

Figure 3.1 (a) Schematic diagram of the roller bottle. (b) Schematic diagram of the computational grid used for the simulations.

To test the sufficiency of the convergence criterion, the simulation was run using convergence criteria ranging from 10^{-5} to 10^{-6} with no significant change in the velocity field when the convergence criteria were varied over this range (<1.0% change in velocity magnitude, on average). Sensitivity analysis (i.e., trajectory closing) showed that such a mesh was sufficiently accurate. These flow data were then used for all subsequent particle tracking simulations. The condition adopted in the base case simulation was $\Omega = 0.25$ rpm, $\mu = 0.125$ Pa s at room temperature, giving a Reynolds number of Re 1.32. At this Re, the flow can be considered to be in the creeping flow regime. The flow was assumed to be sufficiently slow so that it could be assumed that the free surface was always horizontal and no surface tension effects were present at the bottle walls. The free surface was therefore modeled as a frictionless wall (i.e., a surface that exerts no stresses on the fluid). Several assumptions were also made regarding the particles simulated in the post-processing program. The particles in the flow were sufficiently dilute so that they did not interact with each other or affect the fluid flow in any way. The particles were assumed large enough so that Brownian forces could be ignored. All the assumptions used here are valid for the majority of industrial uses of roller bottles.

Validation

Two flow visualization experiments were performed to validate the simulations. The experiments were conducted in a glass roller bottle 10 cm in diameter and 20 cm in length. The working fluid was glycerin, which has a density of 1.26 g/cm^3 and a viscosity of 1.25 Pa s. The bottle was rotated using an apparatus that consisted of a set of rollers whose rotation speed and direction were accurately controlled by a computer. The rotation rate for the two experiments performed was 0.25 rpm, giving a Re of 0.13. This flow condition can be assumed to be representative of all conditions in the creeping flow regime, because in this regime the velocity field becomes independent of Re. The velocity field was first validated via comparison with particle imaging velocimetry (PIV) results. The use of particle imaging techniques to study fluid flow problems is well documented in the literature and has been reviewed by Adrian. Information that is pertinent to our specific system is provided here. The flow was seeded with 12-μm-diameter silver-coated hollow glass spheres that have a specific gravity of 1.17. These particles are sufficiently small as to provide accurate measurements of the instantaneous velocity field. A 10-mJ New Wave Research pulsed mini-YAG laser was used as the illumination source.

The laser beam passed through a series of optical components to produce a sheet of light 1 mm thick (at a focal distance of 1 m). A 32 x 32 grid was analyzed using cross-correlation data acquisition, with 10 ms between laser pulses. Resultant data from PIV was in the form of velocity direction and magnitude at the center of each interrogation area. The experimental velocity field was measured in the vertical plane normal to the bottle axis in the middle of the bottle. The length and color of the arrows represent the velocity magnitude in the plane, ranging from low to high. The velocity fields are qualitatively identical, with the exception of some small experimental scattering near the free surface of the liquid. The computational velocity field was further validated by comparing computational and experimental mixing patterns obtained for passive tracers at the center plane of the bottle. This first step for simulating mixing patterns is to determine the trajectories followed by point particles as a result of the equation of motion:

$$dx/dt = v \qquad \ldots(3)$$

where x is the particle position and v is the fluid velocity. The flow pathlines predicted using 4th-order Runge–Kutta integration of equation 3, which are also the flow streamlines because the flow is steady. A stagnation point is located on the symmetry line at 0.6 unit lengths below the free surface (the unit length is the height of the liquid). A similar approach was used to simulate the evolution of a vertical filament initially placed at the center of the bottle. The "line" consisted of 10,000 particles moving at the flow velocity. As the particles are moved by the flow, the line is twisted around the stagnation point. The line stretch in this steady two-dimensional flow is linear with time.

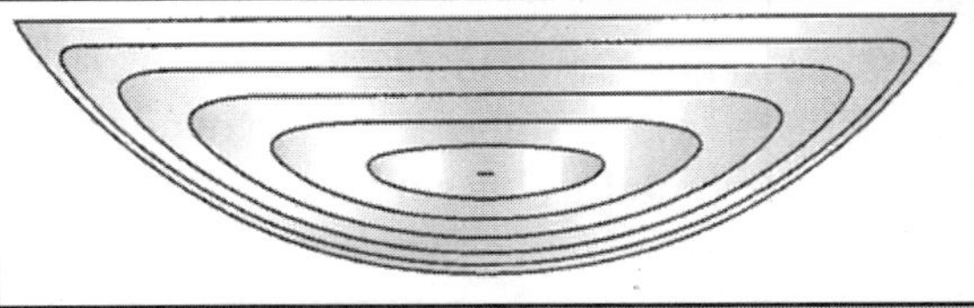

Figure 3.2 Computational stream lines of the flow at the center plane of the bottle under steady flow conditions.

The evolution of a fluid line was also observed experimentally to verify the simulations. Once again, excellent qualitative agreement between experiments and simulations was obtained, and it was concluded that the simulations indeed provide an accurate representation of the velocity field in the roller bottle. Once the computational velocity field was validated, it was used to obtain a detailed characterization of fluid motion in the bottle. To provide a clear illustration of a three-dimensional velocity vector field is not a trivial task. This is accomplished by using two-dimensional cuts of the velocity field, where the in-plane components are shown using a vector and the total magnitude of the velocity field is shown using color contours, with red signifying high velocity and blue indicating low velocity. The velocity field shows both reflection symmetry relative to the center plane (expected from the symmetry of the boundary) and antisymmetry between the left and right sides of the bottle (expected from the creeping flow condition and the rotational symmetry of the bottle). As discussed next, this observation can have important implications because the flow induced by the end walls of the bottle is the only means of axial mixing. Due to the no-slip condition at the bottle wall, all fluid follows the bottle wall in a counterclockwise direction. When the fluid reaches the free surface (at the right side), it detaches from the wall and flows outward in the *z*-direction. At the left side, the fluid immediately adjacent to the wall is drawn downward; this motion is "fed" by an inward flow from the *z*-direction. However, detailed examination of the velocity vectors indicates that the *x*- and *y*-components of the velocity field are largely unaffected by end-wall effects; such effects appear primarily in the axial (*z*) component of the velocity field.

The end-wall effects are more clearly shown by a top view of the bottle. The end of the bottle is at the bottom of this figure, and the symmetry plane is at the top. Flow in the axial (*z*) direction is clearly visible in this figure. The fluid flows upward from the end wall in the lower right-hand corner, travels in a counterclockwise loop, then flows downward at the lower right-hand corner as a result of the end-wall flow. There is a stagnation point 0.46 cm from the end of the bottle, as shown by the dark blue color. Moving toward the center of the bottle, end effects have less influence on the flow; there is less than 2% difference in the velocity field between the two-dimensional symmetry plane and the plane at 5 cm from the end wall. Similar flow patterns were observed for other values of Re of less than 10. An alternative illustration of the three-dimensional flow pattern in the roller bottle is obtained by computing the pathlines of the flow. The center of the bottle is at the left-hand side of the figure, and the end wall is at the right. Examination of the pathlines from left to right reveals the effect of the end wall on the flow field. Toward the end of the bottle, the

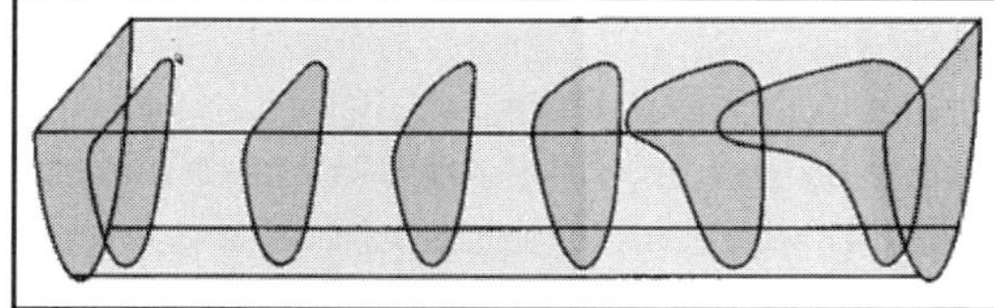

Figure 3.3 Pathlines of passive tracer particles starting from different positions inside the bottle.

pathlines begin to bend near the free surface toward the center of the bottle. All the pathlines shown are closed loops. The pathlines at the center plane of the bottle are two-dimensional loops, and the pathlines away from the symmetry plane are three-dimensional loops. In either situation, however, the same result ensues: fluid particles in the flow remain trapped in the closed periodic orbits and do not contact other portions of the flow. Such regular motion patterns can pose significant barriers to the mixing process. Experimental evidence of this observation and approaches for overcoming mixing limitations follow.

Cellular Motion in a Roller Bottle

The initial point for the modeling of small particles in the flow field is given by the equations of motion proposed by Maxey and Riley:

$$m_p \frac{dv}{dt} = (m_p - m_F)g + m_F \frac{Du}{Dt} - 6\pi a^2 \mu \int [\pi v (t - \tau)]^{-1/2} \left(\frac{dX}{dt}\right) d\tau$$
$$- \left(\frac{1}{2}\right) m_F \frac{d\,[u - v(1/10)a^2 \nabla^2 u]}{dt} 6\pi a \mu X \qquad \ldots(4)$$

where $X = u - v - (1/6)a^2\nabla^2 u$. In these equations, u is the (dimensional) fluid velocity vector, v is the (dimensional) cell velocity vector, a is the cell radius, v is the fluid kinematic viscosity; and m_p and m_F correspond to the weight of the cell and that of the fluid displaced by the cell, respectively. The five terms on the right-hand side of the equation correspond to buoyancy forces, pressure forces, flow history effects, added mass effect, and Stokes drag forces. This equation can be significantly simplified for the roller bottle problem. Because u is known from the numerical solution of the flow, the relative magnitude of each term can be evaluated. Analysis has shown that the leading order terms are inertia, buoyancy, and drag forces, thus the equation becomes:

$$\frac{dv}{dt} = Gg + \frac{1}{S_{tk}}(u - v) \qquad \ldots(5)$$

where $G = 2(\gamma - 1)gL/[(2\gamma + 1)U^2]$, and $S_{tk} = (2\gamma + 1)a^2\, U/(9vL)$; γ is the particle density relative to the fluid density, L is the depth of the fluid, and U is the bottle rotation velocity. If we assume that the particles are cells and the rotation speed and liquid are typical of those used in the biotechnology industry, this equation can be further simplified. Under such conditions, $G \sim O(10^{+3})$ and $S_{tk} \sim O(10^{-6})$, so that the inertial term can be neglected, greatly simplifying the problem. The particle velocity is obtained immediately as

$$v = u + V_s, V_s = \frac{2a^2(\gamma - 1)}{9vU} g \qquad \ldots(6)$$

where v, the total particle velocity, is simply the sum of u, the fluid velocity, and V_S, the particle terminal settling velocity. Once v is known, particle positions are found by numerically integrating the equations

$$\frac{dx}{dt} = v \qquad \ldots(7)$$

using the standard 4th-order Runge–Kutta method. To understand small particle behavior in a steady rotating roller bottle, equations 6 and 7 are used to simulate the motion of particles on the center plane of the bottle. Six particles were initially placed on the vertical center line. The particles showed different behaviors: while the two particles nearest to the free surface reached the rotating wall in a fraction of a bottle rotation, the remaining four particles moved in closed orbits. Such particles will never reach the bottom wall. This simulation indicates that whether a particle settles or not depends not only on its settling velocity, but also on its initial location within the flow field.

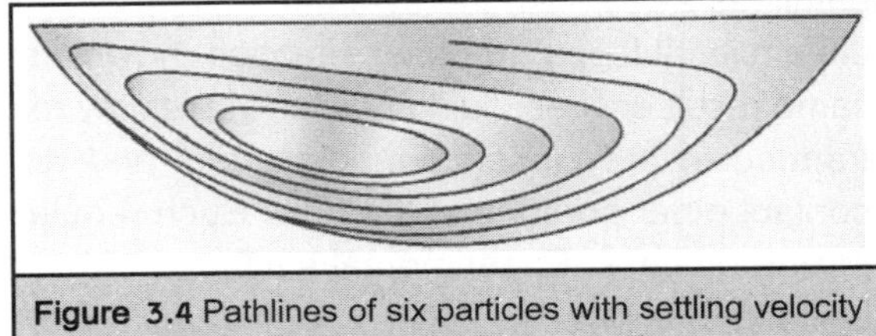

Figure 3.4 Pathlines of six particles with settling velocity of $V_s = 0.05U$ placed in the flow.

Moreover, as the settling velocity used in this simulation is much larger than those corresponding to real systems, data actually indicates that a substantial fraction of the cells in a roller bottle could actually remain trapped indefinitely in recirculating trajectories within the bottle flow, never reaching the bottle wall. This simulation result was investigated experimentally by using long-exposure-time photography to reveal the pathline followed by a fluorescent particle with a settling velocity. A 200-μm fluorescent polystyrene particle was placed in the center plane of the roller bottle. This particle had a terminal settling velocity of approximately 0.5 cm/min, or 0.06 U, where U is the linear rotation velocity of the roller bottle. Using long-exposure photography, the particle pathline was photographed for one bottle revolution under fluorescent light. The original position of the particle was approximately 0.6 length units from the free surface (where the unit length is the height of the fluid). The pathline revealed closely resembles the pattern, which is a simulation for particles of roughly similar settling velocity. When the particle was originally placed in a different location, closer to the bottle center, the particle followed a different closed pathline. When placed close enough to the upper surface, the particle eventually settled and remain attached to the bottle wall.

CELLULAR SEDIMENTATION

Sedimentation of Particles

A more complete picture of cell behavior can be obtained by following a large number of particles over a long period of time. Such simulations were conducted by placing 20,000 particles uniformly distributed across the entire central cross section of the bottle and following them for 20 rotations of the bottle. Whenever a particle touched the bottle wall, it was (arbitrarily) assumed that the particle had "settled" and was removed from the simulation. For these simulations, the bottle was rotating counterclockwise at a nondimensional linear velocity of $U = 1.0$. For $V_s = 0$, the structure of the particle positions is identical to those of the flow streamlines, because such particles are in fact fluid particles. As V_s increases, the particle paths are perturbed away from streamlines, for $V_s = 0.02$ U most particles remain trapped in closed periodic orbits and never reach the bottle walls; only those particles initially very close to the free surface and the bottle wall are able to reach the bottle boundary. As settling velocity increases, more particles settle, but only for large values of V_s (i.e., $V_s = 0.2\ U$) are most particles able to reach the bottle walls. Although such simulations give information about the dynamic behavior of particle motion, the settling rate cannot be inferred from them.

To obtain the spatial distribution of particle settling rate, the settling time of each particle was recorded and plotted as a function of initial particle location. The bottle is rotated clockwise at a dimensionless linear velocity of $U = 1.0$ at the wall. Results correspond to $V_s = 0.02$, 0.05, 0.1, and 0.2. The colors from dark gray to white represent the particle residence time from 0 to the maximum computing time. The smooth core region in dark gray represents the particles that do not settle at all. For small $V_s = 0.02$, only a narrow strip of particles under the top free surface and near the bottom on the right side of the bottle settle. The width of this strip increases as V_s increases and the size of the red core decreases. The tip of this core region is always at the upper-left corner of the flow and leans to the left side of the bottom of the bottle. At $V_s = 0.1$, most of the particles in the bottle will settle. When V_s is increased to 0.2, almost all the particles settle to the wall.

The spatial distribution of particle settling time in the flow is shown clearly by the shades of gray. Short-lived particles are always on the right side of the bottle for this flow direction. If the flow direction were reversed. These results clearly reveal that the particle residence time distribution in suspension is strongly affected by the recirculation of the flow. The fraction of particles suspended in the flow at a given time is plotted against the dimensionless time, where one bottle revolution corresponds to 12.56 time units. All the settling processes are nonlinear (except for $V_s = 0.2$). Particles settle to the wall relatively fast at the initial stage, and then the rate of settling (the slope of the curves) levels off in a relatively short time, about 8 to 12 time units. With large settling velocity ($V_s = 0.2$, the heaviest particles examined), the particles settle at almost a constant rate.

Bottle Rotation Speed on Cell Culture

Previous experiments demonstrated that the growth of anchorage-dependent cells and cell culture products (e.g., varicella virus) in roller bottles was much less efficient than that observed in stationary cultures such as T-flasks. It was hypothesized that these differences were caused in part by decreased cell (uninfected and infected cells) sedimentation rates present in the roller bottle cultures as the result of circulating liquid flows that trapped the particles in circular orbits. The decreased sedimentation times resulted in lower cell plating efficiencies for uninfected cells and increased degradation times for virus-infected cells, that were unable to reach the cell culture surface for attachment or infection. To further investigate this hypothesis, the kinetics of virus-infected cell sedimentation and attachment were studied at different rotation rates. The attachment to and growth of cells on the roller bottle surface were also studied to determine the primary mechanism controlling the differences observed in cell growth rates as a function of rotation rate.

Methodology

Oka strain of varicella was obtained from the Biken Institute, Japan. MRC-5 cells were obtained from ATCC or NIBSC and used to generate a manufacturer's working cell bank (MWCB) at Merck. MEM was manufactured by Merck and supplemented at the time of use with fetal bovine serum

(FBS), neomycin, and glutamine. Eagle's basic medium (BME) was purchased from Gibco. Horseradish peroxidase–conjugated donkey antigoat antibody (affinity-purified F[ab′]2 fragment) was obtained from Jackson Immunoresearch Labs. Goat antivaricella antiserum was obtained from Lampire. 3,3- Diaminobenzidine (DAB) was obtained from Sigma, prepared as a 25 mg/mL solution in water, and stored at –20°C. Immediately before use, DAB stock solution was thawed and diluted with a peroxide/nickel enhancing solution in phosphate buffer. A freshly trypsinized, concentrated cell suspension was placed in roller bottles with MEM growth medium supplemented with 10% FBS, 50 μg/mL neomycin, and 2 mM glutamine. The bottles were incubated at 37°C and rotated at the desired rate. After 4–6 days (depending on the process used), the roller bottles were refed with fresh nutrient medium.

Roller bottle cultures with an established cell monolayer on the roller bottle surface were infected by the addition of freshly trypsinized, varicella virus-infected MRC-5 cells. The infected cells were added to the liquid nutrient medium in the roller bottle 6–8 days after the initial roller bottle cell plant. Particle counting and size analysis utilized an Elzone particle size analyzer model with a 48-μm orifice. An in situ immunostaining procedure was used. The medium was aspirated from the culture vessel (T-flask or roller bottle) using vacuum. The culture vessel was then rinsed three times with PBS with 1% BSA. After the final rinse with PBS, a fixing solution of 5% acetic acid, 5% water, and 90% methanol was added to the culture vessel and exposed to all surfaces for 15–30 min. The methanol is then aspirated from the culture vessel with vacuum. The culture vessels are then rinsed three more times with PBS, and stored with a PBS overlay at 4–8°C until stained. Immunostaining of the fixed bottles was carried out by a dual antibody method. First, 2–3 mL of a 1:500 dilution of goat antivaricella antiserum in PBS with 1% BSA was added to the culture vessels, covering the entire growth surface of the T-flask and a 3- to 4-cm-wide strip along one side of the roller bottle. The vessels were then placed on a rocker platform and rocked gently at room temperature for 1–2 h to allow binding of the primary antibody.

The culture vessels were then rinsed three times with PBS to remove unbound antibody, and 3 ml of a 1:500 dilution of the anti- goat antibody was added. The vessels were again incubated at room temperature with gentle rocking for 30–45 min. Unbound secondary antibody was then removed by three PBS rinses. Then, 10 mL of peroxidase substrate solution was added to each culture vessel and the vessels were incubated at 37°C until foci of infected cells appeared. The peroxidase solution was then aspirated from the vessels with vacuum and the vessels again rinsed three times with PBS. The stained areas were left under a PBS overlay, and the vessels were stored at 4–8°C. Infectious foci were counted using a microscope at 10x magnification. Varicella plaque assay was conducted in the following manner: MRC-5 cells were grown in 60-mm plates with 5 mL/plate of BME with 10% FBS, 0.2% glutamine, and 0.05% neomycin; 10 plates were prepared per sample with 5 plates per dilution. The plates were then incubated at 35.5°C with a 3% CO_2 gas environment for 2 days. Before addition of the samples, the cell growth medium was removed and 5 mL/plate of MEM with 2% FBS, 0.2% glutamine, and 0.05% neomycin was added to each plate. Diluted virus samples of 0.1 mL were added to each plate.

Samples were diluted to ensure a plaque counting range of 15–80 plaques per plate. The plates were mixed well and incubated for 7 days at 35.5°C with a 3% CO_2 gas environment. Following incubation, the medium was removed by aspiration and 2 mL of 0.2% Coomassie blue stain in 100% ethanol containing 1% v/v glacial acetic acid was added to each plate. After 15 min at room temperature, the stain was removed and the plate was rinsed with cold water. The plates were inverted to dry. Plaques were then counted in each plate using a light box to assist in visualizing the plaques.

Laboratory output

In the first study, the kinetics of virus-infected cell sedimentation and attachment were studied at three rotation rates: 0 (T-flask stationary culture), 1/4 rpm (roller bottle culture), and 1/2 rpm (roller bottle). A mono- layer of MRC-5 cells was established on the surface of the T-flasks and roller bottles. Following growth of these cells to confluence, varicella-infected MRC-5 cells in nutrient medium were added to the T-flasks or roller bottles and rotated at 0, 1/4, and 1/2 rpm. Samples of the supernatant were collected at various times from 0 to 36 h and were measured for infected cell number remaining. At the same time intervals, flasks and roller bottles were sacrificed and stained with a varicella-specific monoclonal antibody marker to determine the number of infectious foci on the cell monolayer. Infectious foci appear when an infected cell has attached to, fused with, and infected a cell attached to the roller bottle surface followed by infection of neighboring cells. Thus, the disappearance of infected cells from the supernatant could be correlated to the appearance of infectious foci on the cell monolayer (where it is assumed that one infected cell equals one infectious focus). The infectivity of varicella-infected cells degrades exponentially at culture temperatures. To optimize the use of this material and to provide for as synchronous an infection as possible, the time that the infected cells remain in solution is to be minimized. This finding is further supported by the increased appearance of infectious foci on the cell monolayer as the rotation rate is decreased, thus establishing the correlation between number of infected cells disappearing from the supernatant with the increase in infectious foci.

Additionally, a mass balance was performed to demonstrate that as a function of time the appearance of infectious foci equaled the initial number of cells in the supernatant minus the intrinsic degradation rate in solution (degradation rates were determined from independent experiments). In the second study, lower rotation rates were used to further optimize the roller bottle rotation rate for the infection of the cell monolayer. Rotation rates lower than 1/8 rpm were not selected because the cell monolayer remains without nutrient medium coverage for too long, resulting in poor cell growth and viability. Thus, the following three rotation rates were selected in the second study: 0 (T-flask stationary culture), 1/8 rpm (roller bottle culture), and 1/4 rpm (roller bottle). The disappearance of infected cells correlates with a decrease in rotation rate, further supporting the hypothesis. In the final study performed, cell growth rates and final cell densities were determined as a function of roller bottle rotation rate to evaluate if rotation rate was a critical parameter in achieving high cell densities once the cells attached to the culture vessel surface. Two rotation rates were evaluated in this study: 1/4 and 1/2 rpm. Higher cell numbers were achieved when higher rotation rates were utilized.

The higher cell densities as rotation rate is increased may be attributed to improved transport of nutrients, gases, and waste products to and from the bulk solution to and from the cell monolayer.

The results of this experiment pointed to the need to further optimize the cell proliferation steps of the process. These studies were carried out, and the resulting process has shown much greater productivity. The liquid flow modeling shows good qualitative agreement with the experimental data. In both cases, the settling dynamics are nonlinear, in which the rate at which particles are removed from the suspension starts out high and decreases thereafter. In addition, the model correctly predicted that decreases in rotation rate resulted in improved sedimentation of cells onto the cell monolayer.

REMOVING PRACTICAL PROBLEMS

Unsteady Rotation

Comparison indicates that incomplete settling is caused by the strong recirculation present in the flow, which in two-dimensional cuts of the flow field causes the flow to behave as sets of nested closed loops. When a particle is in the lower part of the flow (below the stagnation point), it falls through loops toward an outer loop. However, if the particle does not reach the wall, it then moves to the upper part of the flow and then falls toward an inner loop. The combination of these two effects causes the particle to move up and down in an closed orbit. In fact, without Brownian motion a particle can exhibit only one of two phenomena: settle to the bottom of the bottle or move in a closed orbit. Intuitively, such closed orbits can be disrupted by perturbing the flow in a time-dependent manner, perhaps resulting in enhanced settling. This hypothesis was tested by examining settling in time-periodic flows, which are the simplest and most easily controllable time-dependent flows. There are different ways of introducing time periodicity into the flow.

A simple approach is to reverse the flow direction periodically. For the creeping flow used in this study, transient effects that occur when the flow reverses direction are negligible because inertial terms in the Navier–Stokes equations are very small; therefore, the periodic flow can be simulated using two identical steady flows that are mirror images of each other, each one acting for a predetermined amount of time in an alternating fashion. Although mixing of passive tracers will not improve at all in such a periodic flow because the tracers would simply move back and forth along the same streamlines, settling of heavy particles is greatly enhanced by periodic reversal of the direction of rotation. The settling behavior of particles depended strongly on the value of the flow period T, defined as the total displacement of the bottle wall (measured in dimensionless units, where one unit is equal to half the depth of the liquid) between a pair of reversals (i.e., a flow period). For example, if the bottle completes a rotation between reversals, the flow period is two total rotations and one obtains a value of $T = 25.13$. For some values of T, complete settling can be achieved in short times even for a much smaller settling velocity than those studied for steady flow.

The particle settling rate in the periodic flows was studied by placing 6,540 particles uniformly in the flow domain. As before, particles that hit the wall were considered to be settled at the wall. Two small values of settling velocity were studied, $V_s = 0.002$ and 0.005. The particles were followed for 30 periods for each case, and the fraction of particles suspended in the flow was plotted against time. The results for $V_s = 0.005$ are shown, which demonstrates that the settling processes for all values of T are linear. At small values of T, $T < 25$, almost all particles settle very fast, and the smaller the value of T, the faster the settling. However, at $T = 25$, almost no settling occurs, and the same behavior is observed for $T = 50$. The particle settling rate increases from small T to $T = 25$ and then decreases until $T = 35$. The settling rate for $T = 35$ and 40 are about the same during most of the flow process, except at the later stage when settling in the $T = 40$ case continues at a linear rate while settling in the $T = 35$ case slows down. Therefore, the fastest settling is achieved at small T or between $T = 35$ and $T = 40$. Of those two conditions, the latter is preferable because for $T < 25$, the bottle does not complete a rotation between reversals, and therefore some regions of the bottle wall never touch the liquid. Such a condition would lead to drying and lysis of the cell monolayer in such regions of the bottle wall and therefore to reduced process efficiency.

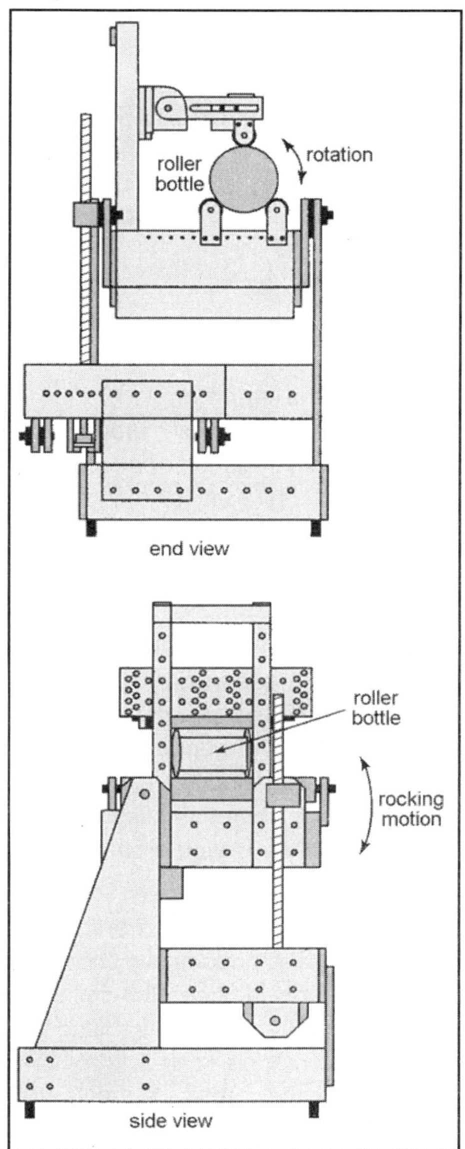

Figure 3.5 Experimental apparatus used to investigate mixing enhancement by introduction of rocking motion.

Rocking Motion

As mentioned, the flow patterns revealed by simulations suggest slow fluid mechanical mixing, especially in the axial direction. This expectation was confirmed experimentally by using a pH indicator (bromothymol blue) and an acid–base neutralization to study mixing behavior. The experiments were carried out on an apparatus that consisted of a set of rollers whose rotation speed and direction were accurately controlled by a computer. To achieve flows within the creeping regime, glycerin was used as the working fluid. The glycerin initially was mixed with the indicator and 0.5 mL 1 M NaOH, causing the glycerin to turn a deep blue color. With the roller bottle on the apparatus, approximately 20 mL of glycerin was then removed from the bottle with a syringe and mixed with 1.0 mL of 1 M HCl, turning the fluid yellow. This fluid was then injected back into the roller bottle near the back wall (the bottom of the bottle when upright) with a syringe and carefully mixed with the surrounding fluid at the base of the bottle using a specially modified spatula. After the acidic glycerin was injected back into the roller bottle, the mixer was turned on and the bottle was rotated at constant speed. After 32 revolutions, mixing within each half of the bottle is complete. This mixing results from the axial motion of fluid caused by the end-wall effects.

However, even after 64 bottle revolutions, mixing is minimal between the two halves of the bottle; neutralization of the base on the right- hand side will occur only through the action of diffusion. Recent experiments dealing with mixing of dry powders in a rotating and rocking cylinder have shown that a small-amplitude rocking motion in the vertical direction can greatly enhance mixing in the axial direction. This approach was implemented here to attempt to enhance mixing in the roller bottle. Two conditions were examined: (i) mixing at a rocking frequency of 1.6 bottle revolutions per rocking cycle and (ii) mixing at a rocking frequency of 3.2 bottle revolutions per rocking cycle. The roller bottle was rocked approximately 12 degrees at a rocking rate of 1.6 bottle revolutions per rock, and mixing improved substantially in the system as a result of the rocking. Similarly, mixing for a rocking rate of 3.2 bottle revolutions per rock; rocking indeed enhanced mixing, which is essentially complete after 64 revolutions. Moreover, because rocking disrupts the recirculating flow patterns described earlier, it is also likely to disrupt cyclical cell motions and enhance settling.

It is important, however, to realize that if rocking were used to enhance mixing in a cell growth process, the choice of rocking frequency and amplitude should take into account the need to keep the cells submerged in the nutrient medium as uniformly as possible. If an integer number of revolutions per rocking cycle were chosen, then the cells at the top of the bottle when the rocking reached its maximum angle of inclination would always be the same and would be exposed to nutrients for a substantially shorter fraction of time than those at the bottom. The rocking frequencies of 1.6 and 3.2 revolutions per rocking cycle were chosen because these frequencies should keep the cells fairly uniformly exposed to the nutrients, although the cells closer to the neck of the bottle will still be exposed less to the nutrients.

CONCLUDING REMARKS

The three-dimensional flow in a roller bottle was simulated using a finite-volume method and used to study the particle settling behavior in the bioreactor. From the steady flow simulations of cell motion in two-dimensional cuts of the velocity field, it was shown that a large number of particles in the flow remain trapped in circulating loops and never reach the walls of the roller bottle. Only for large settling velocities (i.e., $V_s = 0.2\ U$) do most particles reach the bottle walls. Thus, in many situations, poor mixing and settling of cells could occur. It was shown that settling could be enhanced by destroying the closed-orbit motion of the cells.

A simple time-dependent flow that periodically reversed the rotation direction was an effective approach for this purpose. Particles were found to settle faster for small values of the flow period *T*, but such short periods are undesirable because such conditions fail to wet the entire bottle wall, which is essential to keep the cell monolayer intact and viable. Longer periods were also shown to be effective for enhancing settling, and nearly optimal values were identified. Major limitations in mixing performance in the axial direction were caused by the presence of a vertical plane of symmetry at the center of the bottle. Such limitations, however, can be readily overcome by introducing a small- amplitude vertical rocking motion that disrupts the plane of symmetry, leading to much faster and complete axial mixing. The frequency of such motion was shown to have a significant effect on mixing rate. Additional work is needed to identify optimum values of frequency and amplitude.

4

BIOLOGICAL PRODUCTION OF GLUTAMIC ACID

Amino acid fermentation is now the key industry supplying natural amino acids on an industrial scale. The discovery of a glutamic acid producing bacterium by Kinoshita et al. in 1956 was the start of the subsequent development of amino acid production by regulation of biosynthetic metabolism. At that time, a two-step L-glutamic acid production process through reductive amination of α-ketoglutaric acid had been studied. Instead, screening of glutamic acid producing microorganisms was carried out using a glutamic acid requiring strain of *Leuconostoc mesenteroides* to directly produce glutamic acid by fermentation from carbohydrate, and glutamic acid producing bacteria were isolated from nature and some habitats with high frequency. The glutamic acid producing bacterium first discovered was reported as *Micrococcus glutamicus* in 1958, and other glutamic acid producing coryneform species, all patent strains, were subsequently isolated by many scientists. They were subjected to experiments, and industrial production began with various new species of the genera *Brevibacterium, Corynebacterium, Microbacterium,* and *Arthrobacter.*

BIOLOGICAL AGENTS

The genus *Corynebacterium* was proposed by Lehmann and Neumann in 1896 to accommodate the diphtheria bacillus. Thereafter, various isolated strains not only of pathogenic bacteria for animal and plants but also of saprophytic bacteria were identified as members of the genus *Corynebacterium*. In this way, many bacterial species with diverse properties were simultaneously classified into this genus. Under the circumstances, Jehsen introduced a generic term of *coryneform* to clarify this genus definition. The word *coryneform* was coined by Oerskov in 1923 to explain the morphology of ray fungi. Jensen used the word *coryneform* to describe one of the more characteristic morphological features, V formation, particularly on cell division of aerobic and anaerobic gram-positive, nonacid-fast, pleomorphic, nonbranching, rod-shaped, nonsporogenous bacilli. However, it was difficult to distinguish between coryneform bacteria and strains with more or less similar morphological characters, and it was not possible to clearly classify them. In the fifth edition of

Bergey's Manual of Determinative Bacteriology, the genus *Corynebacterium* belonged to the family Mycobacteriaceae of the order Actinomycetales. However, in the sixth edition (1948), the family Corynebacteriaceae was created under the order Eubacterial, and this family included the genera *Corynebacterium, Listeria,* and *Erysipelothrix*. Furthermore, in the seventh edition, the genera *Microbacterium, Cellulomonas,* and *Arthrobacter* were included in the family Corynebacteriaceae, in addition to the previously mentioned three genera.

The generic term *coryneform group of bacteria* first appeared in Part 17, "Actinomycetes and Related Organisms," in the eighth edition (1974), and the genera *Corynebacterium, Arthrobacter, Cellulomonas,* and tentatively, the genus *Kurthia* were included in this group. In addition, the genera *Brevibacterium* and *Microbacterium* were included, though as *incertaesedis*. Most of these genera were heterogeneous in composition. The genera *Listeria* and *Erysipelothrix*, which were not coryneform bacteria, were moved to the gram-positive, asporogenous, rod-shaped bacteria group in the eighth edition and treated as genera of uncertain affiliation.

Table 4.1 Systematic position of the genus *Corynebacterium* and related genera in *Bergey's Manual*

Bergey's Manual Edition (year)	*Position of the genus Corynebacterium and related taxa*
Determinative Bacteriology Fifth (1939)	Order II. Actinomycetales
	Family I. Mycobacteriaceae
	Genus I. *Corynebacterium*
	Genus II. *Mycobacterium*
	Family II. Actinomycetaceae
	Genus I. *Leptotrichia*
	Genus II. *Erysipelothrix*
	Genus III. *Proactinomyces*
	Genus IV. *Actinomyces*
Determinative Bacteriology Sixth (1948)	Order I. Eubacteriales
	Family VIII. Corynebacteriaceae
	Genus I. *Corynebacterium*
	Genus II. *Listeria*
	Genus III. *Erysipelothrix*
Determinative Bacteriology Seventh (1957)	Order IV. Eubacteriales
	Family XII. Corynebacteriaceae
	Genus I. *Corynebacterium*
	Genus II. *Listeria*
	Genus III. *Erysipelothrix*
	Genus IV. *Microbacterium*
	Genus V. *Cellulomonas*
	Genus VI. *Arthrobacter*

Bergey's Manual Edition (year)	*Position of the genus Corynebacterium and related taxa*
Determinative Bacteriology Eighth (1974)	Actinomycetes and related organisms
	Coryneform Group of Bacteria
	Genus I. *Corynebacterium*
	a. Human and animal parasites and pathogens
	b. Plant pathogenic corynebacteria
	c. Nonpathogenic corynebacteria
	Genus II. *Arthrobacter* Genera incertae sedis
	Brevibacterium
	Microbacterium
	Genus III. *Cellulomonas*
	Genus IV. *Kurthia*
Systematic Bacteriology Vol. 2 (1986)	Irregular, nonsporing, gram-positive rods
	Genus *Corynebacterium*
	Plant pathogenic species of Corynebacterium
	Genus *Gardnerella*
	Genus *Arcanobacterium*
	Genus *Arthrobacter*
	Genus *Brevibacterium*
	Genus *Curtobacterium*
	Genus *Caseobacter*
	Genus *Microbacterium*
	Genus *Aureobacterium*
	Genus *Cellulomonas*
	Genus *Agromyces*
	Genus *Arachnia*
	Genus *Rothia*
	Genus *Propionibacterium*
	Genus *Eubacterium*
	Genus *Acetobacterium*
	Genus *Lachnospira*
	Genus *Butyrivibrio*
	Genus *Thermoanaerobacter*
	Genus *Actinomyces*
	Genus *Bifidobacterium*
Determinative Bacteriology Ninth (1994)	Irregular, nonsporing, gram-positive rods
	Genus *Acetobacterium*
	Genus *Acetogenium*

Bergey's Manual Edition (year)	Position of the genus Corynebacterium and related taxa
	Genus *Actinomyces*
	Genus *Acromicrobium*
	Genus *Agromyces*
	Genus *Arachnia*
	Genus *Arcanobacterium*
	Genus *Arthrobacter*
	Genus *Aureobacterium*
	Genus *Bifidobacterium*
	Genus *Brachybacterium*
	Genus *Brevibacterium*
	Genus *Butyrivibrio*
	Genus *Caseobacter*
	Genus *Cellulomonas*
	Genus *Clavibacter*
	Genus *Coriobacterium*
	Genus *Corynebacterium*
	Genus *Curtobacterium*
	Genus *Eubacterium*
	Genus *Exiguobacterium*
	Genus *Falcivibrio*
	Genus *Gardnerella*
	Genus *Jonesia*
	Genus *Lachnospira*
	Genus *Microbacterium*
	Genus *Mobiluncus*
	Genus *Pimerobacter*
	Genus *Propionibacterium*
	Genus *Rarobacter*
	Genus *Rothia*
	Genus *Rubrobacter*
	Genus *Sphaerobacter*
	Genus *Terrabacter*
	Genus *Thermoanaerobacter*

The classification and identification of coryneform bacteria have long been difficult and confusing. As a result, new isolated strains were identified in any genera as new species. However, with the application of modern taxonomic methods, such as cell wall analysis, nucleic acid base composition, and DNA relatedness, the relationships of coryneform bacteria and traditionally associated taxa had been clarified. The definition of the genus *Corynebacterium* should now be restricted to those microorganisms that contain *meso*-A_2pm, acetylated muramic acid, mycolic acid with ca. 22 to 38 carbon atoms, and arabinose and galactose in the cell wall. In addition, these species have dihydrogenated menaquinones with either eight or nine isoprene units. In this respect, the genus *Brevibacterium* was distinguished from the genus *Corynebacterium* in that it does not contain mycolic acid and polysaccharide components such as arabinose and galactose. Furthermore, the genera *Arthrobacter* and *Microbacterium* were defined as a group of microorganisms that are additionally different in the diamino acid and major menaquinone. As a result, the genus *Corynebacterium*, defined in chemotaxonomic characteristics, includes most animal pathogenic and parasitic species belonging to section a, and some saprophytic species (e.g., *Corynebacterium glutamicum*) belonging to section c in the eighth edition. On the other hand, all plant pathogens and many saprophytic species belonging to section b were excluded from this genus. *Bergey's Manual* was newly started in 1986 with a new title, *Bergey's Manual of Systematic Bacteriology*, in which the genus *Corynebacterium* and the related genera are included in section 15, and, consequently, the term *coryneform bacteria* disappeared from the index.

STATUS OF GLUTAMIC ACID PRODUCING BACTERIA

The strain isolated as accumulating a remarkable amount of glutamic acid was initially named *Micrococcus glutamicus* nov. sp. by Kinoshita et al., with ATCC 13032 as a type strain. This strain was later revealed to belong to the genus *Corynebacterium* based on chemotaxonomic characteristics and reclassified to *C. glutamicum* by Abe et al. On the other hand, many other glutamic acid producing species, patent strains, reported as different species than *C. glutamicum*, were found to be so-called coryneform bacteria. They were also found to have very similar chemotaxonomic properties based on studies by Keddie, Bousfield, and Yamada and Komagata, and were revealed to be bacteria belonging to *Corynebacterium* in the strict sense. Furthermore, Komatsu and Kaneko showed in 1980 that eight strains, *C. acetoacidophilum* ATCC 13870, *Brevibacterium lactofermentum* ATCC 13655, *B. divaricatum* 14020^T, *B. sacchalolyticum* ATCC 14066, *B. roseum* ATCC 13825, *B. immariophilum* ATCC 14068, *B. ammoniagenes* ATCC 13745, and *Microbacterium ammoniaphilum* ATCC 15354, exhibited 73 to 102% DNA binding to the type strain of *C. glutamicum*. However, as no measures had been taken to make taxonomical changes for these strains, they were subjected to studies using these names. In 1991, Liebl et al. showed that five strains, *B. divaricatum* 14020^T, *B. flavum* ATCC 14067, *B. lactofermentum* ATCC 13665 and ATCC 13869, and *C. lilium* ATCC 15991^T were closely related to the type strain of *C. glutamicum* based on the high levels of DNA relatedness, and were thus transferred to the same species as *C. glutamicum*. *C. acetoacidophilum*, *C. acetoglutamicum*, and *C. acetophilum* have been reported to produce remarkable amounts of glutamic acid from acetic acid as a carbon source other than carbohydrate. While *C. acetoacidophilum* ATCC 13870 was confirmed to be genomically similar to *C. glutamicum*, the taxonomic positions of the other two strains are unknown. Furthermore, hydrocarbon-assimilating bacteria *Corynebacterium hydrocarboclastus* ATCC 15592, which produce a remarkable amount of glutamic acid from hydrocarbons, exhibited more than 70% DNA binding to the type strain of *Rhodococcus erythropolis*.

BIOLOGICAL CHANGES IN GLUTAMIC ACID PRODUCING BACTERIA

C. glutamicum has some characteristic features, including that it undergoes V-formation by snapping division upon cell division, and requires biotin for growth. However, glutamic acid production was observed only in a very narrow range of biotin content, that is, at suboptimal concentration (2.5–5.0 μg/mL) for cell growth. In 1962, Somerson et al. found that the addition of penicillin was very effective, even in the presence of excessive amounts of biotin. Therefore, it is widely known that in order to accumulate a remarkable amount of glutamic acid extracellularly, low levels of biotin concentration, or penicillin treatment at high levels of biotin concentration, are required. However, the detail of their mechanisms is not known, although there have been reports attributing the accumulation of glutamic acid due to the change in the membrane permeability. At a low concentration of biotin, cells swelled and elongated. According to electron microscopic observation, outer layers changed grossly and lost ordered structures, and many blebs were observed at low concentrations of biotin. It was observed that glutamic acid producing bacteria have an increased sensitivity to lysozyme or SDS treatment when cultured at low concentrations of biotin.

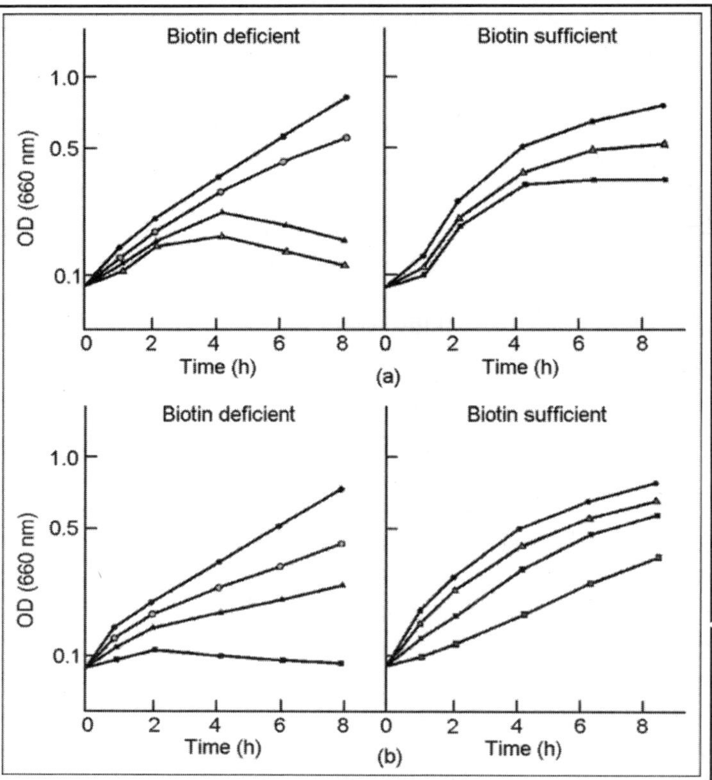

Figure 4.1 Effect of lysozyme and SDS on growth of *Corynebacterium glutamicum* ATCC 13032^T in biotin-sufficient and biotin-defficient cultivation: (a) Effect of lysozyme on the growth of *C. glutamicum* ATCC13032.

5

PROTEIN DENATURATION

The use of enzymes in organic solvents for biotransformations is now commonplace. The nonaqueous milieu permits the advantages of organic solvents as the reaction media, and enzymes provide the opportunity for high selectivity and catalytic rates under mild reaction conditions. Indeed, enzymes are fantastic catalysts in terms of effectiveness—enzyme-catalyzed reactions provide up to 10^{15}-fold rate accelerations as compared with uncatalyzed processes—and reaction specificity (e.g., enantio-, regio-, and chemioselectivity). Unfortunately, exploiting such advantages is often limited by the low stability of the biocatalysts. Even in aqueous environments, enzymes undergo denaturation that is, they loose their unique three-dimensional structure and catalytic activity. The risk of denaturation is even greater when proteins are placed into conditions much less "natural" than the conventional aqueous environment. The goal of this review is to consider the general mechanisms behind the often low stabilities of proteins in the presence of organic solvents, and to evaluate more specific mechanisms responsible for denaturation of proteins in systems with organic solvents. Special attention is given to enzymes as the most important class of proteins in biotransformations.

MECHANISM OF PROTEIN DENATURATION

Mechanisms of protein denaturation in aqueous media have been studied for decades and are mostly understood now. Denaturation of proteins is caused by various factors such as high and low temperatures, high pressures, ultrasound, high-intensity irradiation (including microwaves), addition of organic solvents, certain salts, and specific compounds such as urea. Despite the high structural diversity of proteins, their denaturation is described by very similar features. Denaturation, or more specifically inactivation of enzymes (i.e., loss of their catalytic function) can be either reversible or irreversible. For reversible denaturation, the enzyme completely recovers the native structure and catalytic activity after the denaturing action is eliminated, for example, by reestablishing room temperature after heating or by removing chemical denaturants during dialysis. Reversible denaturation is often a two-state transition between the folded native state, N, and the unfolded denatured state, U. Because structural differences between N and U are significant, concentration of both forms can be followed in the transition zone quite easily by circular dichroism or fluorescence spectroscopy, or other physical methods.

This ability to follow concentrations of both protein forms during denaturation enables researchers to obtain the thermodynamic characteristics of protein stability and quantitatively compare the stability under a variety of external conditions including organic solvents. Often after incubation under denaturing conditions (especially at temperatures higher than 50–60°C), the enzyme does not recover its native structure and catalytic activity; this is irreversible denaturation. The reason may be purely kinetic, for example, heat-denatured enzyme remains in a catalytically inactive conformation after cooling because at room temperature the protein molecule is not flexible enough to surpass the high kinetic barrier that exists between the denatured and the native states. On the other hand, in the unfolded state, the protein becomes more susceptible to aggregation and chemical changes in the structure that also lead to irreversible inactivation.

Reversible and Irreversible Denaturation

When protein is reversibly denatured, the two protein forms, N and U, are in equilibrium, which is governed by the equilibrium constant K:

$$\mathrm{N} \xleftrightarrow{\mathrm{K}} \mathrm{U},\ K = [\mathrm{U}]/[\mathrm{N}]$$

where [N] and [U] represent concentrations of the native and unfolded proteins, respectively, and are determined experimentally under denaturing conditions by measuring catalytic activity or physical characteristics of the protein. Other thermodynamic parameters of denaturation are determined from the value of K by studying the temperature dependence of denaturation:

$$\Delta G_D = -RT\,(\ln K); \qquad \Delta H_D/RT^2 = d(\ln K)/dT$$

$$\Delta S_D = (\Delta G_D - \Delta H_D)/T; \qquad \Delta c_P = d(\Delta H_D)/dT$$

where ΔG_D is the free energy of denaturation, R is the gas constant, and T is the temperature in degrees kelvin. ΔH_D is the enthalpy of denaturation, ΔS_D is the change in entropy of denaturation, which is a characteristic of the degree of protein unfolding, and Δc_p is the change in heat capacity on denaturation and is a measure of the hydration effects in a protein. This formalism also applies to denaturation of proteins in organic solvents, at least in cases when the denaturation is not complicated by irreversible processes such as aggregation. Irreversible inactivation is studied by following the rate, v_{in}, of disappearance of the native form of the protein:

$$\mathrm{N} \xleftrightarrow{k_{in}} \mathrm{I};\ v_{in} = -d[\mathrm{N}]/dt = k_{in}[\mathrm{N}]$$

where I is the inactivated enzyme and k_{in} is the rate constant of inactivation. Aliquots of solution, or suspensions with enzyme in organic solvents are sampled at time intervals, and the residual enzymatic activity is assayed in aqueous solution (or in organic solvent).

Protein Stabilization

Closely related to low stability is the problem of stabilization of enzymes. In the last two decades, numerous approaches have been elaborated to increase stability of proteins in aqueous media.

These approaches employ immobilization and cross-linking of proteins, the methods of site-directed and random mutagenesis, and covalent modification of proteins; also, stabilization is achieved by addition of low molecular weight compounds. Despite the variety of organic solvent systems containing enzymes, two of them are most important in the practical sense and for understanding protein denaturation by organic solvents. First, enzymes are often dissolved in binary mixtures of water and polar organic solvents with concentrations of the solvent as high as several dozens volume percent. Second, lyophilized or immobilized enzymes are studied when suspended in neat organic solvent or in the solvent containing as little as a few volume percent of water.

ROLE OF BIOLOGICAL CATALYSTS

Exploiting enzymes in peptide synthesis and other biotransformations demands the use of mixtures of water with polar organic solvents (i.e., alcohols and formamides, among others) as the reaction media. However, almost all organic solvents in binary mixtures with water have unfavorable denaturing effects on enzymes. Even polyhydric alcohols (glycerol and ethylene glycol, among the others), which at low concentrations are known to stabilize proteins against heat treatment, induce denaturation at concentrations higher than 50 vol %. Catalytic activity of enzymes dissolved in mixtures of water with a polar organic solvent are described by a characteristic profile. Addition of cosolvent to aqueous solution of enzyme up to some critical concentration does not change or increase enzymatic activity. This activation could often be explained by more favorable solvation of the polar transition state of enzymatic reaction by water–cosolvent mixture than by water. However, beginning with this critical concentration of a cosolvent, enzymatic activity drops (often reversibly) in a threshold manner, reaching a very low or even zero value.

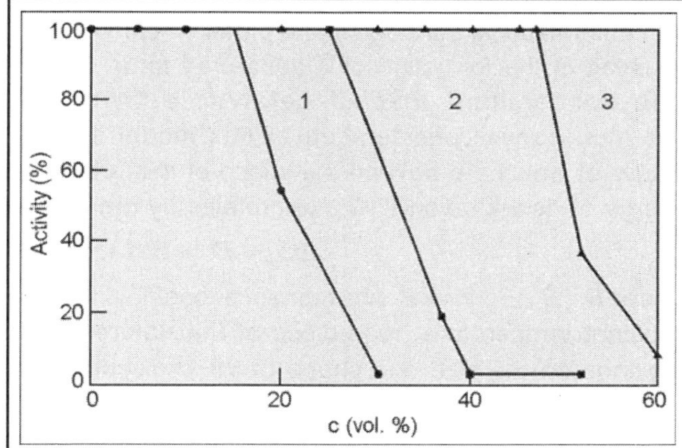

Figure 5.1 Dependence of relative catalytic activity of dissolved enzymes on the concentration of organic solvents in binary mixtures with water. 1, laccase in water-acetone mixture; 2, α-chymotrypsin in water-ethanol mixture; 3, trypsin in water-acetonitrile mixtures.

This threshold decrease in activity is accompanied by an abrupt change in spectroscopic or other physical properties of the dissolved enzyme, and this fact has been explained by a conformational transition in the protein molecule. The fact that the plummet in enzymatic activity occurs at the same concentration as the spectral changes do, implies that it is the denaturation that causes the threshold inactivation of the dissolved enzymes.

Mechanistic description of this inactivation is based on the important prerequisite that the protein molecule in aqueous solution is surrounded by a shell formed by water molecules that are bound by hydrogen bonds to the protein surface. According to a widely accepted notion, this hydration shell, or at least a portion of it, represents an integral part of the protein and is essential for protein structure and function. Consequently, displacement of bound water molecules by an organic solvent results in a dramatic change of the whole structure and leads to protein denaturation. In line with these ideas, reversible inactivation of enzymes in water–cosolvent mixtures consists of three major steps: (i) dehydration of the protein molecule, that is, removal of a certain critical amount of water molecules from the protein hydration shell; (ii) binding of the organic cosolvent by the partially dehydrated protein molecule; and (iii) conformational transition in the protein molecule, resulting in the formation of a denatured form, D, with concomitant loss of catalytic activity. Thermodynamic consideration of this model the details of which are beyond the scope of this article, leads to the following expression for the free energy of denaturation (ΔG_D) of proteins by organic solvents:

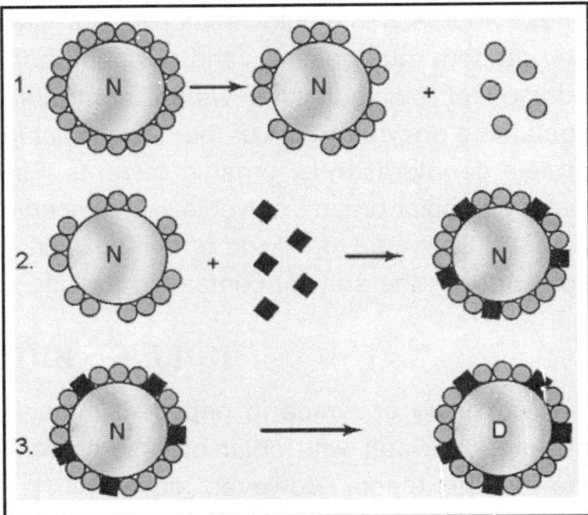

Figure 5.2 Schematic representation of major molecular steps involved in reversible protein denaturation by organic solvents.

$$\Delta G_D = B_0 + B_1 n + B_2 E_T(30) + B_3 n \log P$$

where B_0, B_1, B_2, and B_3 are numerical coefficients, which remain unchanged for a given protein at constant temperature, regardless of the nature of the solvent; n is a numerical parameter that depends on the size and shape of the molecules of organic solvents and gives an idea of how many molecules of water can be displaced from the protein surface by one molecule of the organic solvent; $E_T(30)$ is the Dimroth–Reichardt parameter, which is directly related to the free energy of the solvation process of the organic cosolvent; P, the partition coefficient of the solvent in a water–octanol biphasic system, is a measure of organic solvent hydrophobicity.

The values of $E_T(30)$ and P for different solvents are well documented, and the parameter n can be calculated from the information on the organic solvents contained in comprehensive handbooks on organic and physical chemistry.

Equation 1 predicts that cosolvents having high values of $E_T(30)$ and $\log P$ possess greater denaturing ability. In fact, the solvents such as 1,4-dioxane, tetrahydrofuran, and higher alcohols (e.g., isomers of butanol), which are both hydrophobic and have a high solvation capacity, appear to be strong denaturants and cause enzyme inactivation at concentrations as low as 10–30% (v/v). Inversely, Equation 1 predicts that hydrophilic solvents, such as glycerol, ethylene glycol, and formamide, have a low denaturing capacity.

Indeed, it has been experimentally confirmed that many enzymes are not inactivated by these solvents at concentrations as high as 50–60 vol %. Thus, solutions with high concentrations of these solvents are convenient as a media for enzyme-catalyzed reactions carried out in binary water–cosolvent mixtures.

Role of Organic Cosolvents

Molecular events behind enzyme inactivation in water– cosolvent mixtures also prompt ideas of how to stabilize enzymes against this denaturation. Because protein unfolding is an indispensable step in the inactivation by solvents, one could try the approaches recommended as best for deceleration of protein unfolding in aqueous solutions. By using the method of covalent immobilization, proteolytic enzymes were stabilized against denaturation by alcohols and dimethylformamide. The stabilization effect expressed as an increase in the threshold concentration at which enzymatic activity dropped abruptly, rose with the number of covalent bonds between the enzyme and support. Stabilization against denaturation by polar organic solvents has been also obtained when noncovalent complexes of enzymes with polyelectrolytes were formed due to multiple electrostatic interactions between oppositely charged groups in the interacting macromolecules.

The delay in activity decrease in this case was also accompanied by a shift in the concentration of organic solvents, at which point significant structural changes in the protein molecule (unfolding) occurred. Probably the same reason (i.e., increase in rigidity of the protein molecule in water–cosolvent mixtures) underlies stabilization of enzymes by chemical cross-linking with bifunctional reagents, as was the case with cross-linked crystals of thermolysin in dimethylformamide. Equilibrium between the native and denatured states of a protein can be efficiently shifted toward state N by addition of substrates and specific ligands that are preferentially bound with the native enzyme. Due to this reason, enzymes are often more stable in organic solvents in the presence of saturating concentrations of their substrates. Another productive idea for stabilization of enzymes in water–cosolvent mixtures is to strengthen the binding of the hydration shell to the protein surface. Protein denaturation includes the step of dehydration, namely, dissociation of essential water molecules from the hydration shell. It has been assumed that if water molecules are more tightly bound with the protein, denaturation could be decelerated. Indeed, introduction of polar and charged groups into the protein by covalent modification led to a significant increase in the concentration of organic solvents that caused denaturation of α-chymotrypsin.

Stabilization was observed when enzymes were modified with low molecular weight compounds (pyromellitic anhydride, glyceraldehyde) and poly(ethylene glycol) (PEG). This stabilization was explained by the ability of the newly attached hydrophilic (carboxylic and hydroxylic) groups to hold water, thus preventing protein dehydration by organic cosolvent molecules. Tighter binding of the hydration shell could also be achieved due to replacements of the proteins' amino acid residues by site-directed or random mutagenesis. Recent studies from Arnold's and Wong's groups showed that such replacements increased the stability of proteases in the presence of a strong denaturant, dimethylformamide.

The stabilization was explained by numerous reasons, such as enhancement of internal hydrogen bonds and ionic forces, optimization of internal hydrophobic interactions, reduction of solvation energy, and introduction of conformational restrictions to the protein molecule. Also it is possible that chemical modification or replacement of amino acid residues on the protein surface both change the protein's affinity to organic solvent molecules and, decrease the protein's sensitivity to denaturation. A stabilizing approach recently developed by Arnold and Wong suggests increasing the rigidity of a protein by introducing amino acid residues (e.g., histidines) that are able to specifically bind with metal ions. In fact, in organic solvents, catalytic activity and stability of such metal-chelated proteases was significantly higher than in their nonmodified analogs.

Precipitation Techniques

It has been commonly noticed that the initially reversible denaturation of proteins by organic solvents is further accompanied by irreversible inactivation during long (a dozen minutes to hours) incubation. This irreversible inactivation becomes especially important when the concentration of cosolvent is higher than the denaturation threshold. Proteins may even precipitate if their concentration in water–cosolvent solution is of the order of micromoles per liter or lower; normally these concentrations are used to detect catalytic activities. The general features of this inactivation were described by Singer, who found that proteins lose only a part of their native structure in organic solvents—the denatured protein retains a large portion of secondary structure, although the tertiary structure is destroyed. This conformation, analogous to the recently discovered molten globule state, turns out to be very "sticky," and proteins have a strong tendency to form inactive aggregates due to formation of interprotein hydrogen bonds and hydrophobic interactions. However, even in the insoluble form, enzymes may show a small, though detectable, activity in the presence of high concentrations (60–80 %, v/v) of polar solvents. For example, such an activity was detected when solid α-chymotrypsin was added to a water–cosolvent mixture. On the other hand, the enzyme does not show any activity when the precipitate is formed after an aqueous solution of the enzyme is added to the organic solvent. This difference does not seem too surprising in view of the recent fact that the enzyme directly suspended in water–cosolvent mixture contains a higher percentage of the native secondary structure than the enzyme first dissolved in water and then mixed with the solvent to form a precipitate.

Most likely, in the former case, only the enzyme molecules forming the surface layer of particles are inactivated due to direct contact with the organic solvent, whereas the enzyme molecules inside the particles are protected from denaturation by interprotein contacts. Oppositely, in the latter system, enzyme molecules are first denatured and then aggregate, giving rise to inactive precipitates. This inactivation caused by aggregation has been successfully combated by protein immobilization or by modification with polymers such as PEG; both are highly efficient in preventing interprotein contacts. Aggregate formation and protein precipitation can also be delayed by chemical modification of proteins with low molecular weight compounds (for example, charged reagents) or by replacement of amino acid residues on the protein surface through mutagenesis. Sometimes these modified enzymes remain catalytically active without noticeable precipitation in the significantly wider range of cosolvent concentrations (by 30-40 vol %) than their nonmodified counterparts.

BIOLOGICAL CATALYSTS IN ANHYDROUS MEDIA

Neat organic solvents or solvents containing only a few volume percent of water are very convenient media for many enzymatic reactions. Most proteins are insoluble or sparingly soluble in neat organic solvents and most often are used as suspensions of freeze-dried protein powders. These systems have found numerous applications in chemical processing, food-related conversions, and analyses.

Table 5.1 Potential advantages of enzymatic processes in nonaqueous media

1. Increased solubility of apolar substrates.
2. Enhanced thermostability.
3. Easier product recovery from low-boiling sovents.
4. Immobilization is often unnecessary because the enzyme is insoluble.
5. Easier recovery of the enzyme by simple filtration or centrifugation.
6. If immobilization is required for better performance, simple adsorption onto nonporous surfaces may be sufficient. Enzymes do not desorb from these surfaces in nonaqueous media.
7. Alteration in substrate specificity.
8. Suppression of water-dependent side reactions, e.g., nonenzymatic blank reaction in enzymatic hydrocyanations ofaldehydes, which leads to racemic product.
9. Shifting thermodynamic equilibria to favor synthesis over hydrolysis, e.g., esterification and peptide formation.
10. The possibility of carrying out reactions with other neucleophiles, which would not be feasible in water, e.g., transesterification and ester aminolysis.
11. Easier integration of enzymatic with conventional chemical steps in a multistep synthesis (chemoenzymatic processes).

The main problem that still remains with nonaqueous enzymology is a rather low catalytic activity of enzymes in anhydrous media in comparison with activity in water. Many factors such as diffusional (mass transfer) limitations, sterical shielding of the enzyme's active site from the access by a substrate, and unfavorable energetics of substrate desolvation may contribute to this decrease in activity. Here we focus on aspects of the problem that are related to the stability of enzymes.

Lyophilization

It has been established that many proteins are denatured during lyophilization. This denaturation can be either reversible (lyophilized proteins regain their native three-dimensional structure and function when placed in aqueous solution) or irreversible. Especially sensitive to irreversible inactivation during lyophilization are oligomeric proteins, which often dissociate into individual subunits. Carpenter and coauthors have found that enzymes can lose their activity both at freezing and at drying, which constitute the lyophilization protocol, and that denaturation at these steps follows essentially different mechanisms.

Temperature stresses

This process includes a number of "stresses" such as protein exposure to low temperatures and ice, contact with the test tube surface, concentration of the dissolved compounds, and eutectic crystallization of solution components, and each of these stresses may decrease enzyme activity. For example, eutectic crystallization from a buffer-containing solution may result in dramatic alterations in pH, for example, up to 3 pH units. Such changes largely influence the catalytic activity of enzymes, which is very sensitive to the ionization state of the enzyme's functional groups participating in catalysis. As shown for subtilisin Carlsberg in anhydrous media, the enzyme activity increased more than 300 times upon changing the pH of the aqueous solution (from which the enzyme was lyophilized), from 5 to 11. However, addition of compounds such as polyvinylpyrrolidone and bovine serum albumin (BSA) during freezing is known to prevent such large shifts in pH and may be helpful to increase the activity of enzymes in organic solvents. Inactivation of enzymes in the freezing process is dominated mainly by unfolding of the protein molecule.

Protection of labile proteins during freezing may be achieved by adding a variety of solutes, such as polyols, sugars, amino acids (normally at concentrations of a few hundred millimolar), synthetic polymer PEG, and a protein BSA, effective at concentrations of 1%(w/w) or smaller. This cryoprotection is explained by the mechanism of preferential hydration of proteins. Sugars, polyols, and some other compounds are preferentially excluded from contact with the surface of a protein because their interaction with a protein's functional groups is unfavorable in comparison with interactions of the same compounds with water molecules. Consequently, in the presence of these compounds, the protein is preferentially hydrated, that is, the hydration shell becomes more tightly bound to the protein molecule. This leads to an energetically unfavorable situation as chemical potentials of both the protein and the additive increase. As a result, the compact native structure is stabilized by the additives because denaturation (unfolding) would lead to a greater surface area of contact between the protein and the solvent and would increase this thermodynamically unfavorable tendency.

Dehydration

The frozen protein sample is already significantly dehydrated due to formation of ice from the bulk water. However, an essential amount of water comprising the protein's hydration shell (several hundreds of water molecules per protein of molecular weight of 20–25 kDa) remains unfrozen and is removed only by sublimation under high vacuum (i.e., by lyophilization). How much water remains bound to a protein after lyophilization is a question that is not resolved. Most likely, only a few water molecules are bound by several hydrogen bonds to polar groups in the protein's interior, whereas almost all water on the protein's surface is stripped.

This latter fact leads to the appearance of a great number of polar groups with the noncompensated ability to form hydrogen bonds, which is satisfied by the formation (mainly intermolecularly) of β-sheet structures because the β-structure allows for the highest degree of hydrogen bonding in proteins. Due to low molecular flexibility in anhydrous media, these structural changes in freeze-dried enzymes are conserved, and the catalytically active conformation is not completely recovered after the enzyme is placed in organic solvent. Actually, active site titration of chymotrypsin in organic solvents has shown that about 40% of the active enzyme was destroyed at the drying step.

Excipient effects

Stabilization during the lyophilization procedure demands the protection of proteins against inactivation at both the freezing and drying steps. As discussed in the previous section, many additives are effective as protectors during freezing. Fewer compounds are known to protect enzymes against dehydration stress during drying; sugars are one example. Due to several hydroxyl groups, sugars substitute for molecules of water to form multiple hydrogen bonds with polar groups on the protein surface, thus preventing alterations in the protein's secondary structure. Especially successful are combinations of several additives. For example, PEG was shown to protect lactate dehydrogenase and phosphofructokinase from inactivation during freezing, but not to stabilize these proteins during drying.

This is not surprising because PEG crystallizes during freezing and is then not able to form the hydrogen bonds with proteins that are necessary for protein protection at the drying step. However, in the presence of 1% PEG and small amounts of glucose or trehalose, which alone provide only minimal protection, the activity of both lyophilized enzymes recovers almost completely after rehydration in aqueous solution. In this case, PEG stabilizes the proteins by the mechanism of preferential hydration during freezing, while sugars act as water substitutes during drying because of their hydrogen bonding with the proteins. However, protection during freeze-drying is not the sole mechanism responsible for higher catalytic activity in organic solvents of enzymes lyophilized with excipients. For example, several proteases were shown to be activated a thousandfold in anhydrous media when they were lyophilized with an excess of KCl or other salts.

Although certain salts are known to be good cryoprotectants, the activation of enzymes in that case was shown to be due to some other effect than protection during freeze-drying. The most plausible explanation, suggested by Khmelnitsky et al., is that the activation was due to stabilization of the reaction transition state by a polar matrix microenvironment. Inactivation at the steps of freezing and drying can, however, be avoided if a method other than lyophilization is used for preparing enzyme catalysts. Vacuum evaporation of aqueous solutions of enzymes or precipitation by acetone did not lead to better catalytic activities in anhydrous media. Much higher activities have been obtained with crystals of enzymes.

In a crystalline microenvironment, additional polar and hydrophobic contacts between neighboring protein molecules significantly increase stability by preventing unfolding, aggregation, or dissociation in the case of oligomeric proteins. "Mild" cross- linking with bifunctional chemical reagents makes the crystals more solid (cross-linked enzyme crystals) and also increases protein stability. One more way to avoid lyophilization is to obtain noncovalent complexes of enzymes with surfactants, which exist due to ion-pairing between the molecules of surfactant and enzyme. Such complexes are quite soluble in nonpolar solvents (up to concentrations of several milligrams per milliliter) and possess very high catalytic activities. For example, the value of k_{cat}/K_m for transesterification of *N*-acetyl-L-phenylalanine ethyl ester catalyzed by the complex of subtilisin with Aerosol OT dissolved in octane was equal to 70% of the value for the hydrolysis reaction in water.

Factors Affecting the Activity of Biological Catalysts

In a dry state almost all enzymes are inactive; they need a small quantity of water to display catalytic activity. This water provides enzymes with conformational mobility, which is necessary for catalytic function. Because of this, lyophilized (dry) enzymes are often preequilibrated with small amounts of water before starting reactions in organic solvents. However, contact with anhydrous solvents may lead to significant dehydration of enzymes ("stripping" essential water of the protein) and inactivate them. This tendency is especially strong with polar solvents. Some of them were shown to dehydrate enzymes on a very rapid time scale—5 min or faster. The driving force for dehydration is the higher affinity of water to molecules of polar solvent than to the protein's functional groups. The interaction energy for a hydrogen bond between water and dimethyl sulfoxide was calculated as equal to 8.8 kcal/mol; the solvent interaction with an amide H has the same energy, and both are much stronger than hydrogen bonds between water and a carbonyl O or an amide H and between a backbone carbonyl O and amide H.

These data explain why dimethyl sulfoxide, as well as other polar solvents, has the tendency to strip water molecules from the protein surface and form new hydrogen bonds with protein functional groups. On the other hand, contact with nonpolar solvents leaves the major portion of the protein molecule hydrated and causes only small rearrangements of side chains on the protein surface. Effects of polar solvents on the structure of suspended enzymes depend mainly on the nature of the solvent and on enzyme stability. For example, acetonitrile, butanol, ethyl acetate, and tetrahydrofurane, each containing 1% (v/v) of water, produced few changes in the structure of bovine pancreatic trypsin inhibitor in addition to those already brought about by lyophilization. On the other hand, this protein was strongly denatured by methanol, dimethyl formamide, and dimethyl sulfoxide— solvents known for their ability to dissolve proteins. Contact with polar solvents is accompanied by significant changes in the protein's hydration. For example, 4% (v/v) isopropanol added to crystalline chymotrypsin in hexane nearly doubled the number of water molecules observable in the protein's crystal structure. These additional water molecules did not contact directly with the protein's surface groups but were integrated within the second hydration sphere of the protein.

Structural changes in proteins caused by organic solvents have been successfully inhibited by the simple addition of low molecular weight compounds during lyophilization. For example, lyoprotectors such as ammonium sulfate or sucrose did not only stabilize enzymes during lyophilization but were quite efficient in protecting enzymes from inactivation by organic solvents, probably by maintaining water in the enzyme's active site. Polar solvents may significantly increase molecular mobility of proteins in nonaqueous media by eliminating flexibility constraints imposed by protein–protein contacts. It has been shown that the rates of interconversion between several conformations of a peptide, gramicidin A, in an anhydrous milieu were increased in protic solvents due to a faster exchange between hydrogen bonds.

Molecular dynamic simulations also showed that subtilisin is much more flexible in polar versus nonpolar solvents. Probably, such an increase in molecular flexibility by added polar cosolvents (dimethyl sulfoxide and formamide) is responsible for the more than hundredfold acceleration of the reactions catalyzed by subtilisin in neat nonpolar solvents, although this explanation has been disputed. Recent studies with Fourrier-transform infrared spectroscopy have pointed out that enzymes conserve a much more native-like secondary structure in neat solvents than they do in water-cosolvent solutions with a solvent concentration of 60 vol %.

The explanation is that protein denaturation (unfolding) is significantly inhibited by the more severe conformational restrictions imparted on the enzyme molecules in anhydrous versus water-cosolvent media. This may explain why enzymes are often more catalytically active in neat organic solvents than in water-cosolvent mixtures, that is, why the addition of water is detrimental rather than beneficial for stability of the native enzyme's conformation.

Temperature Resistance in Anhydrous Media

The fact of the enormously high stability of lipase suspended in organic solvents was mentioned in one of the first works in nonaqueous enzymology: this enzyme retained high activity after several hours of incubation in anhydrous media at 100°C. Further studies confirmed that high thermal stability is a hallmark of proteins in nonpolar solvents, as opposed to their low stability in aqueous media; for such comparison, see the recent review by Zaks. For example, the melting temperature for proteins suspended in nonpolar solvents is 110–145°C, which is 50–70°C higher than in aqueous solutions. Thermal stability of ribonuclease A was nearly the same when measured in either a free-of-solvent state or hydrophobic solvents. The explanation was that the conformational flexibility of proteins is significantly retarded in systems with a small water content, as the latter acts as a lubricant in all macromolecular motions, owing to its high propensity to form multiple hydrogen bonds with a protein.

Water removal also increases the strength of intramolecular hydrogen bonds and salt bridges, due to the higher energy of electrostatic interactions in the media with low dielectrics. All this creates a high kinetic barrier between the native and unfolded states and stabilizes native conformations of proteins. Also molecular mobility is suppressed by protein–protein interactions, which exist in multitude in lyophilized and crystalline protein samples, where the packing density of macromolecules is very high. Elimination of water decelerates almost all detrimental chemical reactions, which are also responsible for irreversible inactivation of proteins in nonaqueous media. For example, a very slow inactivation of proteins in nonpolar solvents at temperatures higher than 100°C is caused by the formation of aggregates cross-linked by intermolecular reactions of transamidation and disulfide exchange. On the other hand, proteins are significantly less thermostable in polar versus nonpolar solvents. The reason may be that in hydrophobic solvents a protein has more intraprotein hydrogen bonds, which make the protein molecule more rigid and more resistant to high-temperature unfolding. Thermal stability of proteins in organic solvents often declines on addition of even small amounts of water, which are quite sufficient to revive molecular dynamics in proteins. An almost linear decrease in the melting temperature has been observed as a function of the amount water in anhydrous media. This stability decrease was explained by stabilization of the unfolded protein state, rather than destabilization of the folded native state. High-temperature inactivation in anhydrous media is accompanied by migration into the solvent of water bound to protein; it was suggested that this dehydration could be the rate-limiting step of the whole inactivation process.

The tendency to dehydrate proteins essentially depends on the hydrophobicity of organic solvents. Molecular dynamic simulation has shown that more water molecules leave the surface of a protein placed in dimethyl sulfoxide than in the acetonitrile simulation, whereas no water at all leaves the protein in carbon tetrachloride. Finally, protein thermal stability in anhydrous solvents significantly depends on conditions of lyophilization, such as pH of the buffer or presence of excipients. For example, the complex of chymotrypsin with sorbitol denatures in organic solvent at much higher temperatures than the enzyme in the absence of sorbitol.

CONCLUDING REMARKS

The problem of low activity and stability of enzymes in organic solvents led biochemists and biochemical engineers to the necessity of protein stabilization, and some recent advances were discussed in this review. A variety of methods have been applied, including familiar immobilization and chemical modification, and recently designed genetic procedures, which have become routine in the last decade. Essentially, not only protein stability is increased by methods of enzyme engineering, but also a variety of other important enzyme properties are changed, such as pH profiles of catalytic activity and stability, substrate specificity, including chemio-, regio- and enantioselectivity, and others. It is a task of the next decade to learn how to tailor in the most rational way the whole combination of enzyme properties that are necessary for successful utilization of enzymes in biotransformations.

6

BIOPROCESSING OF PULP

The use of enzymes in the pulp and paper industry started in the late 1980s and has grown rapidly since then. This growth has been fueled by several factors, including (i) an improved understanding of the interactions between enzymes and the constituents of pulp and paper processing, (ii) an increased need for the industry to adopt environmentally benign technology, and (iii) development of cost-effective production technology for the relevant enzymes. The pulp and paper industry processes wood and other cellulose-containing fiber, such as straw. Typically, pulp is made by processing the fiber feedstock, and paper is made by combining one or more types of pulp with various non-pulp additives designed to achieve the desired properties, such as smoothness, opacity, or color. Enzymes are used to aid pulp and paper manufacturing by modifying the following types of substrates:

1. The cellulose fiber
2. Noncellulosic constituents of the fiber, such as hemicellulose and lipids
3. Paper additives, such as starch
4. Contaminants, such as slime

This article provides general background on pulp and paper processing and describes the use of enzymes in each of the processing areas.

MECHANISM OF PULPING

Although these processes have common elements, the differences are sufficiently large that enzymes are used for different purposes.

Mechanical Process

Newsprint and other bulk papers are made primarily by mechanical means, that is, the raw fiber is processed primarily by mechanical action. First, the fiber feedstock is cleaned and chipped to uniform size. The raw fiber is then torn apart by mechanical means in the pulping. Closely associated with the pulping is refining, which improves the strength of the pulp. Enzymes are used to improve the de- watering and refining of mechanical pulps. After refining, the pulp is mildly bleached, usually with peroxide. Enzymes are not used to aid in the bleaching of mechanical pulps, although enzymes are used in chemical pulp bleaching. After bleaching, the pulp is ready for papermaking. Mechanical pulp is produced at lower cost than chemical pulp, but the strength and surface properties are not suitable for more demanding applications, such as writing paper or wrapping paper.

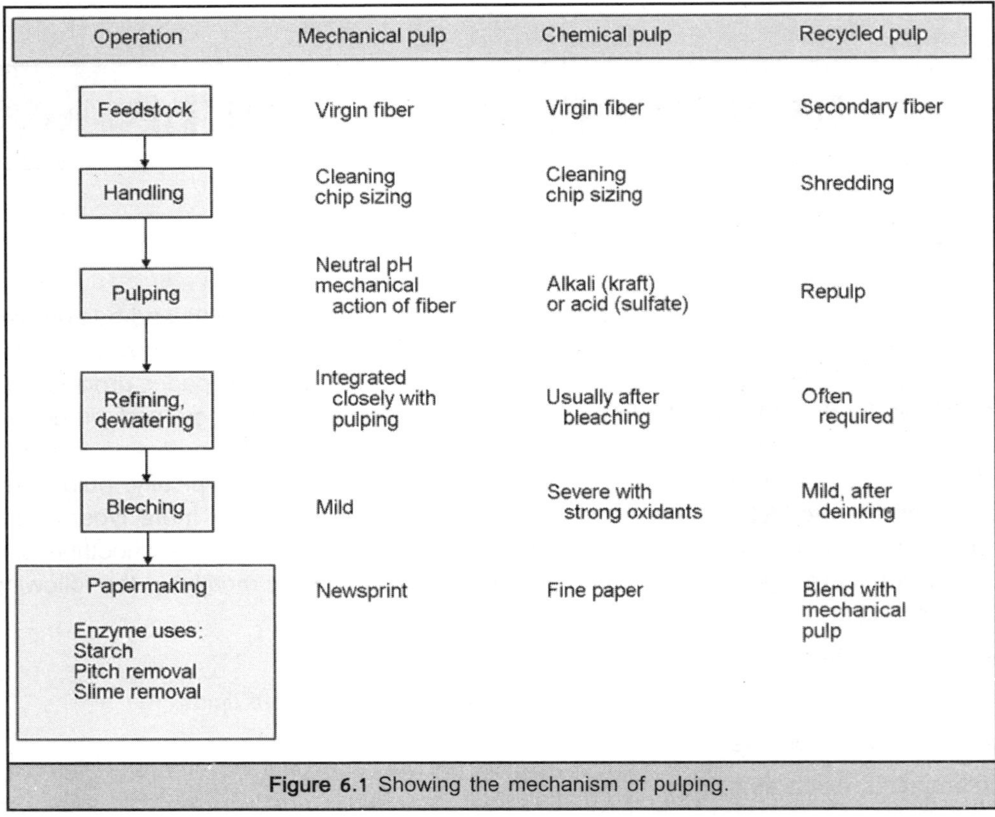

Figure 6.1 Showing the mechanism of pulping.

Chemical Process

These higher performance papers are made from chemical pulps, most notably by cooking in alkali (kraft pulp) or in acid (sulfite pulp). For chemical pulps, the fiber source and initial handling are similar as for mechanical pulps. However, the chemical pulping processes remove impurities, principally lignin, from the fiber, which improves the strength and makes the surface more uniform. After pulping, bleaching is carried out at harsher conditions and with more aggressive oxidizing chemicals than used for bleaching mechanical pulps; the resulting product is very white. Enzymes are widely used to decrease the usage of these bleaching chemicals, particularly chlorine gas and other chlorine-based oxidizing chemicals. After bleaching, the pulp is refined (the reverse of the order for mechanical pulp). Data suggest that enzymes can enhance the refining of chemical pulps, but this is not as widely reported as with mechanical pulps.

Recycling Process

A third source of pulp that is becoming more prominent involves the use of recycled paper as the feedstock. This is known as secondary fiber processing. After mechanical shredding and repulping, the ink is removed from the pulp, a process known as deinking. There are laboratory data and limited mill results that show benefits of enzyme use in deinking. The deinked stock is then lightly refined and blended with pure mechanical pulps for paper production. Enzymes are also reported to be beneficial in the refining of secondary fiber.

Production of Papers

Regardless of the pulping process, the finished pulp is combined with inorganic fillers, paper additives to improve properties, and process additives to improve the performance of the paper machine. The pulp and additives are formed into sheets and dried on a paper machine. Enzymes are used in papermaking to modify the starches used in paper sizing, to remove pitch, and to prevent slime accumulation. Although an important benefit of enzyme use in the industry is a decreased amount and toxicity of effluent discharges, there is little use of enzymes for direct treatment of waste streams. The environmental benefits of enzyme use in the industry are mentioned in the following relevant process sections, and each processing area and the associated use of enzymes is described in more detail.

ROLE OF BIOLOGICAL CATALYSTS IN PRODUCTION OF PULP

Woody and Non-woody Fibers

Historically, wood has been by far the most prominent cellulose-containing feedstock for the industry. Numerous species of hardwood and softwood are used around the world. Non woody fibers such as bamboo, kenaf, and straw are also widely used, particularly in China and tropical areas. Regardless of the fiber source, a certain amount of physical handling is carried out to obtain cleanliness and particle size suitable for pulping. This processing takes the form of debarking, washing, and screening, to remove the foreign matter, and mechanical chipping for most pulping processes. Nonwoody fibers require removal of seeds and other nonfiber constituents.

In virtually all cases, the intact fiber structure is not very susceptible to enzyme attack. Consequently, there have been few laboratory scale, let alone commercial, enzyme processes involving treating intact fiber. There has been much research focused on the use of whole cells to digest wood chips. This is targeted particularly to remove lignin to improve the subsequent pulping and bleaching and to remove pitch by digesting lipids, of which at least one commercial product is available. Whole cell applications on intact fiber are plagued by the long (several day) reaction times required, the extreme variability in day-to-day chip storage times, and the vulnerability of the microbes to subfreezing temperatures. Whole cell applications will not be discussed further in this article.

Process of Pulping

Pulping is the process by which the macroscopic structure of raw wood fiber is broken apart, rendering the fiber much more pliable and thereby suitable for papermaking. Two general classes of pulping operations are chemical pulping and mechanical pulping. These operations will be discussed in terms of wood processing, although the same operations apply to the processing of nonwoody fibers.

Chemical pulping

In *chemical pulping*, the fiber is heated in the presence of chemicals that dissolve a portion of the fiber, particularly the lignin. Today, most of the chemical pulp is from the alkali or kraft process, in which wood chips are cooked at 160 to 190°C (320–374°F) for about 3 h in a concentrated liquor of sodium hydroxide and sodium sulfide. After the alkali cook, the pulp is a slurry with the liquor; the cooking liquor is reused through an extensive recovery process. The kraft cooking treatment greatly increases the cellulose content of the fibers.

Table 6.1 Most common chemical pulp composition

Constituent	*Feedstock*	*Pulp*
Cellulose	60	85
Hemicellulose	15	10
Lignin	20	4
Miscellaneous: resin, ash, etc.	5	1

The other significant chemical pulping process is acid (sulfite) pulping. Sulfite pulping is particularly prevalent in central Europe. In sulfite pulping, the wood chips are cooked in sulfurous acid analogous to the alkali liquor of kraft pulp. The cellulose content is higher than that of kraft pulp, but the resulting pulp is not as strong. Sulfite pulp is used primarily as a cellulose chemical feedstock. The conditions of chemical pulping are too harsh to be tolerated by commercial enzymes, so there has not been significant attention paid to enzymes in this area. The most significant use of enzymes in chemical pulp has been in conjunction with bleaching rather than pulping.

Mechanical pulping

In mechanical pulping, the fiber is refined by rotating plates; today, this is usually carried out at temperatures of about 140°C (284°F) and designated as thermomechanical pulping. The power input onto the fiber and the spacing between the plates that the pulp is forced through determine the severity of the pulping. Refining is a series of compressions and decompressions that gently separate and fibrillate individual fibers. This enhances pulp strength relative to the older stone ground-wood process, which involved coarse grinding on a large stone. As with chemical pulp, the conditions of mechanical pulping are too harsh for enzyme treatment. However, enzyme treatment of mechanical pulp has important implications in the pulping and refining, as described in the next section.

Dehydration and Refining

One process parameter that is of primary interest in mechanical pulping and in secondary fiber processing is the rate of drainage of water from the pulp. This is important because the subsequent rate of water removal on the paper machine is often rate limiting in mill production throughput. The presence of fine particles in the pulp markedly decreases the rate of drainage. Typically, pulps consist of about 12% to 15% fines that pass through a 200-mesh screen; particle size is usually based on a Bauer-McNett classifier. Secondary fiber tends to have particularly high levels offines, and the level offines increases with repeated recycling of stock. Refining, which is required in many cases to increase the strength of the pulp, also creates fines that decrease the rate of drainage. Mixtures of cellulase enzymes and hemicellulase enzymes are used to increase the drainage rate of mechanical pulp and secondary fiber. The speed of the paper machine can then be increased. Alternatively, additional refining or pulping can be carried out to obtain pulp of increased strength with the original rate of drainage.

Table 6.2 Pulp dehydrating by enzyme

Parameter	*Control*	*Enzyme-treated*	*Refine + enzyme*
Freeness (mL CSF)	192	274	156
Burst (kPa m^2/g)	1.83	1.84	2.07
Corrugation (newtons force)	129	124	141

The fines consist largely of cellulose and hemicellulose and, relative to the remainder of the pulp, have a high surface area for enzyme attack. There is some evidence for three distinct actions by the enzymes: (i) a portion of the fines are solubilized, (ii) there is a flocculation of the particles, and (iii) there is a cleaning of the small fibrils on the fiber surface. All these would tend to increase the drainage of the pulp. A commercial product of Nalco (Naperville, Ill.) that is used to aid pulp dewatering is a mixture of cellulase and hemicellulases from the fungus *Trichoderma*. The enzyme is supplied in liquid solution and is pumped into the storage chest before the pulp machine headbox or, in the case of secondary fiber, just after repulping.

The pulp is diluted in water (2 to 4% solids consistency) in a stock chest, and the enzyme is simply added to this slurry; the flow and mixing are sufficient to disperse the enzyme into the pulp. The process pH of 5 to 7 is in the optimal range for the enzyme; the process temperature of 35 to 50°C (95 to 122°F) is adequate. The retention time is at least 30 mins, and the enzyme dosage is 1.5 L (–1/2 U.S. gallon) per ton of pulp. The drying or refining of the sheet destroys the enzyme, preventing further action on the pulp. Careful control of the enzyme dosage is important, excessive enzyme treatment can weaken the pulp because of cellulose hydrolysis. In addition, excessive fines creation or destruction can affect the opacity of the sheet.

Oxidizing and Bleaching

The process by which pulp is brightened or made completely white with oxidizing chemicals is bleaching. Enzymes are used to aid in the bleaching of kraft pulp. In most cases, mechanical pulp is only mildly bleached with hydrogen peroxide and sodium hydrosulfite. There are few effluent problems with these bleaching chemicals. The major usage of bleaching chemicals and associated effluent problems are with kraft pulp. In bleaching kraft pulp to produce bright white finished pulp, the residual lignin is virtually removed. Bleaching is carried out as a continuous process. The pulp is reacted with oxidizing compounds that preferentially attack the aromatic rings on the lignin while damaging the cellulose as little as possible.

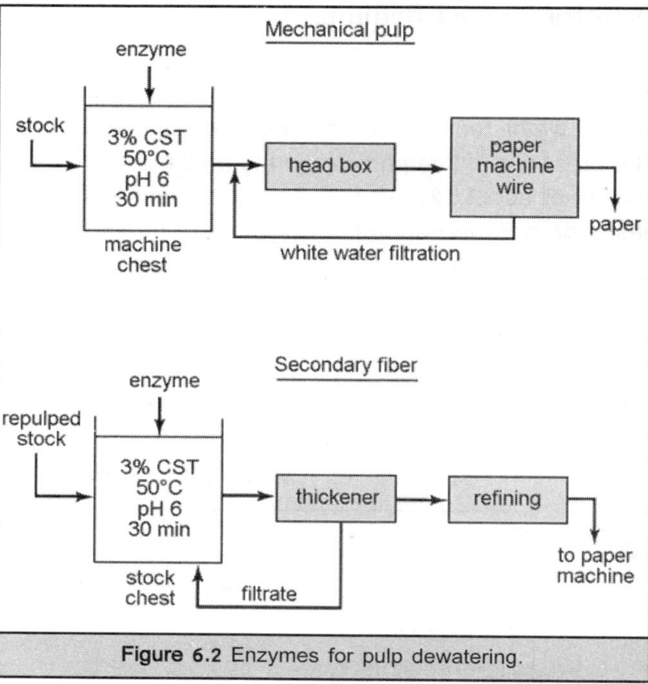

Figure 6.2 Enzymes for pulp dewatering.

Bleaching is carried out in 3 to 6 discrete stages, which taken together are known as a bleaching sequence. A classical three-stage sequence uses chlorine gas in the first stage, followed by an alkali extraction, then finally reaction with chlorine dioxide, a chemical that is particularly suited to brightening kraft pulp. In recent years, the bleached pulp industry has been motivated to decrease the use of chlorine-based bleaching compounds for environmental, regulatory, and market reasons. There are several options for mills to do this, but most require the expenditure of large sums of capital. Enzymes can be used to eliminate some of the chlorine-based compounds with little capital investment and are thus attractive to the industry.

Action of xylanase enzyme

Xylanase enzymes, which act on the xylan in the pulp, have been used commercially by the industry and are the example of the growing use of enzymes to aid in the bleaching of kraft pulp. At this writing (late 1997), the percentage of the bleached kraft pulp that is treated with xylanase enzymes, by country, is: Canada, 12%; Finland, 6%; Sweden, 3%; United States, 2%. These enzymes are marketed to mills in North America by Iogen Corporation of Ottawa, Canada; ICI of Toronto, Canada; and Clariant of Charlotte, N.C. The beneficial effect of xylanase enzymes in bleaching was reported by Viikari et al. The commercial enzymes used for enhancing the bleaching of kraft pulp all contain β-1,4 endo-xylanase activity that performs most of the action on the pulp.

These enzymes may also contain other hemicellulase enzymes. Xylanase enzymes act on the xylan in the pulp; this is a portion of the hemicellulose with a β-1,4-linked xylose backbone, with arabinosyl and other side chains. The enzymes do not bleach the pulp, but rather change the pulp structure such that the oxidizing bleaching chemicals (whether chlorine-based or otherwise) work more effectively on the pulp. Xylanase enzymes can decrease the use of oxidizing chemicals by 15 to 20%; more typically, a decrease of 10 to 15% is observed. The enzyme treatment can also increase the brightness of the bleached pulp and decrease the quantity of off-white fiber particles, known variously as shives, sclereids, and vessels. The unbleached pulp, known as brownstock, is pumped from the final washer into a storage tank before it enters the bleach plant. The pulp is a mat about 2 inches thick, of 8 to 14% solids consistency. Although the pulp is well washed, it has sufficient residual alkali soda that its pH is 9 to 12. This is higher than the effective pH range of the commercial enzymes, so the pulp must be acidified, usually with sulfuric acid, to the appropriate pH. The temperature of the pulp is typically 50 to 70°C (122 to 158°F). The capacity in the storage tank usually is sufficient for at least 1 to 2 hours of stock retention, which is sufficient time for the enzyme to act on the pulp.

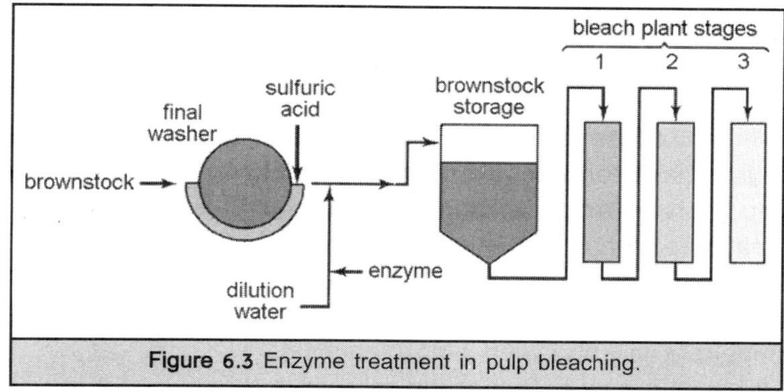

Figure 6.3 Enzyme treatment in pulp bleaching.

The pulp is pumped from the storage tank into the bleach plant, where contact with the bleaching chemicals destroys the enzyme. The enzyme- treated pulp then proceeds as normal through the bleach plant, except with a decreased requirement for bleaching chemicals. The decrease in demand for bleaching chemicals requires changes to be made in the bleach plant control strategy. The details of the how this is done are covered elsewhere. The enzymes are supplied in liquid solution, often in 1,000-L (275 gallon) totes. At the typical enzyme usage of 0.2 to 0.6 L per ton of dry pulp, a mill that processes 1,000 tons of pulp per day requires 2 to 5 totes per week of enzyme. The temperature on the operating floor can reach 35°C (95°F), so the enzyme must withstand this temperature during storage or be maintained in a refrigerated tank.

It is imperative that the enzyme, which is added to pulp in such small quantities, be evenly dispersed into the pulp mat to obtain uniform treatment and bleaching response. The enzyme solution is usually added to the pulp with some dilution water to increase the bulk liquid dispersion. The combined stream is added before a stock pump to take advantage of the mixing provided by these pumps. The equipment that is used to deliver this stream to the pulp varies from mill to mill but has included simple hoses, shower bars, and other devices. The pH of the pulp must be decreased for effective enzyme treatment. This is accomplished by adding acid to the pulp, usually well in advance of the point of enzyme addition. The most common acid used is sulfuric, but sulfurous acid, carbon dioxide, and waste acid from various mill process streams have been used. When sulfuric acid is used, it is stored as a 93 to 98% concentrate. The typical acid requirement is 3 to 5 kg/ton (6.6 to 11 lb/ton) pulp, which is an acid flow of 1 L of concentrate per minute on a 1,000 ton per day bleaching line. As with the enzyme, the acid must be dispersed uniformly throughout the pulp. The acid stream is pumped through a hose or shower bar and onto the pulp. The choice of location and equipment to handle the acid and add it to the pulp takes into account the mill's equipment configuration and desires for safe acid handling and avoidance of equipment corrosion.

Several properties of xylanase enzymes are important in bleaching applications. These include:

Bleaching benefit

The precise mechanism of the enzyme's action on pulp is not known. Hence, no single enzyme activity has been identified that characterizes the enhanced bleaching of pulp. The bleaching benefit for a given enzyme is measured by treating pulp and bleaching. That having been said, xylanase activity is present in all the commercial enzymes and is important in enhancing pulp bleaching. However, it is by no means the only active enzyme present; enzymes of the same xylanase activity can have very different effects in bleaching.

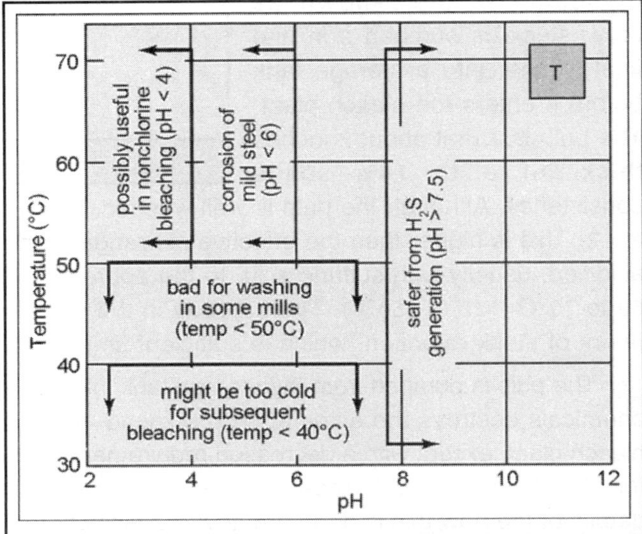

Figure 6.4 pH and temperature for enzymes in pulp bleaching: Effects on mill operations. The box labeled T indicates typical brownstock conditions.

Conditions of use

All the commercial xylanases are characterized by a range of pH and temperature that must be maintained for full effectiveness. The pH range is typically 1 unit broad, and the temperature range spans about 10°C (18°F). Figure 6.4 shows the implications of various pH and temperature ranges on mill operation; different enzymes are preferred for different mills. Enzymes that can operate at pH 10 to 12 and at temperatures up to 70°C (158°F) are the subject of much research.

Absence of deleterious effects on the pulp

This includes degradation of the pulp by cellulase; the enzymes must have little or no cellulase activity.

Enzyme manufacturers have used several different technologies to produce xylanase enzymes that comply with these requirements. These technologies are as follows:

1. *Produce xylanase using a natural microbe that does not make cellulase.* This seemingly straightforward approach has not generally been successful, because most such microbes make only a limited amount of protein. Examples of such microbes are species of *Bacillus* and *Streptomyces* bacteria and *Aspergillus* fungi.
2. *Clone and express xylanasein a microbethatdoes not make cellulase.* This approach has been somewhat more successful than the first, with host organisms including *Aspergillus oryzae, Bacillus pumilis,* and *Streptomyces lividans.*
3. *Produce xylanase along with cellulase and remove the cellulase using downstream processing.* This approach has worked well using *Trichoderma* fungi, which are well known as high protein-

producing microbes. Methods of cellulase removal include size-differentiated filtration, polyethylene glycol precipitation, precipitation with metal ions, ion exchange chromatography, and salting out.

4. *Delete cellulase genes from* Trichoderma. This approach has also worked well for commercial products.

In addition to xylanase, many other hemicellulase enzymes have been evaluated for their ability to enhance the bleaching of pulp. Of these, only mannanase appears to deliver benefits comparable to xylanase. Mannanase enzymes are not used commercially for bleaching at present.

Role of ligninase enzyme

Research over 3 decades focused on the use of ligninase enzymes to oxidize and remove the lignin from the pulp. The motivation for this research was straightforward: lignin removal is the objective of bleaching. However, for many years ligninase technology was not suitable for mill applications because reaction times of several days were required. A breakthrough discovery decreased the time required to bleach pulp with ligninase enzymes to 1 hour, which has motivated much recent effort in the area. Four proteins have been identified that participate in enzymatic delignification: lignin peroxidase, cellobiose dehydrogenase, manganese peroxidase, and laccase. Lignin peroxidase is a manganese-containing enzyme that requires hydrogen peroxide to oxidize the aromatic rings and side chains of lignin.

Cellobiose dehydrogenase is a multifunctional enzyme that has recently been shown to be able to participate in enzymatic delignification. From the point of view of commercial-scale use, laccase is the furthest along in development of the ligninase enzymes. Laccase is a copper-containing enzyme that catalyzes the oxidation of phenolic groups on lignin. The most prominent microbe for laccase production is the white-rot fungus *Trametes versicolor,* and production of the enzyme is induced by phenolic compounds. The copper that forms the redox-active center of laccase is part of the functional enzyme as it is produced and does not need to be added to the system for lignin degradation to occur. Call determined that inclusion of a mediator compound dramatically increases the rate of lignin oxidation. The mediator reported by Call is 1-hydroxybenzotriazole; many other effective mediators have since been reported.

Laccase and mediator are proposed to be used in a mill as follows. Pulp at an intermediate brownstock washer is adjusted to pH 4.5 and 45°C (113°F), then exposed to oxygen at moderate pressure. The enzyme and mediator are then added and act on the pulp for 1 hour. After treatment, the pulp is washed to remove the soluble lignin. The laccase-mediator process is reported to remove 50% of the lignin in a single-stage reaction and a further 25% in a second reaction. This magnitude of effect is much greater than that possible with xylanase and would be an extremely important step forward for mills to eliminate chlorine-based oxidizing chemicals entirely. The laccase-mediator bleaching has been demonstrated in pilot plant trials, but it has not been used commercially because of high costs of production of the enzyme and the mediator. The fourth ligninase enzyme, manganese peroxidase, offers the potential advantage over laccase in that it does not require a mediator. Manganese peroxidase is an iron-containing enzyme that uses hydrogen peroxide to oxidize chelated manganese, which then acts on lignin. The enzyme is most effective at 45°C

(113°F), pH 4.5. The rate and extent of lignin removal of manganese peroxidase is not as great as with a laccase-mediator system, but it is still very impressive (i.e., 40% delignification in 4 h). The areas for development of this technology are the need for careful control of the levels of peroxide, manganese, chelating agent, and enzyme and the costs.

Removal of Ink (Deinking)

The removal of ink is an important part of secondary fiber processing. Large-scale deinking is new and evolving rapidly. Deinking is carried out by repulping at 3 to 4% solids consistency, then diluting to 1% consistency and aerating the slurry while adding flocculating surfactants and ink solvents. The ink particles float to the surface and are collected and removed. Several reports indicate that cellulase enzymes can increase the efficiency of deinking processes. It appears that cellulase can release ink particles, both bound to fiber fines and separately from fiber, and thereby enhance the removal of ink by flotation. This in turn decreases the ink count, allows a higher brightness to be achieved, or alternatively decreases the usage of the deinking chemicals. There is some question, however, as to whether the enzyme treatment decreases the strength of the pulp. Some reports state that the enzymatic deinking process is best carried out under alkaline conditions. There has been one published report of a mill scale trial using this enzymatic deinking; the results confirmed the laboratory data. EDT of Norcross, Ga., has marketed cellulase and xylanase enzymes for mill-scale deinking applications.

BIOLOGICAL CATALYSTS IN PAPERMAKING

Most of the content of paper is finished pulp (which can be blends of both mechanical and chemical pulps) and inorganic fillers such as clay. The remainder of the paper, perhaps up to 10% of the total weight, consists of a variety of additives designed to achieve certain properties in the paper and other additives to improve the operation of the paper machine in the mill. Among the additives to influence paper properties is modified starch for coated paper. Amylase enzymes are used in the starch modification. Enzymes are also used to remove pitch from the paper and to eliminate slime buildup on the paper machine.

Modification of Starch

The oldest use of enzymes in the pulp and paper industry is in the modification of starches for coated paper. This practice dates back to the early 1970s and is now carried out more frequently within the starch processing plants than in the paper mills. Starch imparts many beneficial properties to paper, including strength, stiffness, and erasability. The starches used in papermaking are often modifications of the natural starch polymers found in the locale of the mill, such as corn starch in the United States, potato starch in Britain, and rice starch in Japan. Of particular interest in enzyme use is starch for sizing. Sizing refers to an increased stiffness of the paper that is useful as a coating or for further paper processing. The starch is added to the paper partway through the drying section of a paper machine. The starch is added at a size press, which is two rollers through which the paper picks up starch solution while passing. The presses run best at 45 to 60°C (113 to 140°F), so the viscosity of the starch solution must be in an acceptable range at this temperature.

The standard starch products for the paper industry are suitable for some grades of paper, but in other cases the viscosity is too high for good spreading through the press. In other cases, a mill desires the flexibility to vary the starch viscosity for different paper grades. More recent work has suggested that amylase enzymes can form unique starch products for the paper industry. The generic process for decreasing starch viscosity is to decrease its molecular weight by hydrolyzing the polymer to shorter chains. The first method to do this involved a dilute acid cooking of the starch; this is still practiced in some mills today. The second method, hydrolysis with α-amylase enzymes, was originally carried out in the late 1960s. These enzymes catalyze the addition of a water molecule across the linkages between two glucose units in the starch polymer. The desired reactions for paper starch is to shorten the starch polymer.

Complete conversion to glucose, which would be achieved using glucoamylase and pullulanase enzymes, is undesirable. The two types of reactor systems used for enzymatic starch hydrolysis in paper mills are batch and continuous. In batch systems, roughly 1,000 kg (2,200 lb) of starch is diluted in 4,000 L (1,000 gallons) of water in an agitated tank. The temperature is maintained at, depending on the enzyme used, 25 to 90°C (77 to 194°F). The pH is controlled to 4.5 to 7. The treatment time is typically 30 minutes, after which the system is heated to a suitable temperature to destroy the enzyme's activity, or, for the thermostable enzymes, chemical denaturants are added. Batch systems offer the advantages of simplicity of control and the ability to run at solids levels of up to 40 %. In continuous systems, the starch and water are mixed in a slurry tank and pumped through a steam-heated tube. Enzyme is injected at the mouth of the tube. The running temperature is 125°C (257°F) for 2 min at pH 6. At this point, the target viscosity is reached, and the enzyme, which is added to a concentration of 0.5 g/L, is fully denatured. This system offers advantages of easy automation and high volumetric productivity. The continuous systems are generally for the larger starch users.

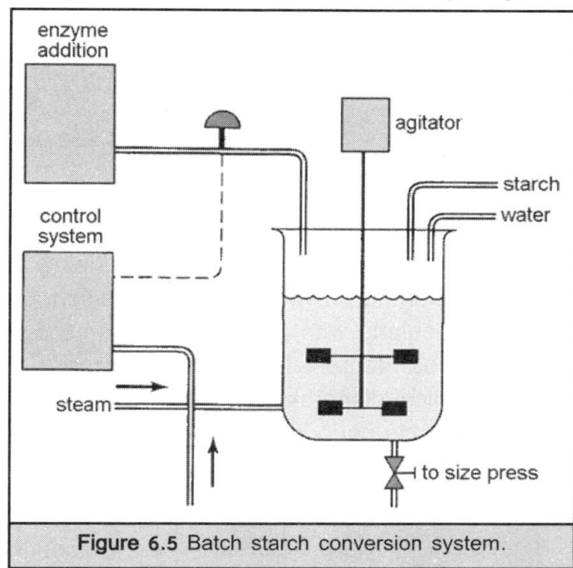

Figure 6.5 Batch starch conversion system.

Removal of Pitch

Pitch is the sticky substance that is prevalent in softwoods, especially pine. Pitch is a collection of compounds known as lipids, which are long-chain fatty acids connected by glycerol linkages. These compounds are analogous to the fatty tissue of mammals and have both hydrophobic and hydrophilic sites on the molecule. This character contributes to the processing difficulties caused by pitch; the compounds are slightly soluble in some points of the process, and they can form a difficult precipitate in other areas. The most troublesome pitch formation is from Japanese Red Pine. Pitch comprises typically less than 1% of the weight of wood, but can reach 3% of the total on this species.

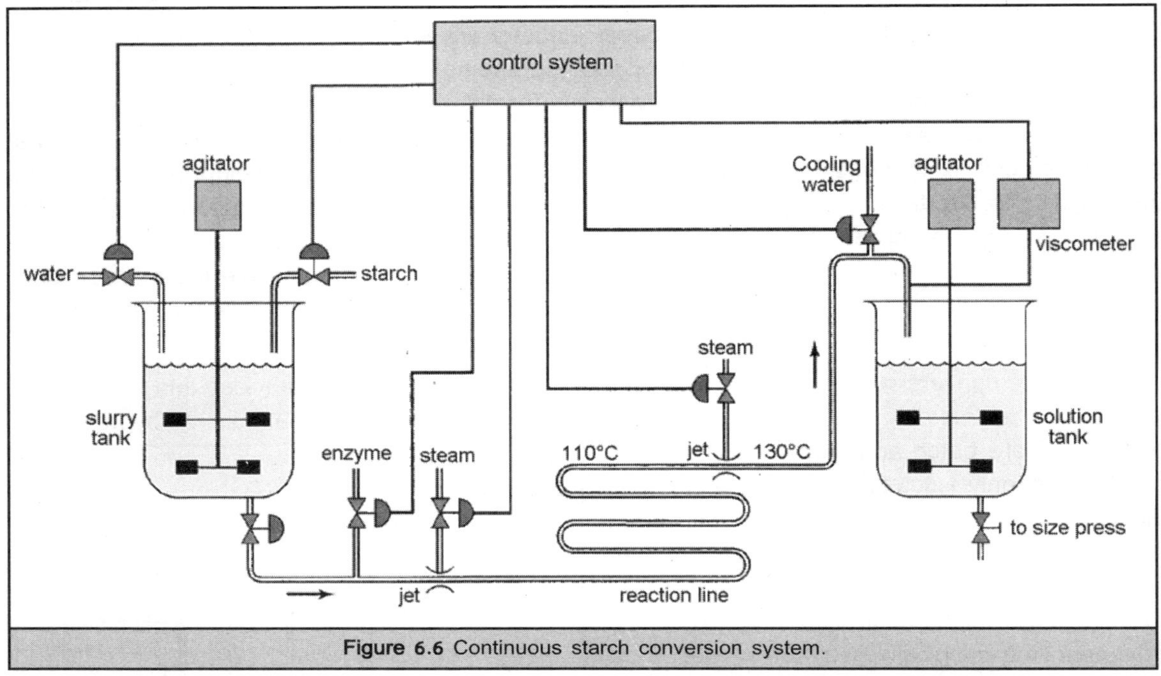

Figure 6.6 Continuous starch conversion system.

The removal of pitch by chemical pulping and bleaching is not particularly efficient, but it is usually sufficient to prevent severe deposition. In mechanical pulping, the pitch problems are more severe, particularly with pitch exposed on the paper machine. At this point, pitch interacts with other additives, decreasing their effectiveness as well as decreasing the rate of drainage of the sheet. Chemical dispersants have been used for pitch control, and the pitch content is decreased by longer storage periods in the wood-yard, which is limited by fungal growth and chip deterioration. A third method of decreasing pitch deposition is by using lipase enzymes to digest the pitch. Lipases hydrolyze the lipids to their constituent fatty acids and substituted polyol.

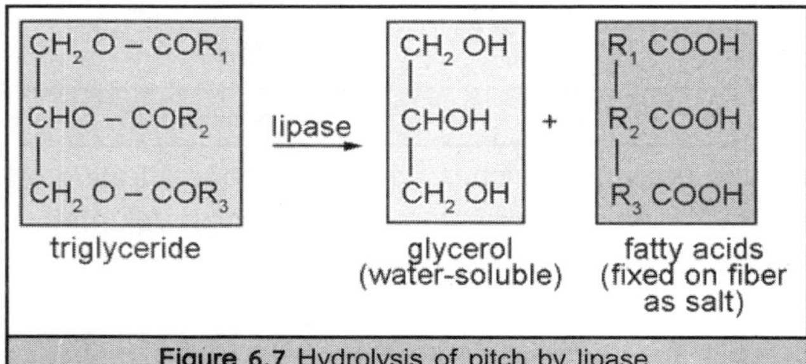

Figure 6.7 Hydrolysis of pitch by lipase.

The polyol (glycerol for triglycerides) is water soluble, and the fatty acids can be bound with talc or other charged compounds. Lipase for pitch removal has been particularly successful in Japan, where 120,000 to 140,000 tons of ground wood pulp are treated annually with lipase, representing 20% of the total groundwood pulp produced in Japan. Lipase has been reported to be added to the groundwood pulping line and to the storage chest before the paper machine. When added to the pulping line (just before the post refiner), a mixture of two lipases added at levels of about 100 ppm and about 600 ppm reduced the triglyceride content of the pulp by 70% over a 1-month mill trial, with an enzyme reaction time of 3 to 3.5 h. This corresponded to a decease in pitch deposits on the paper and equipment and defects in the resulting paper sheets. However, one disadvantage of this location for lipase treatment is the frequent use of alum, particularly in American mills, which can inactivate the enzyme. In another mill trial, lipase was added at 1 to 5 ppm in the storage chest before the paper machine, for a retention time of at least 1 h. The process temperature and pH of 40°C (104°C) and 6.0, respectively, were compatible with the enzyme treatment. The enzyme was added to the dilute slurry in the chest without additional mixing, although in at least one mill lack of good enzyme dispersion caused substandard results. Results of a mill trial using lipase for pitch removal are shown in Table 3. Lipase does not affect the strength or brightness of the pulp. The yeasts *Candida cylindracae* and *Candida rugosa* and *Aspergillus* fungi have been reported to make lipase in high productivity and are used commercially to make lipase.

Removal of Slime

Like pitch, slime deposition causes significant operating problems around a paper machine. *Slime* is the generic name for deposits of a microbial origin in a paper mill. It is impractical to run a paper mill as a sterile system. As a result, a vast array of microbes contaminate the mills, and many of the resulting slime compounds have not been characterized. When confronted with a slime, often the strategy is to try every available biocide until one is found that targets the microbe and destroys the source of the slime. In some cases, however, specific slime compounds have been characterized, and efficient strategies for their removal can be used. One such case is with levan, which is a β-2,6-linked polymer of fructose that forms a slime film. This compound is secreted by several species of *Bacillus* and *Pseudomonas* bacteria that can grow in the recirculated water around the paper machine, especially for fine paper, where the level of inhibiting compounds is low. The enzyme levan hydrolase can hydrolyze this polymer to low molecular weight polymers that are water soluble, thereby cleaning the slime out of the system.

Another advantage of the enzymatic treatment is that it destroys the ability of the microbe to bind to the surface, thereby increasing the effectiveness of chemical biocides. Commercial levan hydrolase is supplied as the product paper machine, although in some cases it has been added at the dryer discharge. The enzyme is effective at pH 4 to 8 and runs best at 50°C. The enzyme can be used alone or in combination with chemical biocides. Other enzymes reported to be effective for slime removal are proteases, cellulase, amylase, and pentosanase. These enzymes are used more typically to treat the process waters of the mill to prevent slime buildup in cooling towers, heat exchanger surfaces, and other surfaces exposed to water.

7

TRANSFORMATION

A variety of methods are available for gene transfer in gram-positive bacteria, and the choice of technique depends on the nature of the experiment, recipient strain, and donor DNA. For naturally transformable species such as *Bacillus,* transformation of competent cells is usually used for the introduction of chromosomal or plasmid DNA. For strains lacking satisfactory levels of natural competence methods such as polyethylene glycol-mediated protoplast transformation have been successfully used, although these are currently being replaced by the more straightforward technique of electroporation. In addition, transduction and conjugation can be used for DNA transfer in gram-positive bacteria. We outline here the major principles of each mode of gene transfer and list particular applications with different gram-positive examples.

Transformation is defined as the transfer of free exogenous DNA into a cell to permanently alter its heredity. Four modes of transformation exist in gram-positive bacteria: (i) transformation of naturally competent bacteria; (ii) transformation of cells with artificially induced competence, (iii) polyethylene glycol-mediated protoplast transformation, and (iv) electrotransformation (electroporation).

ROLE OF EXTRACELLULAR DNA

Transformation of naturally competent bacteria involves the active uptake by a cell of free extracellular DNA and the heritable incorporation of its genetic information. Natural transformation was first detected in *Streptococcus pneumoniae,* where the ability of this bacterium to be transformed by exogenous DNA was used to prove that DNA is the genetic material. Natural competence is a physiologically and genetically determined property of a particular strain. This physiological state is dependent on the expression of specific genes whose products provide the necessary state for competence and subsequent DNA uptake. *Bacillus subtilis* is the most intensively studied naturally transformable species. Almost 40 genes (*com* genes) required for competence have now been identified in *B. subtilis.* About half of them regulate the development of the competent state, and the remainder are involved in DNA uptake and the transformation process (late *com* genes). In *B. subtilis,* the competent state develops postexponentially during the transition from exponential growth to stationary phase. Culture density is one of the physiological conditions controlling the development of competence.

As cells grow to high density, two extracellular peptide pheromones accumulate in the culture medium; they cause increased transcription of the *comS* gene. Synthesis of ComS, in turn, leads to activation of the ComK transcription factor, which directs expression of the late *com* genes. In addition to these cell density signals, there are also nutritional signals that control activation of ComK. Similar mechanisms exist in *Streptococcus*: in *S. pneumoniae,* competence develops in exponentially growing cells at a critical cell density (10^7–10^8 cells/mL). Here, cellular secretion of a small 17-residue peptide is responsible for inducing cells to develop competence. In *B. subtilis,* only 1–10% of the cell population becomes competent whereas in *Streptococcus* spp. (e.g., *S. pneumoniae* and *S. sanguis*) up to 100% of the cells in a culture become competent. DNA uptake mechanisms in *B. subtilis* and *S. pneumoniae* are also similar. In brief, double-stranded DNA noncovalently binds to the surface of a competent cell (approximately 50 [*B. subtilis*] and 30–80 [*S. pneumoniae*] DNA-binding sites are used per competent cell).

The transfer of DNA into the cell initiates with the generation of double-strand breaks. One strand is degraded while the other enters the cell; thus, DNA enters in a single-stranded form. In the case of transformation by homologous DNA, single-stranded DNA can integrate into the recipient chromosome (or resident plasmid) by homologous recombination. In *B. subtilis,* if plasmid DNA contains regions homologous to the recipient genome, the efficiency of transformation will be greater than 10^6 transformants per microgram of DNA. In the case of transformation by a nonhomologous plasmid, following entry of single-stranded plasmid DNA the plasmid must first resume a double-stranded form to replicate. If the transforming plasmid is monomeric, two complementary strands should independently enter the cell to produce a double-stranded plasmid. For this reason transformation by plasmid monomers is very inefficient in *B. subtilis*.

Most plasmid preparations from gram-positive (or gram-negative) bacteria contain plasmid multimers and can yield transformants in *B. subtilis* at a level of 10^3 to 10^5 transformants per microgram of DNA. Plasmid DNA isolated from Rec^- bacteria contains a much lower amount of multimers and will lead to 100 to 1000-fold lower transformation frequencies. To bypass the requirement for plasmid multimers in *B. subtilis,* a so-called plasmid marker rescue transformation system has been developed. This system involves the uptake of donor DNA by competent cells and the subsequent recombination with a partially homologous resident plasmid. A simple procedure can be used for the preparation of competent cells of *S. pneumoniae*. Methods for transformation of other naturally competent streptococci have also been developed, for example, *S. sanguis* Challis, *S. milleri,* and *S. mutans*. For *B. subtilis,* the most commonly used method of preparation of competent cells remains the classical two-step procedure of Dubnau and Davidoff-Abelson.

Laboratory techniques for *Bacillus* transformation (as well as other molecular biological methods for *Bacillus*) can be found in Harwood and Cutting's work. Transformation of naturally competent cells is the method of choice for introduction of DNA into *B. subtilis*. Applications include allele replacement and the introduction of chromosomal markers and plasmids. Transformation has been used for fine structure genetic mapping in *B. subtilis*. This method, however, is limited by the length of DNA taken up by a competent cell. The average size of transforming DNA (about 30 kb) limits the recombinational distance between two markers and hence the genetic distance that can be established. Nonsaturating concentrations of transforming DNA (less than 50 ng/mL in *B. subtilis*) should be used in mapping experiments to avoid congression (i.e., the transformation of the same competent cell by two genetically unlinked markers as a result of independent DNA uptake events). However, congression can be used as a strain construction tool, in that a genetic marker can be introduced into a recipient chromosome by selection for an unlinked marker. For this purpose, however, much higher DNA concentrations (e.g., 1–5 μg/mL in *B. subtilis*) must be used (to get 2–3% congression).

Role of Artificially Induced Competence

Some bacterial species can be artificially manipulated to allow the uptake of exogenous DNA. This so-called chemically induced competence can be achieved typically by treatment of cells with calcium or some other cation and is used in the standard *Escherichia coli* transformation protocol. This is the method of choice for transformation of *E. coli*. Among gram-positive bacteria, *Staphylococcus aureus* has been transformed by this technique. In this procedure, cells collected at early exponential growth phase are washed with a NaCl solution, suspended in a $CaCl_2$-containing buffer, and incubated with DNA. The transformation efficiency has been shown to be dependent on the pH of the buffer, the presence of divalent cations, and the growth phase of the culture. Transformation efficiency is generally higher in nuclease-minus mutants. Also, efficient development of artificial competence requires that the cells are lysogenic for phage φ11.

Certain phage-expressed products are able to enhance competence; for example, both phage tails and the product of early gene 31 have been reported to enhance competence. Efficiencies of 10^4 transform ants per microgram of DNA can be achieved with this system. The main advantage of the artificial competence procedure for *S. aureus* is that it is effective for transformation of chromosomal DNA whereas protoplast transformation is not. This technique has been used extensively for genetic mapping where linkages representing large distances (100–150 kb) of the chromosome can be defined by the uptake of very large chromosomal fragments by the cell.

Generation of Protoplast

Protoplasts are generated by treatment of bacteria with muralytic enzymes (lysozyme or other, depending on the species) to degrade the peptidoglycan cell wall in the presence of an osmotic stabilizer. When exposed to high concentrations of polyethylene glycol (PEG) in the presence of DNA, protoplasts take up the DNA and if allowed to recover on suitably rich media can regenerate the cell wall. Addition of PEG can also cause protoplast fusion. Use of protoplast techniques has advantages and disadvantages. Protoplasts allow the uptake of double- stranded plasmid DNA, enabling plasmid monomers to be used. However, protoplasts are very fragile, burst easily, and are extremely sensitive to traces of detergent on glassware.

In addition, the method is generally considered laborious; protoplast regeneration can be extremely slow (2–6 days) and occurs at a low frequency. Finally, complete media are required, precluding the direct selection for prototrophic markers. Before the introduction of electroporation methods, protoplast transformation had been widely used for transformation of gram-positive species that are not naturally competent.

Protoplast transformation techniques for gram-positive bacteria were first developed for *Streptomyces* and *Bacillus*. PEG-induced transformation of *Streptomyces* protoplasts by plasmid DNA was first reported in 1978 and for *B. subtilis* in 1979. The *B. subtilis* protoplast transformation technique is still used for many *Bacillus* species, such as *B. anthracis, B. amyloliquefaciens, B. stearothermophilus,* and *B. licheniformis*.

In *Streptomyces* spp., transformation and transfection of protoplasts are important tools for gene transfer, being very efficient, even with large constructs up to at least 40 kb, and more than 10^7 transformants per microgram of DNA can be obtained, as has been described for *S. lividans* and *S. coelicolor*. Numerous protoplast transformation techniques have been developed for other *Streptomyces* species: *S. peucetius,* 10^6 transformants/μg DNA; *S. tendae,* 10^6 to 10^7 transformants/μg DNA; *S. clavuligerus,* 10^5 transformants/μg DNA; *S. venezuelae,* 10^7 transformants/μg DNA; *S. niveus,* 10^6 transformants/μg DNA; and *S. lincolnensis,* 10^3 transformants/μg DNA. *Streptomyces* protoplasts are more easily manipulated than those of other bacterial genera and exhibit reproducibly high levels of regeneration. Indeed, this ease of use has enabled protoplast fusion to be used not only for strain construction but also for genetic mapping in *Streptomyces*. Protocols have been developed for protoplast transformation in a wide range of other gram-positive species.

PEG-mediated protoplast transformation methods (giving up to 10^6 transformants/μg DNA) are available for different Actinomycetes such as *Micromonospora olivasterospora, Amycolatopsis (Nocardia) orientalis,* and *Rhodococcus* spp. A method for PEG-mediated transformation of the mycelium has been developed for *Amycolatopsis (Nocardia) mediterranei* that yields 10^6 transformants/μg DNA without the need for protoplast isolation. *Staphylococcus* protoplasts prepared with lysostaphin and stabilized with sucrose and Mg^{2+} are currently being used for PEG-mediated plasmid transformation and protoplast fusion.

Protoplast fusion has also been used for chromosomal mapping in *Staphylococcus aureus* and also for plasmid transfer. There has also been reported a method for PEG-mediated plasmid transformation of a temperature-sensitive peptidoglycan-deficient mutant of *S. aureus,* giving transformation frequencies comparable with those obtained with protoplasts prepared with lysostaphin treatment. PEG-mediated protoplast transformation has been developed for different streptococci, including *S. lactis, S. faecalis, S. faecium, S. thermophilus,* and *S. pyogenes*.

Early techniques used protoplast transformation for introduction of plasmid DNA into mycobacteria. Electroporation techniques have subsequently proved to be more effective, although an efficient method for transformation of *Mycobacterium smegmatis* protoplasts has been described. As with mycobacteria, electroporation has now replaced protoplast transformation techniques in the lactic acid bacteria and *Listeria* spp. PEG-mediated protoplast transformation was the first method of transformation reported for *Clostridium perfringens*. Before the development of electroporation techniques, protoplast transformation was used extensively for the introduction of DNA into *Clostridium acetobutylicum,* the most recent method reported giving up to 10^6 transformants/μg DNA.

Reversible Pore Formation

Application of a high-voltage electric field for a very short period of time (microseconds) causes transient structural changes in lipid bilayers and biological membranes, leading to reversible pore formation (*electroporation*) and allowing the passive influx of DNA molecules. Electroporation as a tool for cell transformation was first used in the early 1980s for the transformation of animal cells. Subsequently, the technique was utilized for prokaryotes (e.g., for transformation of *Bacillus cereus* protoplasts). Transformation of intact bacterial cells by electroporation was first reported in 1987 for *Lactococcus lactis* and *Lactobacillus casei*.

A number of reports on electroporation of different gram-positive bacteria appeared shortly thereafter: *Enterococcus faecalis, Streptococcus thermophilus, Mycobacterium smegmatis,* and BCG, *Clostridium perfringens,* and a number of other gram-positive species—*Lactobacillus acidophilus, Bacillus cereus, Leuconostoc* spp., and *Listeria* spp. The first attempts to transform gram-positive bacteria using electroporation were not very efficient, and for a large number of gram-positive species the technique has been extensively refined. The most important parameters affecting electroporation efficiency appear to be field strength and duration and shape of the pulse. Other parameters involve the cell culture growth conditions (e.g., media composition and phase of cell growth at the time of harvesting for electroporation) and transformation conditions (e.g., final concentration of cells and DNA in the transformation mixture, ionic strength, and the pH of electroporation buffer). Optimal conditions of course, will differ from strain to strain. Some strains can give frequencies of 10^7 to 10^8 transformants/μg DNA, and in these strains library construction is a distinct possibility.

Most strains, however, give frequencies suitable for the exchange of shuttle vectors between strains. However, frequency of electrotransformation of gram-positive bacteria is generally lower than that observed with *E. coli*. Possibly, the dense composition of the gram-positive cell wall presents a more complex barrier to entry of DNA. Several protocols were developed to overcome this problem, such as incorporation of low concentrations of the cell wall synthesis inhibitor penicillin-G in the growth medium substantially increased the efficiency of transformation of *Listeria monocytogenes* (to 10^6 transformants/μg DNA). Gentle treatment of lactococcal cells with muralytic enzymes before electroporation and pretreatment of *Clostridium perfringens* cells with a peptidase lysostaphin also raised the efficiency of electrotransformation.

In the last two cases, muralytic enzymes partially digested the cell wall, providing access of plasmid DNA to the cell membrane pores subsequently created by electroporation. Electroporation has now become the method of choice for the transformation of a wide range of gram-positive bacteria. For example, in *Clostridium* electroporation has superseded the standard method of protoplast transformation used previously. Electroporation methods have been reported for *Clostridium perfringens* (10^7 transformants/μg DNA), *C. beijerinckii, C. acetobutylicum,* and *Clostridium thermosaccharolyticum*. Electroporation is also a method of choice for lactic acid bacteria (*Lactobacillus, Lactococcus*) and has replaced the more time-consuming and frequently unreliable protoplast transformation techniques.

In *Mycobacterium smegmatis,* electroporation is relatively efficient, yielding little more than 10^5 transformants/μg DNA. In other mycobacterial species similar yields have been reported: *Mycobacterium intracellulare* (10^6 transformants/μg DNA), *M. vaccae* (10^3–10^5 transformants/μg

DNA), and *M. bovis* BCG. It was shown that elevation of electroporation temperature to 37°C improves the efficiency of transformation of slow-growing *Mycobacterium* spp. For *Listeria*, electroporation yields as high as 10^6 transformants/μg DNA have been reported for *L. monocytogenes, L. innocua*, and *L. ivanovii*, and somewhat lower yields for *L. seeligeri*.

Efficient electroporation techniques have also been developed for some actinomycetes, for example, *Actinomyces viscosus* and *Actinomyces naeslundii* (10^2–10^7 transformants/μg DNA), *Brevibacterium flavum* and *Corynebacterium glutamicum* (up to 5×10^7 transformants/μg DNA), *Brevibacterium methylicum* (10^5 transformants/μg DNA), and *Rhodococcus fascians* (10^5 transformants/μg DNA). A method for electroporation of *Streptomyces parvulus* and *S. vinaceus* has been reported (10^5 transformants/μg DNA). Electroporation has also been used for transformation in the *Streptococcus* species: *Streptococcus pyogenes* (10^7 transformants/μg DNA), *S. thermophilus* (7×10^5 transformants/μg DNA), *Enterococcus faecalis* (10^5 transformants/μg DNA), *S. parasanguis, S. agalactiae* (10^4 transformants/μg DNA), and *S. suis* type 2 (10^7 transformants/μg DNA). Employed later than for other gram-positive bacteria, electroporation techniques have been developed for the staphylococci, such as *Staphylococcus aureus* (4×10^8 transformants/μg DNA), *S. staphylolyticus*, and *S. carnosus* (10^5 transformants/μg DNA).

Electroporation has been applied to many *Bacillus* species, such as *Bacillus thuringiensis, B. amyloliquefaciens, B. cereus, B. subtilis*, and *B. anthraecis*. However, electroporation is relatively inefficient for the *Bacillus* species and transformation frequencies are low (no more than 10^5 transformants/μg DNA) for most of the species including *B. subtilis*. For *B. subtilis* the maximum electrotrans formation frequency reported was 9×10^4 transformants/μg DNA. *B. thuringiensis*, however, is an exception, with transformation frequencies up to 10^7 transform ants/μg DNA being reported. Recently, an efficient method for electroporation of *B. brevis* cells has been developed. It gives 10^7 up to 10^8 transformants/μg DNA. Electroporation techniques have also been reported for transformation of *Leuconostoc paramesenteroides* and species of Bifidobacterium. Electroporation is rapidly becoming the method of choice for the majority of gram-positive bacteria.

TRANSDUCTION

Gene transfer mediated by a bacteriophage is termed *transduction*; two types of transduction can be defined, generalized and specialized. In the process of generalized transduction, any gene has a more or less equal probability of being transferred. During phage infection of the donor bacterium some phage heads incorporate a "headful" of bacterial DNA rather than phage DNA, and upon lysis the bacteria release phage particles carrying bacterial DNA. The phage particle adsorbs to the recipient bacterium and injects the bacterial DNA fragment, which can recombine into the recipient genome by homologous recombination. During specialized transduction ("restricted transduction"), only genes adjacent to the lysogenic prophage can be transduced as a result of erroneous excision of the prophage.

During excision, adjacent bacterial DNA can be excised together with the prophage (it is also possible for some prophage DNA to remain unexcised). After packaging the newly formed phage particle contains excised bacterial DNA, which can be transferred to other bacteria during the life cycle of the virus.

Generalized Transduction

Several transducing phages have been described for *B. subtilis*. The *generalized transducing phage* PBS1 is the most widely used for genetic analysis. The phage is pseudolysogenic, however, and although it generates turbid plaques on a plate seeded with *B. subtilis* it does not enter a true lysogenic state. PBS1 transduction is used for long-range genetic mapping because a single transducing particle encompasses more than 200 kb (about 10% of the host genome) and can be used to obtain an approximate position of a gene on the chromosome. Following this, DNA-mediated transformation of naturally competent cells is used to obtain a more accurate position of the gene of interest. For rapid mapping of genes, a kit of nine auxotrophic strains has been created. Together, these strains contain 22 auxotrophic markers spaced at relatively even intervals around the chromosome. A PBS1-transducing lysate is prepared from the strain containing the mutation to be mapped (which should have the wild- type allele of the auxotrophic markers carried by the kit strains) and used to infect each kit strain with selection for Aux and screening for the mutant phenotype.

An alternative and perhaps easier approach is to use the large collection of "silent" chromosomal insertions of Tn*917*. Here, a transducing lysate is made from the Tn*917* insertion strain and used to infect the strain with the unknown mutation, followed by selection for MLS resistance (encoded by the transposon) and screening for correction of the mutant phenotype. A PBS1 lysate prepared from the donor strain (donor lysate) is used to infect a motile recipient strain. PBS1 infects cells by adsorption to the flagella, so it can only infect motile cells. Generalized transduction is a useful tool for genetic analysis of closely linked markers in *Staphylococcus aureus*.

Transduction is limited by the size of the phage genome (45 kb for φ11); linkages representing large (100–150 kb) distances are being mapped by transformation of competent cells. Phage-based gene transfer systems serve as efficient tools of gene transfer in mycobacteria. Generalized transduction by the phage 13 is used for marker transfer to *Mycobacterium smegmatis*. Generalized transducing streptomycete bacteriophage FP43 has been used to transduce plasmid pRHB 101 into species of *Streptomyces*, *Streptoverticillium*, and *Saccharapolyspora*. pRHB101 contains a 7.7-kb segment of FP43 DNA (*hft* segment) cloned into plasmid PIJ702. The transducing particles contain linear concatemers of pRHB101, which form circular monomers in the transductants.

Specialized Transduction

Two specialized transducing phages have been extensively used as cloning vectors in *B. subtilis*. φ105 and SPβ. Both phages were designed for cloning *B. subtilis* genes by *trans* complementation. SPβ is a large temperate phage (126 kb) that can tolerate up to 10 kb of extra DNA. It has been used for complementation studies and for transferring *lacZ* fusions between strains. Methods have been developed for integration of foreign DNA into the SPβ prophage by homologous recombination, after which DNA can be transferred to other strains by transduction.

Phage vectors derived from the temperate phage φC31 have been developed for *Streptomyces*. Integration may involve *att* sites of the phage and chromosome, or, for the vectors in which the *att* site has been deleted, the phage can integrate by homologous recombination, either using a preexisting prophage to form a double lysogen or at the chromosomal site homologous to a DNA fragment cloned into the vector.

CONJUGATION

Conjugation requires cell–cell contact and provides a useful method for the interspecies and intergenera plasmid transfer within gram-positive bacteria, especially for strains lacking efficient transformation systems. Conjugative plasmids are very common among gram-positive bacteria. Usually, the conjugative plasmid itself is transferred, and in addition small plasmids can also be transferred. Some gram-positive conjugational plasmids are known to promote transfer of mobilizable plasmids across genus barriers. The conjugative streptococcal plasmid pAMβ1 is efficient in the mobilization of small plasmids in streptococci and lactococci. Large conjugative *Bacillus* plasmids, such as pLS20, can mobilize pUB110 for interspecies transfer. The majority of *Streptomyces* plasmids are self-transmissible. Conjugational crosses are useful means of strain construction in *Streptomyces*. Conjugative *Staphylococcus aureus* plasmids can efficiently mobilize those small plasmids that can form relaxation complexes.

The most efficient conjugation system exists in *Enterococcus faecalis,* termed *pheromone-mediated plasmid exchange*. Here, plasmid-free enterococci excrete short oligopeptide pheromones (clump-inducing agents) that induce plasmid-containing cells to express cell surface proteins (aggregation substance) that allow the donor to bind to a receptor on the surface of the recipient and promote cell–cell contact. The plasmid is then transferred to the recipient cell and represses further pheromone synthesis. This transfer system is extremely efficient, even in matings conducted in liquid. Other conjugation systems are not so efficient and can promote DNA transfer only when the cells are forced into close contact by being placed on a filter or an agar surface. Rare examples of conjugative mobilization of chromosomal markers exist in gram-positive bacteria. Conjugation in *Streptomyces* has been used for mapping of chromosomal genes.

A conjugative transposon Tn*925* from *E. faecalis* not only can mediate its own transfer in *B. subtilis* but also can mobilize the transfer of chromosomal genes. Until recently, it was believed that none of the gram-positive conjugation systems are capable of Hfr-like transfer of chromosomal genes. However, such a mechanism was recently proposed to occur in mycobacterium smegmatis. Conjugative transposons (the elements transposing from a DNA molecule in one cell to a DNA molecule in another cell by a conjugational mechanism but that are incapable of autonomous replication) were first isolated from streptococcal strains associated with human disease. The most carefully studied are Tn*916* from *E. faecalis,* Tn*3 701* from *Streptococcus pyogenes,* and Tn*1545* and Tn*5253* from *S. pneumoniae.*

Conjugative transposons are useful shuttle vectors, especially when the gene to be transferred has to be maintained at the same copy number as the chromosome. The cloned DNA can be inserted into Tn916 at a *Bst*X1 site without disrupting the function of the transposon. The chimeric transposon is then introduced into the gram-positive cell by electroporation or protoplast transformation (the method of transfer should not lead to the rearrangement of large incoming DNA molecules).

The chimeric transposon will then insert into the host chromosome. If it subsequently proves problematic to transform DNA directly into the strain of interest, an intermediate host (e.g., *B. subtilis*) can be used. This host is transformed with DNA of the *E. coli* plasmid containing Tn*916* with the gene of interest cloned into it, and the strain will then act as a conjugative donor to any other gram-positive strain. Thus, the chimeric transposon will be transferred from the intermediate host to the strain of interest by conjugational mating on a solid surface (agar or filter). As an example of its recent application, Tn*916* was reported to be transferred to *Leuconostoc lactis* LM2301, which opens the possibility of genetic studies in this *Leuconostoc* species for which no efficient conjugal or transformation system has been described previously. A series of Tn916-based shuttle vectors for analysis of transcriptional regulation in gram-positive bacteria has also been developed. Conjugative transfer can occur even between gram-negative (*E. coli*) and gram-positive bacteria and can be exploited for transfer to gram-positive bacteria ofthe DNA cloned in *E. coli*.

The first shuttle vector of this type (pAT187) was constructed by Trieu-Cuot et al. It contains replication functions for both *E. coli* (from pBR322) and gram-positive bacteria (from pAMβ1) and an origin of transfer (*oriT*) from the gram-negative broad host range plasmid pRK2. Donor strains of *E. coli* must also harbor pRK2 or some other IncP-type plasmid or chromosomally inserted DNA fragments carrying IncP *tra* functions. Subsequently, other versatile mobilizable vectors have been developed. Such shuttle vectors can be transferred in filter matings to a wide variety of gram-positive bacteria, including *Bacillus, Clostridium, Streptomyces, Enterococcus, Streptococcus, Staphylococcus, Listeria, Nocardia,* and *Rhodococcus*. Improved procedures for conjugative transfer from *E. coli* to *Staphylococcus aureus* and *Listeria monocytogenes* have also been developed involving subinhibitory concentrations of cell wall synthesis inhibitors (penicillins) in the mating medium to increase transfer frequency. Conjugation has proved to be extremely valuable to gram-positive strains that lack a natural competence system and have proved refractory to protoplast transformation and electroporation.

8

ROBUST EXPRESSION OF GENES

Escherichia coli is a widely used organism for the production of recombinant proteins for both basic research and therapeutic applications. In spite of the extensive knowledge of the genetics and molecular biology of *E. coli*, the robust expression of genes in this prokaryote is not necessarily a routine matter. This may be due to the unique structural features of the gene sequence, the stability and translational efficiency of mRNA, the ease of protein folding, degradation of the protein by host cell proteases, major differences in codon usage between the foreign gene and native *E. coli*, and the potential toxicity of the protein to the host. Progress in our understanding of many of these processes is certain to broaden the usefulness of *E. coli* as a tool for gene expression. The major drawbacks of *E. coli* as an expression system are the lack of a secretion mechanism for the efficient release of protein into the culture medium, and the limited ability to facilitate extensive disulfide bond formation. In addition, although bacteria are capable of carrying out a variety of posttranslational modifications, including some types of glycosylation, these are not as complex as those performed by eukaryotic cells. On the other hand, many eukaryotic proteins retain their full biological activity in a nonglycosylated form and, therefore, can be produced in *E. coli*. The objectives here are to focus on expression systems and methodological approaches useful for the high-level production of proteins in *E. coli*. Nutrient composition and fermentation variables are not discussed here.

EXPRESSION PLASMID

The construction of an expression plasmid requires several elements whose configuration must be carefully considered in order to ensure the highest levels of protein synthesis. The promoter is positioned approximately 10–100 bp upstream of the ribosome binding site (RBS) and is under the control of a regulatory gene that may be present on the vector itself or integrated in the host chromosome. Downstream of the promoter is the Shine and Dalgarno (SD) site, which interacts with the 3′ end of 16S rRNA during translation initiation. The distance between the SD region and the start codon ranges from 5 to 13 bases, and its sequence should eliminate the potential for secondary structure formation in the mRNA transcript, which can reduce the efficiency of translation initiation. Both 5′ and 3′ regions of the RBS exhibit a bias toward a high adenine content.

The transcription terminator is located downstream of the coding sequence and serves both as a signal to terminate transcription as well as a protective element comprising stem–loop structures thereby protecting the mRNA from exonucleolytic degradation. In addition, vectors contain a gene that confers antibiotic resistance on the host to aid in plasmid selection and propagation. Finally, the copy number of plasmids is determined by the origin of replication. Although the use of runaway replicons has been reported to result in higher yields of plasmid-encoded protein, in general, higher copy number vectors are not advantageous.

Transcription

Promoters

There are many promoters available for gene expression in *E. coli,* including those derived from Gram-positive bacteria and bacteriophages. A promoter for use in *E. coli* should have certain characteristics to render it suitable for high level protein synthesis. First, it must be strong, resulting in the accumulation of protein comprising 10–30% or more of the total cellular protein. Second, it should exhibit a minimum level of basal transcriptional activity.

Table 8.1. Promoters used for the high-level expression of genes in *E. coli*

Promoter (source)	*Regulation*	*Induction*
lac (E. coli)	*lacI, lacI^{Q}*	IPTG
	lacIts, lacI^{Q}ts	Thermal
	lacIts	Thermal
lpp (E. coli)		IPTG, lactose
trp (E. coli)		Tryptophan depletion, indole acrylic acid
tnaA (E. coli)		Tryptophan, glucose depletion
Mn-SOD *(E. coli)*		Oxygen, paraquat
phoA (E. coli)	*phoB* (positive)	Phosphate depletion
	phoR (negative)	
recA (E. coli)	*lexA*	Nalidixic acid
araBAD (E. coli)	AraC	L-Arabinose
proU (E. coli)		Osmolarity
cst-1 (E. coli)		Glucose depletion
tetA (E. coli)		Tetracycline
cadA (E. coli)	*cadR*	pH
nar (E. coli)	*fnr*	Anaerobic conditions, nitrate ion
nirB (E. coli)		Anaerobic conditions

Continue...

Promoter (source)	*Regulation*	*Induction*
tac (hybrid) *(E. coli)*	*lacI, lacI^Q^*	IPTG
	lacI	Thermal
trc (hybrid) *(E. coli)*	*lacI, lacI^Q^*	IPTG
	*lacI*ts, *lacI^Q^*ts	Thermal
lpp-lac (hybrid) *(E. coli)*	*lacI*	IPTG
$P_{syn(synthetic)}$ *(E. coli)*	*lacI, lacI^Q^*	IPTG
$P_{LtetO-I}$ *(E. coli)*		Anhydrotetracycline
$P_{lac/ara-1}$ *(E. coli)*		IPTG, arabinose
Starvation promoters *(E. coli)*		
P_L (bacteriophage λ)	*lacI*ts857	Thermal
P_L-9G-50 (mutant) (λ)		Reduced temperature (<20°C)
cspA (E. coli)		Reduced temperature (<20°C)
P_R, P_L (tandem) (λ)	*lacI*ts857	Thermal
T7 (T7)	cascaded system	IPTG
T7-*lac* operator (T7)	*lacI^Q^*	IPTG
λP_L, P_{T7} (tandem) (λ, T7)	*lacI*ts857, *lacI^Q^*	Thermal, IPTG
T3-*lac* operator (T3)	*lacI^Q^*	IPTG
T5-*lac* operator (T5)	*lacI^Q^, lacI*	IPTG
T4 gene 32 (T4)		T4 infection
nprM-lac operator (Bacillus)	*lacI^Q^*	IPTG
vhb (Vitreoscilla)	*fnr*	Oxygen, cAMP-CAP
protein A (*Staphylococcus aureus*)		
Pu (Pseudomonasputida)	*xylR*	*m*-Toluic acid
Pm (P. putida)	*xylS*	3-Methylbenzylacohol

Large-scale gene expression preferably employs cell growth to high density and minimal promoter activity, followed by induction or derepression of the promoter. The tight regulation of a promoter is essential for the synthesis of proteins that may be detrimental to the host cell. Furthermore, incompletely repressed expression systems can cause plasmid instability, a decrease in cell growth rate, and loss of recombinant protein production. A third important characteristic of a promoter is its inducibility in a simple and cost-effective manner. The most widely used promoters for large-scale protein production use thermal induction (λP_L) or chemical inducers (tac, trp). IPTG-inducible promoters are widely used in basic research. However, the use of IPTG for the large-scale production of human therapeutic proteins is undesirable due to its toxicity and cost.

Moreover, IPTG is detrimental to the physiology of *E. coli*. The availability of a mutant *lacI*ts gene that encodes a thermosensitive *lac* repressor now permits the thermal induction of these promoters. Cold-responsive promoters have been shown to facilitate efficient gene expression at reduced temperatures. The rationale behind the use of cold-responsive promoters for gene expression is based on the proposition that the rate of protein folding will be only slightly affected at about 15–20°C, whereas the rates of transcription and translation, being biochemical reactions, will be substantially decreased. This, in turn, will provide sufficient time for protein folding, yielding active proteins and avoiding the formation of inactive protein aggregates (inclusion bodies) without reducing the final yield of the target protein.

Terminators

Transcription through a promoter may inhibit its function, a phenomenon known as promoter occlusion. This interference can be prevented by the proper placement of a transcription terminator downstream of the coding sequence in order to prevent continued transcription through another promoter. Efficient transcription terminators serve several important functions: they minimize background transcription, stabilize plasmids, enhance mRNA stability, and can increase substantially the level of protein production. The advantages of tightly regulated promoters have led to the design of completely repressible expression systems that are particularly useful for the expression of genes whose products are detrimental to host growth. Most of these systems have not yet been used for the high-level production of proteins on a large scale.

Translation

Initiation

The wide range of efficiencies in the translation of different mRNAs is predominantly due to the unique structural features at the 5′ end of each mRNA species. To date no universal sequence for the efficient initiation of translation has been devised. The translational initiation region of most sequenced *E. coli* genes contains the initiation codon AUG (91%). GUG is used by about 8% of the genes, and UUG (1%) is rarely used as a start site. The spacing between the SD site and the initiating AUG codon can vary from 5 to 13 nucleotides, and it influences the efficiency of translational initiation.

The SD sequence UAAGGAGG is optimal for protein production. The secondary structure at the translation initiation region of mRNA plays a crucial role in the efficiency of gene expression. It is believed that the occlusion of the SD region and/or the AUG codon by a stem–loop structure prevents accessibility to the 30S ribosomal subunits and inhibits translation.

Several different strategies have been devised to minimize mRNA secondary structure and enhance the expression of genes. These include the enrichment of the RBS with adenine and thymidine residues, the mutation of specific nucleotides upstream or downstream of the SD region, and the use of translational coupling or dicistronic systems.

The insertion of a single amino acid residue (preferably arginine or lysine) between the N-terminal methionine and the protein of interest was reported to increase expression levels from 0.4% to 22% of the total cell protein. This strategy was effective for several types of proteins and different promoters. The recombinant proteins were converted to their native sequences by removal of dipeptides with cathepsin C. The precise mechanism for the observed increase in protein yields is unclear.

Termination

The design of expression vectors frequently includes the insertion of all three stop codons in order to prevent possible ribosome "skipping." In *E. coli* there is a preference for the UAA stop codon. A statistical analysis of more than 2,000 *E. coli* genes revealed local nonrandomness both in the stop codon and in the nucleotide immediately following the triplet and concluded that UAAU is the most efficient translational termination sequence in *E. coli*.

Table 8.2 Codon usage by *E. coli*

UUU	F	21.7	UCU	S	9.1	UAU	Y	16.2	UGU	C	5.1
UUC	F	16.7	UCC	S	8.9	UAC	Y	12.4	UGC	C	6.4
UUA	L	13.4	UCA	S	7.6	UAA	*	2.0	UGA	*	0.9
UUG	L	13.2	UCG	S	8.7	UAG	*	0.3	UGG	W	14.4
CUU	L	11.1	CCU	P	7.0	CAu	H	12.6	CGU	R	21.1
CUC	L	10.7	CCC	P	5.3	CAC	H	9.8	CGC	R	21.5
CUA	L	3.9	CCA	P	8.4	CAA	Q	14.7	CGA	R	3.6
CUG	L	51.9	CCG	P	22.8	CAG	Q	29.0	CGG	R	5.5
AUU	I	29.8	ACU	T	9.4	AAU	N	18.0	AGU	S	8.8
AUC	I	25.1	ACC	T	23.1	AAC	N	21.8	AGC	S	15.7
AUA	I	4.9	ACA	T	7.5	AAA	K	34.4	AGA	R	2.5
AUG	M	27.4	ACG	T	14.0	AAG	K	11.2	AGG	R	1.5
GUU	V	19.0	GCU	A	16.1	GAU	D	32.1	GGU	G	25.3
GUC	V	14.9	GCC	A	25.2	GAC	D	19.5	GGC	G	29.2
GUA	V	11.2	GCA	A	20.4	GAA	E	39.9	GGA	G	8.3
GUG	V	25.8	GCG	A	32.8	GAG	E	18.4	GGG	G	11.0

Synonymous codons

Genes in both prokaryotes and eukaryotes show a nonrandom usage of synonymous codons. These observations imply that heterologous genes enriched with codons that are rarely used by *E. coli* may not be expressed efficiently in *E. coli*. The minor arginine $tRNA^{Arg(AGG/AGA)}$ has been shown to be a limiting factor in the bacterial expression of several mammalian genes because the codons AGA and AGG are infrequently used in *E. coli*. Many studies have demonstrated that the substitution of high-usage for low-usage codons or the coexpression of the cognate tRNA genes can ameliorate the detrimental effect of the rare codons on gene expression levels. The location of low-usage codons in the transcript may adversely impact translation. Thus, the production of β-galactosidase decreased when AGG codons were inserted before the 10th codon from the initiation codon of the *lacZ* gene. Similarly, translational inhibition of a test mRNA was much stronger when consecutive low-usage arginine codons were located near the 5′ end of the mRNA. Moreover, 5′ translational blockage was also observed with a different sequence of consecutive low-usage leucine codons (CUA), making it unlikely that structural effects on the mRNA were responsible. In another case tandem AGG triplets caused a substantial inhibition of gene expression independently of their localization in mRNA, a phenomenon that was attributed to competition of the tandem AGGAGG codons with the natural SD sequence.

In contrast, other workers reported efficient expression of genes that contained low-usage codons. To date, it has not been possible to formulate precise "rules" to predict whether the content of low-usage codons in a specific gene might adversely affect the efficiency of its expression in *E. coli*. The experimental results may possibly be confounded by several variables such as positional effects, the clustering or interspersion of the rarely used codons, secondary structure of the message, and other effects. Nevertheless, it is clear that the codon context of specific genes *can* have adverse effects on both the quantity and quality of the expressed protein. Usually, this problem can be rectified by the alteration of the codons in question, or by the coexpression of the cognate tRNA genes.

Longevity of mRNA

mRNA degradation is not effected randomly by nonspecific endonucleolytic cleavage, since there is no inverse correlation between mRNA length and half-life. Two classes of protective elements are known to stabilize mRNAs in *E. coli*. One class consists of sequences in the 5′ untranslated regions (UTRs) of mRNAs, and the other class includes stem–loop structures from the 3′ UTRs and intercistronic regions. Some of these elements act as stabilizers when fused to heterologous mRNAs, but only under restricted conditions. In contrast, the 5′ UTR of the *E. coli ompA* transcript prolongs the half-life of a number of heterologous mRNAs in *E. coli* under normal conditions of rapid cell growth.

Belasco and colleagues showed that the presence of a stem–loop at or very near the extreme 5′ terminus of the *ompA* 5′ UTR is essential for its stabilizing effect. Furthermore, the half-life of a normally labile mRNA was prolonged in *E. coli* by adding a hairpin structure at its 5′ terminus, and it was proposed that *E. coli* mRNAs beginning with a long single-stranded segment are preferentially targeted by a ribonuclease that interacts with the 5′ terminus before cleaving the mRNA at internal sites. It appears, therefore, that the addition of the *ompA* 5′ stabilizer to heterologous genes might enhance gene expression in *E. coli*. However, it is possible that mRNAs that contain internal RNA processing sites may not be protected by the presence of the 5′ stabilizer.

The other class of mRNA protective elements consists of 3′ UTR sequences that can form stem–loop structures, thereby blocking exonucleolytic degradation of the transcript from the 3′ terminus. Such an element was identified within the transcription terminator of the crystal protein gene of *Bacillus thuringiensis*. Its fusion to the 3′ termini of the penicillinase *(penP)* gene of *B. licheniformis* and the human interleukin 2 cDNA increased the half-life of the mRNAs and enhanced the production of the corresponding polypeptides in both *B. subtilis* and *E. coli*. However, as in the case of some of the 5′ stabilizers, this 3′ "retroregulator" is unlikely to act as a universal mRNA stabilizer.

For example, the replacement of the 3′-terminal hairpins of labile mRNAs with those from stable messages did not enhance the expression of the labile transcripts. It has been suggested that gene expression might be enhanced by the use of host strains that are deficient in specific ribonucleases, such as RNase II or PNPase. This too is unlikely to be an effective strategy, because the absence of RNase II or PNPase, as well as the overproduction of RNase II, had no effect on the average half-life of *E. coli* bulk mRNA. Moreover, strains that were deficient in both RNase II and PNPase were inviable.

The deletion of 110 bp in the 3′ untranslated region of a cDNA encoding the human TNF-α was reported to enhance the expression level of this cytokine. The authors suggested that a UUUAUUA sequence at the 3′ end of the mRNA may have an inhibitory effect on the expression of TNF-α in *E. coli*. It is well known that 3′ AU-rich sequences in cytokines, growth factors, and oncoproteins have an inhibitory effect on translation and may destabilize the mRNA. However, AU-binding proteins are not known to exist in bacteria. Therefore, the observed increase in TNF-α yield is probably due to some other mechanism.

COMPARTMENTS OF *E. COLI*

Cytoplasm

Each of the four "compartments" of *E. coli*, i.e. the cytoplasm, periplasm, membranes/cell surface, and the extra-cellular medium, offers advantages as well as disadvantages for protein production. The formation of inclusion bodies remains a significant barrier to gene expression in the cytosol. Inclusion bodies do offer several advantages, particularly the production of proteins that are toxic to *E. coli*. For example, the highly insoluble bovine prochymosin has recently been used as a fusion partner for the production of antimicrobial peptides as inclusion bodies in *E. coli*. On the other hand, the advantages of inclusion bodies are counterbalanced by the arduous task of refolding the aggregated protein, the uncertainty of whether the refolded protein retained its biological activity, and the reduction in yield of the refolded and purified protein.

To date the precise physicochemical parameters that contribute to the formation of inclusion bodies remain unclear. A statistical analysis of the composition of 81 proteins that do and do not form inclusion bodies in *E. coli* concluded that charge average and turn-forming residue fraction are strongly correlated with inclusion body formation. These observations led to the development of a model to predict the probability of inclusion body formation solely based on the amino acid composition of a protein. The model was used to predict accurately the insolubility of the human T-cell receptor Vβ5.3 in *E. coli*, as well as to enhance the solubility of a protein by adding negatively charged residues. Several experimental approaches have been utilized to minimize the formation of inclusion bodies and improve protein folding.

These include the growth of bacterial cultures at lower temperatures, selection of different *E. coli* strains, substitution of selected amino acid residues, co-production of chaperones, use of *E. coli* DsbA or thioredoxin either as fusion partners or coproduced with the protein of interest, growth and induction of the cells under osmotic stress in the presence of sorbitol and glycyl betaine, addition of nonmetabolizable sugars to the growth medium, alteration of the pH of the culture medium, and the use of strains deficient in thioredoxin reductase. The reducing potential of the cytoplasmic redox state presents still another problem. Bacterial cytoplasmic proteins contain few cysteine residues and few disulfide bonds. Most proteins that contain stable disulfide bonds are exported from the cytoplasm. Thus, mammalian proteins whose complex tertiary structure depends in part on disulfide bond formation may not be produced in their correct conformation in the bacterial cytoplasm.

Table 8.3 Relative merits of different compartments for gene expression in *E. coli*

Advantages	*Disadvantages*
	Cytoplasm
Inclusion bodies. Facile isolation of protein in high purity and concentration, target protein protected from proteases, desirable for production of proteins that, if active, are lethal to host cell.	*Inclusion bodies.* Protein insolubility, refolding to regain protein activity, refolded protein may not regain its biological activity, reduction in final protein yield, increase in cost of goods.
Higher protein yields	Reducing environment: does not facilitate disulfide bond formation.
Simpler plasmid constructs.	N-terminus authenticity.
	Proteolysis.
	Purification is more complex (more protein types).
	Periplasm
Purification is simpler (fewer protein types).	Signal peptide does not always facilitate transport.
Proteolysis is less extensive.	Inclusion bodies may form.
Improved disulfide bond formation/folding.	
N-terminus authenticity.	
	Membranes/cell surface
Surface display may facilitate drug screening, biocatalysis and other applications.	
Some progress made in overexpression of membrane proteins.	
	Extracellular
Least level ofproteolysis.	No secretion, usually.
Purification is simpler (fewest protein types).	Protein diluted, more difficult to purify.
Improved protein folding.	
N-terminus authenticity.	

Mutant *E. coli* strains were isolated that allow the formation of disulfide bonds in the cytoplasm. These mutations inactivate the *trxB* gene that encodes thioredoxin reductase and contributes to the sulfhydryl reducing potential of the cytoplasm. Thioredoxin itself was unnecessary for disulfide bond formation. The authors suggested that the cytoplasm may contain another thioredoxin-like protein that can be reduced by thioredoxin reductase; in the absence of thioredoxin reductase, the oxidized form of this protein facilitates formation of disulfide bonds in the cytoplasm. Other workers have utilized *E. coli* strains carrying null mutations in the *trxB* gene, and observed significant amounts of functional disulfide-containing protein in the cytoplasm. Proteins produced in the cytoplasm may not possess an authentic N-terminus. The cytoplasmic expression of a gene without a leader requires the presence of an initiation codon, the most common one encoding methionine. Although this amino acid might have no adverse effect on the protein synthesized, there are specific cases in which the extraneous methionine residue may have profound consequences.

Bacterial translation is initiated by *N*-formyl-methionine, which is deformylated during synthesis but not necessarily removed. The N-terminal methionine might be cleaved off by an endogenous methionine aminopeptidase (MAP), depending on the side-chain length of the second amino acid residue. Thus, residues with small side chains such as Gly, Ala, Pro, Ser, Thr, Val, Cys, and to a lesser degree, Asn, Asp, Leu, and Ile facilitate the MAP-catalyzed removal of the N-terminal methionine. Several strategies have been successfully used to remove the extra methionine residue from recombinant proteins in vivo, including the coexpression of the *E. coli* methionine aminopeptidase gene and the use of exopeptidases. Finally, cytosolic gene expression necessitates purification of the target protein from the large pool of intracellular proteins.

Advantages of Periplasm

The periplasm offers several advantages for protein targeting. In contrast to the cytosolic compartment, the *periplasm* contains only 4% of the total cell protein. The target protein is thus effectively concentrated, and its purification is considerably less onerous. The oxidizing environment of the periplasm facilitates the proper folding of proteins, and the cleaving in vivo of the signal peptide during translocation to the periplasm is more likely to yield the authentic N-terminus of the target protein. Protein degradation in the periplasm may be less extensive than in the cytosol, but it is still a significant issue. The transport of a protein through the inner membrane to the periplasm normally requires a signal sequence. A wide variety of signal peptides have been employed successfully in *E. coli* for protein translocation to the periplasm. These include prokaryotic signal sequences, such as the *E. coli* PhoA signal, OmpA, OmpT, LamB, OmpF, β-lactamase, enterotoxins ST-II, LT-A, LT-B, protein A from *Staphylococcus aureus*, endoglucanase from *B. subtilis*, PelB from *Erwinia carotovora*, a degenerate PelB signal sequence, the murine ribonuclease, and the human growth hormone signal. However, the presence of a signal peptide does not always ensure efficient protein translocation through the inner membrane.

It is clear that besides the signal peptide, other structural features in proteins are involved in membrane transport. Strategies for the improved translocation of proteins to the periplasm include the supply of components involved in protein transport and processing, such as overproduction of signal peptidase I, use of *prlF* mutant strains, coexpression of the *prlA4* and *secE* genes, coexpression of the *prlF* gene, and expression of the *pspA* gene. The intracellular accumulation of unprocessed proteins has often been interpreted as a possible saturation of the *E. coli* translocation system. For example, Hsiung et al. observed the intracellular accumulation of a greater amount of human growth hormone precursor after IPTG induction, but no increase in the amount of translocated growth hormone. Similarly, the *Chlamydia trachomatis* major outer membrane protein (MOMP) was successfully targeted to the *E. coli* surface, but upon prolonged induction, the unprocessed higher molecular weight MOMP accumulated. Although this may have been caused by saturation of the transport mechanisms, the authors point out that other explanations are also possible, such as differences in the carboxyl terminus of the chlamydial leader peptide, or formation of inclusion bodies, making a large proportion of MOMP unavailable for processing and resulting in the accumulation of unprocessed MOMP.

Other workers dismiss the possibility of jamming of the protein export machinery, pointing out that under proper conditions correctly processed and fully active proteins can accumulate in the periplasm at a level 20–50% of the total cell protein. In agreement with this, insulin-like growth factor I (IGF-I) was expressed in the periplasm at up to 30% of the total cell protein. Moreover, there was no unprocessed IGF-I detected in the cytosol. However, IGF-I expression was achieved in the presence of the thiol oxidant proteins DsbA or DsbC, which might have increased protein yields by possibly assisting translocation of IGF-I. Interestingly, the overexpression of the disulfide oxidoreductases increased the fraction of aggregated IGF-I for reasons not understood. The basic protein Skp has been identified as a periplasmic molecular chaperone. Overexpression of Skp was shown to increase the amount of fusion proteins incorporated into phage.

The effect was most notable on scFv fragments that are aggregation prone, suggesting that Skp acts by increasing the amount of correctly folded proteins, which can then get incorporated into the phage coat. The chaperone composition of the periplasm is not as well characterized as that of the cytoplasm, and the mechanism of action of Skp is unclear. Further studies should demonstrate whether Skp and other periplasmic molecular chaperones have wide applicability in periplasmic protein expression.

Cell Boundaries

The cell surface of *E. coli* may facilitate several applications including vaccine development, biocatalysis, protein-protein interactions, and drug screening. In many cases production of membrane proteins is limited by toxicity and bacterial cell death. A procedure has recently been described for the selection of *E. coli* strains that facilitate the overproduction of proteins and their incorporation into the membrane.

Secreted Proteins

The targeting of synthesized proteins for secretion to the culture medium presents significant advantages. Unfortunately, *E. coli* normally secretes very few proteins, and the manipulation of the various transport pathways to facilitate secretion of foreign proteins remains a formidable task. The methodological approaches to protein secretion in *E. coli* may be conveniently divided into two categories: (i) the exploitation of existing pathways for "truly" secreted proteins; and (ii) the use of signal sequences, fusion partners, permeabilizing proteins, nutrients, or other agents that effect protein secretion as a result of "leakage" or selective and limited permeability of the outer membrane. The first approach offers the advantage of specific secretion of the protein of interest, hence minimum contamination by nontarget proteins. For example, the hemolysin gene has been used for construction of secreted hybrid proteins. Secretion, however, is not a particularly efficient process.

The second approach relies on the induction of limited leakage of the outer membrane to render protein secretion, albeit with modest yields. Examples are the use of the *pelB* leader, *ompA* leader, protein A leader, the coexpression of bacteriocin release protein along with the addition of glycine to the culture medium, and the coexpression of the *kil* gene for membrane permeabilization. A mutated bacteriocin release protein was reported to provide more efficient extracellular production of proteins by *E. coli.*

LOCALIZING PROTEINS

Proteolysis is a selective, highly regulated process that plays an important role in cellular physiology. *E. coli* contains at least 40 enzymes that catalyze the hydrolysis of peptide bonds. These enzymes participate in a wide range of metabolic processes including the removal of N- or C-terminal amino acid residues, catabolism of dipeptides, oligopeptides, and proteins, the selective removal of abnormal proteins, and the inactivation of regulatory proteins. Protein damage or alteration may result from a variety of conditions, such as incomplete polypeptides, mutations caused by amino acid substitutions, excessive synthesis of subunits from multimeric complexes, posttranslational damage through oxidation or free radical attack, and genetic engineering. Such abnormal proteins are efficiently removed by the bacterial proteolytic machine. *E. coli* proteases are localized in the cytoplasm, the periplasm, and the inner and outer membranes.

Table 8.4 Localization of proteases in *E. coli*

Cytoplasm	*Inner membrane*	*Periplasm*	*Outer membrane*
La (Lon)	SppA (protease IV)	Deg P (HtrA, Do)	OmpT (protease VII)
ClpP	DegS (HhoB)	DegQ (HhoA)	OmpP
ClpX	Protease VI	Ptr (Pi, III, pitrilysin)	OmpX
ClpA (Ti)	LepB (signal peptidase I)	Prc (Tsp, Re)	
ClpQ	LspA (signal peptidase II)	Mi	
ClpY	SohB		
HflB (FtsH, TolZ)	Protease V		
Protease II			
Ci			
Fa			
In			
So			

Some determinants of protein instability have been elucidated. The "N-end rule" relates the metabolic stability of a protein to its amino-terminal residue. Thus, in *E. coli,* N-terminal Arg, Lys, Leu, Phe, Tyr, and Trp conferred 2-min half-lives on a test protein, whereas all the other amino acids, except proline, conferred greater than 10-h half-lives on the same protein. As already discussed in "Cellular Compartments, Cytoplasm", amino acids with small side chains in the second position of the polypeptide facilitate the MAP-catalyzed removal of the N-terminal methionine. Therefore, these studies suggest that Leu in the second position would probably be exposed by the removal of the methionine residue and destabilize the protein. The second determinant of protein instability is a specific internal lysine residue located near the amino terminus. This residue is the acceptor of a multiubiquitin chain that facilitates protein degradation by a ubiquitin-dependent protease in eukaryotes.

Another correlation between amino acid content and protein instability is presented in the PEST hypothesis, which proposes that proteins are destabilized by regions enriched in Pro, Glu, Ser, and Thr, flanked by certain amino acid residues. It was suggested that PEST-rich proteins maybe produced efficiently in *E. coli,* which apparently lacks the PEST proteolytic system. Strategies for minimizing proteolysis of recombinant proteins in *E. coli* have been reviewed. These include protein targeting to the periplasm or the culture medium, use of protease-deficient host strains, growth of the host cells at low temperature, construction of N- and/or C-terminal fusion proteins, tandem fusion of multiple copies of the target gene, coexpression of molecular chaperones, coexpression of the T4 *pin* gene, replacement of specific amino acid residues in order to eliminate protease cleavage sites, modification of the hydrophobicity of the target protein, and optimization of fermentation conditions.

The usefulness of some of these methods maybe limited, depending on the intended use of the recombinant protein. Thus, for example, the presence of fusion moieties on the target protein may possibly interfere with functional or structural properties or therapeutic applications of the product. The engineering of enzymatic or chemical cleavage sites for the subsequent removal of the fusion partners involves many considerations: the accessibility of the cleavage sites to enzyme digestion, the purity, specificity and cost of the commercially available enzymes, the authenticity of the N- or C-termini upon enzymatic digestion, the possible modification of the target protein upon chemical treatment, and so forth. For the large-scale production of fusion proteins, some of these difficulties are amplified. Similarly, the fusion of multiple copies of the target gene to effect multidomain polypeptides requires the subsequent conversion to monomeric protein units by cyanogen bromide cleavage. In this case, the target protein must not contain internal methionine residues and must be able to withstand harsh reaction conditions. Moreover, a limited extend of amino acid side-chain modification may occur, and the toxicity of CNBr presents a significant issue for large-scale cleavage reactions.

Similarly, the “rational” modification of a protein sequence requires structural information that may not be available. Molecular chaperones (see later) have been used successfully to stabilize specific proteins, but this approach to date remains a hit or miss affair. The cytoplasm of *E. coli* appears to contain a greater number of proteases than the periplasm. Therefore, proteins located in the periplasm are less likely to be degraded. However, proteolytic activity in the periplasm is substantial. Secretion into the culture medium would provide a better alternative in terms of protein stability. Unfortunately, the technology for secretion of proteins from *E. coli* into the culture medium is still in its infancy.

A major catalyst of protein degradation in bacteria is the induction of heat shock proteins in response to a variety of stress conditions, such as the thermal induction of gene expression or the accumulation of abnormal or heterologous proteins in the cytoplasm. Under these conditions the production of the *lon* gene product, protease La, and other proteases, is enhanced. This problem is minimized by the use of host strains deficient in the *rpoH (htpR)* locus. The *rpoH* gene encodes the RNA polymerase σ^{32} subunit, which regulates several proteolytic activities in *E. coli* Hosts that carry the *rpoH* mutation have been demonstrated to increase substantially the production of foreign proteins in *E. coli.* A large number of protease-deficient hosts exists, including those that are deficient in all known protease loci that affect the stability of secreted proteins.

COMPLICATED FOLDING IN PROTEIN

The efficient posttranslational folding of proteins, the assembly of polypeptides into oligomeric structures, and the localization of proteins are mediated by specialized proteins termed *molecular chaperones*. The demonstration that the production and assembly of prokaryotic ribulose bisphosphate carboxylase in *E. coli* requires both groES and groEL proteins, led to an increasing interest in the use of molecular chaperones for high-level gene expression in *E. coli*. The experimental results from the use of chaperones, however, have been inconsistent, and thus far the effects of chaperone coproduction on gene expression in *E. coli* appear to be protein specific. It is also unclear whether the in vivo levels of different chaperone species are limiting under conditions of gene overexpression. Contradictory results may be due to several reasons. Proteins that are drastically destabilized will probably fold incorrectly, even in the presence of chaperones.

Thus, the truncation of polypeptides, the production of single domains from multisubunit protein complexes, lack of formation of disulfide bonds (which ordinarily contribute to protein structure), or absence of posttranslational modifications such as glycosylation may make it impossible to attain thermodynamic stability. It was shown that calmodulin coexpression was essential to generate active inducible nitric oxide synthase (iNOS) in *E. coli,* and it was suggested that calmodulin binds to iNOS after translation and aids in folding and stabilizing the enzyme. It is now clear that different types of chaperones normally act in concert. Therefore, the overproduction of a single chaperone may be ineffective. Another variable to consider is growth temperature. For example, GroES/EL coexpression increased the production of β-galactosidase at 30°C, but not at 37°C or 42°C, whereas DnaK and DnaJ were effective at all temperatures tested.

Finally, the overexpression of chaperones can lead to phenotypic changes, such as cell filamentation, that can be detrimental to cell viability and protein production. The coexpression of the human or rat protein disulfide isomerase (PDI) enhanced the yield of correctly folded protein in the *E. coli* periplasm. Disulfide bond formation in the *E. coli* periplasm is facilitated by a group of proteins that maintain the correct redox potential. It is thought that DsbA, a soluble periplasmic protein, directly catalyzes disulfide bond formation in proteins, whereas DsbB, an inner membrane protein, is involved in the reoxidation of DsbA. Eukaryotic PDI was capable of complementing the phenotypes of *dsbA* null mutants, but its function was virtually abolished in *dsbB* mutants. In addition, the ability of PDI to enhance the yield of target proteins was increased in the presence of exogenously added glutathione. These observations suggest that PDI depends on the presence of bacterial redox proteins for its reoxidation.

The coexpression of rat PDI has also been reported to enhance the correct folding of tissue plasminogen activator. Protein misfolding can be attributed to the intracellular concentration of aggregation-prone intermediates, suggesting that reducing the rate of protein synthesis should disfavor protein misfolding. Indeed, the use of weaker promoters or conditions of partial induction from stronger promoters can result in higher amounts of soluble protein. The ability of *E. coli* to secrete a large amount of alkaline phosphatase into the periplasm was due to a lower synthetic rate of the *phoA* gene product, and a reduction in promoter strength enhanced the extracellular production of exoglucanase by *E. coli*.

INDUSTRIAL FUSION OF PROTEINS

In recent years there has been a remarkable increase in the variety of fusion proteins for biological research. Other studies have addressed the design and engineering of excision sites for the chemical or enzymatic cleavage and removal of fusion partners. This section briefly summarizes the use of selected fusion systems that have a direct impact on high-level production and, in some cases, secretion of target proteins. Uhlen and colleagues developed a multifunctional fusion partner based on staphylococcal protein A or synthetic derivatives thereof. In addition to its utility as an affinity tag for purification, the protein A moiety acts as a solubilizing partner to improve folding, and the presence of the protein A signal peptide causes the secretion of the gene product to the culture medium. An alternative fusion partner is derived from streptococcal protein G (SPG), a bacterial cell wall protein that has separate binding regions for albumin within the amino-terminal domains, and for IgG within the carboxyl-terminal domain.

The combination of both protein A and SPG domains formed a tripartite fusion protein, thus providing an additional purification option and further protecting the target protein from proteolytic degradation. Another fusion partner, thioredoxin, has been shown to dramatically increase the solubility of fusion proteins produced in the *E. coli* cytoplasm, and to prevent the formation of inclusion bodies. Similarly, the thioredoxin homolog DsbA has been used as a fusion partner to direct the transport of proteins to the periplasm. More recently, a mutant form of DsbA ($DsbA^{mut}$) was used as a fusion partner to confer higher solubility to 10 different eukaryotic proteins expressed in the cytoplasm at 37°C. The consistent performance of $DsbA^{mut}$ led the authors to conclude that this protein may be used as a generic fusion partner for enhancing the solubility of a variety of heterologous eukaryotic proteins.

An additional advantage of $DsbA^{mut}$ may be its ability to exert its solubilizing effect at 37°C rather than at the lower growth temperatures often employed for enhancing solubility of proteins: at reduced temperatures the cell growth cycle becomes longer, with a significant impact on fermentation cycle time and a reduced protein yield. The fusion of genes to the ubiquitin sequence significantly increased the yield of proteins. The remarkable increase in protein yield was thought to be due to protection of the target protein from proteolysis, improved folding, and efficient mRNA translation. Ubiquitin or the ubiquitin metabolic pathway is absent in prokaryotic organisms. In order to remove the ubiquitin moiety from fusion proteins, Baker et al. coexpressed the ubiquitin-specific protease Ubp2 in *E. coli*, thus effecting the cotranslational cleavage of ubiquitin from the fusion protein.

CONCLUDING REMARKS

An efficient prokaryotic expression vector contains a strong and tightly regulated promoter, a Shine–Dalgarno site that is positioned approximately 9 bp 5′ to the translation initiation codon and is complementary to the 3′ end of 16S rRNA, and a transcription terminator positioned 3′ to the protein-coding sequence. In addition, the vector requires an origin of replication, a selection marker, and a gene that facilitates the stringent regulation of promoter activity. The translational initiation region of a gene must be free of secondary structures that may occlude the initiation codon and/or block ribosome binding. UAAU is the most efficient translation termination sequence in *E. coli*. Each of the four "compartments" for targeted protein production, namely, the cytoplasm, periplasm, membranes/cell surface, and the growth medium, offers advantages and disadvantages for gene expression.

The formation of inclusion bodies can be minimized using a variety of techniques, but it remains a significant barrier to high- level protein production in the cytoplasm. The effectiveness of molecular chaperones has been protein specific so far. Codon usage can have adverse effects on the synthesis and yield of recombinant proteins. Positional effects appear to play an important role in protein synthesis. Thus, the presence of rare codons near the 5′ end of a transcript probably affects translational efficiency. This problem may be rectified by the alteration of the rare codons, and/or the coexpression of the cognate tRNA genes. To date there is no generally applicable strategy to prevent the degradation of a wide variety of mRNA species in *E. coli*. Although certain 5′ and 3′ stem–loop structures have been shown to block mRNA degradation, these seem to stabilize only specific mRNAs, under restricted conditions. One exception appears to be the 5′ UTR of the *E. coli ompA* transcript, which prolongs the half-life of a number of heterologous mRNAs in *E. coli*. The wide variety of existing fusion partners have utility in the production, detection, and purification of recombinant proteins.

Specific fusion moieties can increase the folding, solubility, resistance to proteolysis, and secretion of recombinant proteins into the growth medium. Protein misfolding, attributed to the intracellular concentration of aggregation-prone intermediates, may be minimized by a combination of experimental approaches: replacement of amino acid residues that cause aggregation, coexpression of molecular chaperones and foldases, reduction of the rate of protein synthesis, the use of solubilizing fusion partners, and the careful optimization of growth conditions. Proteolysis in *E. coli* may be minimized by the use of protease-deficient host strains, the construction of fusion proteins, the coexpression of molecular chaperones, the elimination of protease cleavage sites through genetic engineering, and the optimization of fermentation conditions. Host strains that are deficient in the *rpoH (htpR)* locus are effective, particularly for thermally induced expression systems.

9

MICROBIAL PRODUCTION OF L-DOPA

Tyrosine phenol-lyase (TPL) (deaminating) catalyzes 3,4-dihydroxyphenyl-L-alanine (L-DOPA) synthesis from pyrocatechol, pyruvate, and ammonia in a one- step reaction. In the investigation to produce L-DOPA by this enzyme reaction, a bacterial strain (*Erwinia herbicola*) has been chosen as the best producer of TPL out of more than 1,000 others tested. Culture conditions for the preparation of cells containing high TPL activity and reaction condition for the synthesis of L-DOPA were optimized with this bacterium to support L-DOPA, appearing as the crystals in the reaction mixture, at final amounts reaching 110 g/L. L-DOPA is the precursor of the neurotransmitter dopamine and is an effective agent in the treatment of Parkinson's disease. In this disease, the amount of dopamine in the brain of the patient is insufficient. L-DOPA had been produced mainly by a synthetic method that contains many chemical reaction steps, including a step for optical resolution. TPL is a pyridoxal phosphate dependent multifunctional enzyme obtained from bacteria that catalyzes the synthesis of L-DOPA from pyruvate, ammonia, and pyrocatechol in a one-step reaction. The L-DOPA from *E. herbicola,* is produced in cells containing high amount of TPL. This biocatalytic method is one of the most economical processes to date for the production of L-DOPA.

ORGANISMS WITH HIGH TPL ACTIVITY

The activity of TPL in microorganisms was investigated with 646 strains of bacteria, 140 strains of yeast, 138 strains of fungi, and 117 strains of actinomycetes. TPL activity was determined by measuring the amount of L-tyrosine or L-DOPA synthesized through the β-replacement reaction between L-serine and phenol or pyrocatechol, respectively. Cells of microorganisms were directly added to the reaction mixture as the enzyme for practical applications. TPL activity occurred widely in a variety of bacteria, most of which belong to the Enterobacteriaceae, especially to the genera *Escherichia, Aerobacter, Proteus,* and *Erwinia*. No activity was observed in yeasts, fungi, and actinomycetes. A large number of bacterial strains isolated from rice and corn plants showed high TPL activity. Cells of *E. herbicola* ATCC 21434, which showed the highest activity, were selected as a promising source of enzyme for the preparation of L-DOPA. Aeration, pH, and temperature affect the formation of TPL. Cell growth requires oxygen transfer, but the condition for TPL synthesis may have a different optimum. For example, the optimum volume of the medium for the formation of TPL was 40 to 60 mL in a 500-mL shaking flask, but cell growth was rather repressed in this volume. The optimum pH for enzyme formation in this example was in the range of 7.0 to 7.5, and the optimum temperature for enzyme formation was in the range of 27 to 32°C.

TPL activity occurs during cultivation of *E. herbicola,* and continues beyond the active growth phase. Cells were grown at 31.5°C for 32 h with reciprocal shaking. The medium contained 0.2% tyrosine, 0.2% KH_2PO_4, 0.1% $MgSO_4 \cdot 7H_2O$, 1.0% yeast extract, 0.5% polypeptone, and 0.5% meat extract (pH 7.5) and supported maximum formation of the enzyme during the early stationary phase (cultivated for 26 to 30 h); thereafter, the enzyme gradually disappeared with the consumption of L-tyrosine added to the medium. Additions of yeast extract, meat extract, polypeptone, and the hydrolyzate of soybean protein to the basal medium enhance cell growth as well as the formation of TPL. DL-Methionine, DL-alanine, and glycine stimulate enzyme formation. However, L-serine, L-ornithine, and L-cystine inhibit enzyme formation. Pyridoxine is added as a required precursor of coenzyme for enzyme formation. Catabolite repression of formation of TPL by *E. herbicola* has been observed on adding glucose, pyruvate, and α-ketoglutarate to the medium at high concentrations. Glycerol is a suitable carbon source for cell growth as well as for the accumulation of the enzyme in growing cells. A marked increase in enzyme formation has been seen when glycerol is added together with succinate, fumarate or malate.

Among the metallic ions tested, Fe^{2+} ion was the only one effective for the formation of TPL. TPL is an inducible enzyme, and the addition of L- tyrosine to the medium is essential for formation of the enzyme. However, when large amounts of L-tyrosine were added, inhibition of enzyme formation and repression of cell growth were observed. The phenomena seem to be caused by the phenol that is liberated from the L-tyrosine. Various analogs of L-tyrosine that were not decomposed during cultivation were investigated in the search for an inducer of TPL, but no effective, nondecomposable inducer was found among the L-tyrosine analogs tested. L-Phenylalanine, D-phenylalanine, and phenylpyruvate did not induce this enzyme by themselves, but they show a synergistic effect on the induction of the enzyme by L- tyrosine. L- and D-phenylalanine are competitive inhibitors of TPL. Cells of *E. herbicola* with high TPL activity are typically prepared by growing them at 28°C for 28 h in a medium containing 0.2% KH_2PO_4, 0.1% $MgSO_4$ $7H_2O$, 2 ppm Fe^{2+} ($FeSO_4 \cdot 7H_2O$), 0.01% pyridoxine • HCl, 0.6% glycerol, 0.5% succinic acid, 0.1% DL-methionine, 0.2% DL-alanine, 0.05% glycine, 0.1% L-phenylalanine, and 12 mL of hydrolyzed soybean protein in 100 mL of tap water, with the pH controlled at 7.5 throughout cultivation. Under these conditions, TPL is efficiently accumulated in the cells of *E. herbicola* and makes up about 10% of the total soluble cellular protein.

RESTRICTED SYNTHESIS OF L-DOPA

Pyruvate, Ammonia, and Pyrocatechol in Synthesis

The optimum pH for this reaction is around 8.0, and the optimum temperature is 16°C. Because pyruvate is unstable in the reaction mixture at a high temperature, low temperature is favorable for the synthesis of L-DOPA. The reaction is carried out at 16°C for 48 h in a reaction mixture containing various amounts of sodium pyruvate, 5 g of ammonium acetate, 0.6 g of pyrocatechol, 0.2 g of sodium sulfite, 0.1 g of EDTA, and cells harvested from 100 mL of broth, in a total volume of 100 mL. The pH was adjusted to 8.0 by the addition of ammonia. At 2-h intervals, sodium pyruvate and pyrocatechol were added to the reaction mixture to maintain the initial concentrations. The maximum synthesis of L-DOPA was obtained when the concentration of sodium pyruvate was kept at 0.5%. L-DOPA appears as the crystals in the reaction mixture.

When a higher concentration of pyruvate was added to the reaction mixture during the synthesis of L-DOPA, two by-products, 1-methyl-1,3-dicarboxy-6,7-dihydroxy-1,2,3,4-tetrahydroisoquinoline and 1-methyl-3-carboxy-6,7-dihydroxy-3,4-dihydroisoquinoline, were formed. The by-products have been shown to be formed through a spontaneous condensation reaction. TPL catalyzes a variety of α,β-elimination, β-replacement, and racemization reactions and the synthesis of L-tyrosine and its related amino acids. The mechanism for these reactions has been studied in some detail, by adopting the general mechanism proposed for pyridoxal-dependent reactions.

Table 9.1 Comparison of chemical and enzymatic methods of L-DOPA production

	Chemical method	*Enzymatic method*
Starting materials	Vanillin, hydantoin, acetic anhydride, hydrogen	Pyrocatechol, pyruvate, ammonia
Total number of unit-reaction	8	1
By-products	Ammonia, CO_2, acetate	H_2O
Necessity of optical resolution	Yes	No
Equipment	Specific plant	Commonly usable fermenter
Period (day)	15	3
Contaminating amino acids	L-Tyrosine, 3-methoxytyrosine, trihydroxy-phenylalanine	L-Tyrosine

An enzymatic method for preparing L-DOPA was developed using bacterial cells with high TPL activity. This method is simple and is one of the most economical processes to date for preparing L-DOPA. L-DOPA is currently produced by this method with 60-m^3 scale fermenters. The yield of L-DOPA compared to pyrocatechol is 98% and to pyruvate, 90%, with the final accumulation of L-DOPA reaching 110 g/L. Certain L-tyrosine-related amino acids can be synthesized from the starting materials of pyruvic acid and appropriate derivatives of phenol, such as pyrocatechol.

10

MICROBIAL PRODUCTION OF β-GALACTOSIDASE

β-Galactosidase catalyzes the hydrolysis and transgalactosylation reactions of β-D-galactopyranosides, such as lactose. The enzyme occurs in a wide variety of organisms including microorganisms, plants, and animals. The application of β-galactosidase in bioprocess technology has been achieved exclusively with microbial enzymes, which have long been used for the hydrolysis of lactose for increasing the digestibility of milk or for improving the functional properties of dairy products. During the past decade, another potential application of the enzyme has also been developed; the β-galactosidase-catalyzed transgalactosylation has proved to be useful for structural and functional modifications of food materials, medicines, and other biologically active compounds. In this article, we survey the enzymology and application in bioprocess technology of the microbial β-galactosidases with special emphasis on application of their transgalactosylation.

METHODOLOGY

β-Galactosidase occurs in a variety of microorganisms including yeasts, molds, bacteria, actinomycetes, and archaebacteria. β-Galactosidase is known as a marker enzyme for coliform bacteria, which are indicators of the fecal pollution of water. Microbial β-galactosidases have been purified by a combination of several conventional techniques, such as salting-out fractionation, ion exchange, gel filtration, hydroxyapatite, and hydrophobic interaction chromatographies. The affinity chromatography of β-galactosidase uses *p*-aminophenyl-(or *p*-aminobenzyl) β-D-thiogalactopyranoside agarose, which is commercially available and has been used for the purification of the *Escherichia coli lacZ* β-galactosidase, its recombinant derivatives, and the other β-galactosidases.

Chemicals Required

The specificity and molecular properties of the microbial β-galactosidases differ significantly with the source of the enzyme. For example, the *Saccharopolyspora rectivirgula* β-galactosidase is highly specific for β-D-galactosides, whereas the *Sulfolobus solfataricus* enzyme shows broad substrate specificity and can efficiently hydrolyze β-glycosides other than β-galactosides as well. The catalytic competence and specificity of transgalactosylation also depend on the enzyme. The reported native molecular weights and subunit structures of β-galactosidases range, respectively, from 19,000 to 630,000 (the *Kluyveromyces lactis* enzyme) and from monomer (e.g., the *S. rectivirgula* enzyme) to heterooctamer (the *E. coli ebg* enzyme).

The bacterial β-galactosidases generally show optimum pHs at neutral regions whereas most mold enzymes show them at acidic regions, which in some cases reach pH 2. Some β-galactosidases of the *lacZ* family require essential monovalent and divalent ions for their activities; the *S. rectivirgula* β-galactosidase is a multimetal enzyme that requires multiple divalent ions for its maximum thermostability and activity. However, some other β-galactosidases (e.g., the *Rhizobium meliloti* enzyme) do not show such metal ion requirements. These observations strongly suggest the structural diversity of the microbial β-galactosidases. Many chromogenic, fluorogenic, and luminogenic substrates that are specific for β-galactosidase have been designed. D-Galactose, D-galactosylamines, β-D-thiogalactosides, and their analogs and derivatives serve as specific inhibitors of some microbial β-galactosidases.

Taxonomic Status

Henrissat has compared all the available sequences of glycosyl hydrolases using hydrophobic cluster analysis and classified these enzymes into families. Many microbial β-galactosidases have thus far been cloned and sequenced and have been classified, on the basis of sequence similarities, into at least four categories according to Henrissat's classification.

Glycosyl hydrolase family 2

β-Galactosidases belonging to this glycosyl hydrolase family, typified by the *E. coli lacZ* β-galactosidase, consist of a large subunit protein of approximately 1,000 amino acids and show very high sequence similarities to the animal β-glucuronidases, suggesting that the β-galactosidases of family 2 have a close evolutionary relationship with the β-glucuronidases. Many β-galactosidases of this family require monovalent and divalent metal ions for maximum activity (e.g., the *E. coli lacZ, Kluyveromyces lactis, Streptococcus thermophilus,* and *Saccharopolyspora rectivirgula* enzymes).

Glycosyl hydrolase family 1

The family 1 β-galactosidases include the enzyme of an archaeon, *Sulfolobus solfataricus*. This enzyme is essentially a β-glycosidase that can efficiently hydrolyze β-galactosides as well as the other β-glycosides. The *S. solfataricus* enzyme, consisting of 489 amino acid residues, differs in subunit size from the family 2 of enzymes and exhibits no sequence similarity to that family. Interestingly, the lactase/phlorizin hydrolase, a mammalian β-galactosidase playing a central role in lactose digestion in the small intestine, belongs to family 1. Thus, it might be possible that family 2 β-galactosidases evolved to the animal β-glucuronidases whereas family 1 enzymes evolved to the digestive lactase enzymes functioning in the mammalian intestines.

A comparison of stereo- and primary structures suggested that, along with β-glucosidases, cellulases, xylanases, and glycanases, the β-galactosidases belonging to families 1 and 2 form a superfamily, the 4/7 superfamily, which is characteristic of a common eightfold β/α architecture with two conserved, catalytically important glutamates near the C- terminal ends of β-strands four and seven.

The lacG β-galactosidases

Strains of *Arthrobacter* sp. and *Bacillus stearothermophilus* produce β-galactosidases belonging to a recently proposed β-galactosidase family, the *lacG* family, which shows no sequence similarity to enzyme families 1 and 2.

Glycosyl hydrolase family 35

Xanthomonas manihotis, which is a gram-negative phytopathogenic bacterium, *Arthrobacter* sp., *Bacillus circulans,* and *Aspergillus niger* produce β-galactosidases with strong sequence similarities to animal β-galactosidases belonging to family 35 of glycosyl hydrolases. Domain 1 of the *Xanthomonas manihotis* β-galactosidase shows similarity to regions within catalytic domains of cellulases and xylanases belonging to family 10 of glycosyl hydrolases. It should be mentioned that the strain of *Arthrobacter* sp. can produce β-galactosidases of different enzyme families, that is, families 1 and 35 as well as the *lacG* family. Cumulative knowledge of amino acid sequences and stereostructures of the microbial β-galactosidases will provide important information for further insights into the classification and evolutionary and structure–function relationships of the microbial β-galactosidases.

STATUS OF SOME MICROBIAL β-GALACTOSIDASES

Although many β-galactosidase-producing microorganisms are known, only a limited number of microorganisms have been selected as sources of the enzymes used in industry. Most of these are of fungal origin, such as *Kluyveromyces lactis, K. fragilis, Aspergillus niger,* and *A. oryzae.* These fungi were chosen mainly because they can inexpensively produce the β-galactosidase and are generally recognized as safe (GRAS) as food additives. Recent extensive studies led to discovering many other fungal and bacterial enzymes that are applicable in bioprocess technology. In this section, we describe the molecular properties and current status in bioprocess technology of some microbial β-galactosidases.

Fungal β-Galactosidase

The *Kluyveromyces lactis, K. fragilis, Aspergillus niger,* and *A. oryzae* β-galactosidases, the commercially available GRAS enzymes, are currently used in industry for manufacturing milk and dairy products. The *Paecilomyces varioti* β-galactosidase is a thermostable extracellular enzyme with a pH optimum at the acidic region and is potentially applicable to the processing of acid whey. The *A. oryzae, Cryptococcus laurentii,* and *Sterigmatomyces elviae* enzymes exhibit high transgalactosylation activities and are used for the production of the "galactooligosaccharides," which are derived from lactose.

Kluyveromyces lactis

Among the fungal enzymes, biochemical and genetic studies have been the most intensive with the *K. lactis* enzyme. The *K. lactis* β-galactosidase is encoded by the *LAC4* gene, whose regulation involves the galactose/lactose induction and catabolite repression systems. The *K. lactis* enzyme consists of a subunit protein with an M_r of 117,618 and is probably a dimer judging from its sedimentation coefficient under nondenaturing conditions, although higher aggregates are also known. The deduced amino acid sequence of the β-galactosidase is very similar to that of the *E. coli lacZ* enzyme; thus, the enzyme belongs to the family 2. However, Maxilact, a commercial preparation of the *K. lactis* enzyme, was shown to contain four enzyme forms of β-galactosidase, whose native molecular weights were estimated to be 19,000, 41,000, 550,000, and 630,000, respectively. The biochemical relevance of these observations to the reported subunit size and structure remains to be clarified.

Aspergillus niger

A. niger produces three forms of β-galactosidases with M_r of 124,000, 150,000, and 173,000, respectively, which can be separated by ion exchange, gel filtration, and hydrophobic interaction chromatographies. These isoforms share very similar isoelectric points, amino acid compositions, and kinetic parameters. The observed multiplicity seems to vary with the culture conditions and is probably related to differences in the carbohydrate content. The amino acid sequence of the *A. niger* β-galactosidase exhibits similarity to animal β-galactosidases belonging to the glycosyl hydrolase family 35.

Crypotococcus laurentii and Strerigmatomyces elviae

C. laurentii and *S. elviae* have been found to produce β-galactosidases that have high galactosyl transfer activities. The *C. laurentii* cells entrapped in the calcium alginate gels are currently used for industrial-scale production of 4'-galactosyllactose-containing "galactooligosaccharides". The *C. laurentii* and *S. elviae* β-galactosidases share the following similar characteristics: (i) occurrence in the cell wall fractions of the yeast cells; (ii) homodimer of a subunit protein with an M_r of 86,000–100,000; (iii) optimum pH for hydrolysis reaction at pH 4.3–5.0; (iv) high transgalactosylation activity that yields 4'-galactosyllactose as a main product from lactose; and (v) strong inhibition by Hg^{2+} but not by *p*-chloromercuribenzoic acid. However, the *S. elviae* enzyme is distinct from the *C. laurentii* enzyme in terms of thermostability; the *S. elviae* enzyme is stable up to 80°C for 1 h although *S. elviae* is a mesophile; in contrast, the *C. laurentii* enzyme is inactivated after incubation at 58°C for 10 min. The *S. elviae* enzyme exhibits sevenfold higher β-D-fucosidase activity than β-D-galactosidase activity.

β-Galactosidase of Bacteria and Actinomycetes

Escherichia coli

The *E. coli lacZ* β-galactosidase is the best characterized β-galactosidase; its three-dimensional structure was already known. The regulation mechanism of the transcription of the *lacZ* β-

galactosidase gene was established as the operon model. The extensive structural and mechanistic studies of the *E. coli lacZ* enzyme have provided the basis for the detailed insights into β-galactosidase catalysis. However, the usefulness of this enzyme in bioprocess technology has been examined only on a laboratory scale; the *E.coli lacZ* enzyme stops short of making a contribution in the food industry, mainly because of the unacceptability of the bacterium in uses related to food. Rather, the *E. coli lacZ* β-galactosidase and its gene are essential tools in genetic engineering, immunochemical, and molecular biology studies. The *E. coli lacZ* β-galactosidase is a tetrameric protein consisting of identical subunits of a 1,023-amino acid polypeptide chain, which folds into five sequential domains with an extended segment at the amino terminus.

The enzyme has four active sites per tetramer, each of which is made up of elements from two different subunits. All the catalytically important amino acid residues are located in domain 3 of the subunit. The enzyme requires the essential divalent and monovalent cations for its maximum activity, but the role of these metal activators in the enzyme mechanism is not yet known in detail. The proposed mechanism of the hydrolysis and transgalactosylation reactions catalyzed by the *E. coli lacZ* β-galactosidase are reviewed in another section. The "α-complementation" phenomenon would be worth noting as a unique characteristic of this enzyme. A fragment of amino-terminal peptide, which consists of approximately 50 amino acids encoded by a plasmid, is capable of intraallelic (α-) complementation with a defective form of the β-galactosidase encoded by the host. The α-complementation has been used for designing plasmid vectors that permit histochemical identification of recombinant clones in genetic engineering studies. The *E. coli* cells produce a second type of β-galactosidase, the *ebg* β-galactosidase ("*ebg*" denotes "evolved β-galactosidase"), which was found as a gene product that confers a lactose-positive phenotype to *lacZ*⁻ mutants.

The enzyme is encoded by two distinct structural genes of the *ebg* operon, *ebgA* and *ebgC* genes that encode polypeptide chains of M_r 110,000 and 20,000, respectively. The sequence of the *ebgA* gene shows 50% nucleotide identity with that of the *lacZ* β-galactosidase gene, indicating that the *ebg* β-galactosidase has a strong evolutionary relationship with the *lacZ* β-galactosidase. The active form of the *ebg* β-galactosidase is an $\alpha_4\beta_4$ heterooctamer, which is a 1:1 complex of the *ebgA* and *ebgC* gene products. Because the wild-type *ebg* β-galactosidase is too catalytically feeble to allow the *E. coli* cells to grow on lactose, spontaneous mutations must occur in the *ebg* β-galactosidase genes before growth can be sustained; when the *lacZ*⁻ mutants are placed under selection pressure with lactose as a sole carbon source, strains producing mutant *ebg* enzyme with enhanced catalytic competence are selected. Thus, although the *ebg* β-galactosidase has made no significant contribution to bioprocess technology to date, the *ebg* system has been extensively studied as an interesting model for the biology of acquisitive evolution and the chemistry of the evolution of catalytic function.

Bacillus circulans

The commercial preparation of *B. circulans* β-galactosidase contains two isozymes, isozymes 1 and 2, which are reported to be monomeric with M_r of 240,000 and 160,000, respectively. The enzyme preparation could be applicable to enzymatic hydrolysis of lactose in skim milk. The isozymes are different from each other in terms of substrate specificity and the ability to catalyze transgalactosylation. Isozyme 2 has an extremely high transgalactosylation activity; the reaction of the enzyme with 4.56% (w/v) lactose yields the transgalactosylation products (galactooligosaccharides) consisting of di-, tri-, tetra-, and pentasaccharides, and the amount of the oligosaccharides reaches the maximum, which is 41% (w/w) of the total sugar in the reaction mixture.

In contrast, isozyme 1 produces oligosaccharides whose amount is only 6% (w/w) of the total sugar at maximum under the same conditions. Thus, isozyme 2 of *B. circulans* β-galactosidase is used for the production of "galactooligosaccharides" as well as for the functional modification of food materials, medicines, and other bioactive compounds by transgalactosylation. Two β-galactosidase-encoding genes, *bgaA* and *bgaB*, have been identified in *B. circulans* and are potential candidates that code for these isozymes. Relevance of these genes to the isozymes, however, still remains to be established. Recently, *B. circulans* has also been found to have a third β-galactosidase, *bgaC*, which is an *exo*-β-1,3-galactosidase belonging to glycosyl hydrolase family 35. The commercial preparation of *B. circulans* β-galactosidase does not contain the activity of this third β-galactosidase. The recombinant *bgaC* β-galactosidase also showed transgalactosylation activity, which is useful for the synthesis of β-1,3-linked galactosyl oligosaccharides.

Streptococcus thermophilus

S. thermophilus is one of the dairy lactic streptococci that are used as starter cultures for a variety of industrial dairy fermentations such as those during the yogurt-making processes. The primary function of these dairy lactic streptococci is the rapid conversion of lactose into lactic acid, an essential process in dairy fermentation. Two distinct biochemical pathways involved in lactose utilization by the lactic streptococci have been established. One of these pathways uses the lactose permease system for the transport of unmodified lactose, which is then hydrolyzed by an intracellular β-galactosidase. The other pathway is mediated by the phosphoenolpyruvate-dependent phosphotransferase system, in which lactose is phosphorylated during translocation and the resultant lactose-6-phosphate is hydrolyzed by phospho-β-galactosidase. In *S. thermophilus* cells, the lactose permease system plays a central role in lactose transport across the bacterial cell membrane, and the β-galactosidase functions as a key enzyme in the metabolism of the internalized lactose in dairy lactic fermentation.

The *S. thermophilus* β-galactosidase is promising for industrial hydrolysis of milk lactose, judging from its thermostability and established safety as a food additive. Also, the enzyme is used as a catalyst for the industrial production of "galactooligosaccharides" from lactose because of its high transgalactosylation activity. The *S. thermophilus* β-galactosidase consists of a subunit protein of M_r 116,860. However, its quaternary structure is not yet firmly established. The enzyme exists in three distinct forms, whose apparent native molecular weights are 204,000, 186,000, and 282,000, and one of which is converted to another in a time-, temperature-, and concentration-dependent manner. The native molecular weight of 500,000–600,000 has also been reported. Structural analysis of the *S. thermophilus* β-galactosidase genes showed that the β-galactosidase is a family 2 enzyme with 48%, 35%, and 32.5% amino acid sequence identity to the *Lactobacillus bulgaricus, Escherichia coli,* and *Klebsiella pneumoniae* enzymes, respectively.

The enzyme requires monovalent and divalent metal ions for its maximum activity, and the specificity of the metal activation varies with the substrate. The enzyme shows high transgalactosylation activity to produce a variety of disaccharides as main products from lactose, in contrast to the *Aaspergillus oryzae* β-galactosidase, for example, producing tri-, tetra-, and pentasaccharides as main transfer products from lactose.

Saccharopolyspora rectivirgula

A strain of *S. rectivirgula*, a thermophilic actinomycete, produces an extracellular thermostable β-galactosidase with high transgalactosylation activity. The *S. rectivirgula* β-galactosidase is a monomeric enzyme with a molecular weight of 145,000 and $S^{\circ}_{20,w}$ of 7.1*S*. When the enzyme reacts with 1.75 M lactose at 70°C and pH 7.0 for 22 h, it yields oligosaccharides in the maximum yield (other than lactose) of 41% (w/w). The enzyme is stable at pH 7.2 up to 60°C (for 4 h in the presence of 10 μM $MnCl_2$) or 70°C (for 22 h in the presence of 1.75 M lactose and 10 μM $MnCl_2$). Thus the enzyme is applicable to an immobilized enzyme system at high temperatures (>60°C) for efficient production of the oligosaccharides from lactose.

This monomeric enzyme has eight specific binding sites for divalent metal ions. These sites are classified as follows: a very tight (class I) site for Ca^{2+}, three tight (class II) sites consisting of two Ca^{2+}-specific sites (class II_{Ca}) and one Mn^{2+}-specific site (class II_{Mn}; K_d for Mn^{2+}, 2.0 nM), and four loose (class III) sites for Mn^{2+} (K_d, 1.2 μM) and Mg^{2+} (K_d, 2 μM). The class II Mn^{2+} is catalytically important in maintaining the native structure essential for activity. Occupation of class III sites by Mg^{2+} or Mn^{2+} is of physiological importance to attain sufficient thermostability by which this extracellular β-galactosidase remains active for a prolonged time at elevated temperatures, as was observed during the growth of *S. rectivirgula*.

ESCHERICHIA COLI LACZ β-GALACTOSIDASE-CATALYZED REACTIONS

The mechanism of the hydrolysis and transgalactosylation reactions has been the most intensively studied with the *Escherichia coli lacZ* β-galactosidase. In this mechanism, hydrolysis and transgalactosylation occur at the same catalytic site and share a common galactosyl enzyme intermediate. Extensive kinetic studies have also suggested the presence of two distinct subsites near the active site of the *lacZ* enzyme, the "galactose site" and "glucose site," that respectively bind galactose and glucose portions of the lactose molecule. The glucose site shows affinities to a variety of nucleophilic compounds, such as sugars, alcohols, and thiols. The presence of such a glucose site is important in the kinetic behaviors of hydrolysis and transgalactosylation of lactose.

Reactions with ONPG

The kinetic mechanism of the *Escerichia coli lacZ* β-galactosidase-catalyzed reactions can be most simply shown by the reaction with 2-nitrophenyl β-D-galactopyranoside (ONPG), which is a synthetic substrate whose aglycone serves as a good leaving group. The ONPG (termed S) binds to the enzyme to form a Michaelis complex ES. The release of the aglycone, 2-nitrophenolate, occurs very rapidly (k_2 = 2,100 s^{-1} at 25°C and pH 7.0) to yield a galactosyl enzyme intermediate, which partitions into subsequent hydrolysis and transgalactosylation. For the hydrolysis reaction, the water molecule reacts with the galactosyl enzyme intermediate with k_3 of 1,200 s^{-1}, yielding galactose and free enzyme (degalactosylation). For the transgalactosylation, a galactosyl acceptor (termed Nu) present in the reaction mixture binds to a glucose site of the enzyme with an affinity of K_i'' to form a ternary complex, followed by the formation of the galactosyl adduct (Gal-Nu).

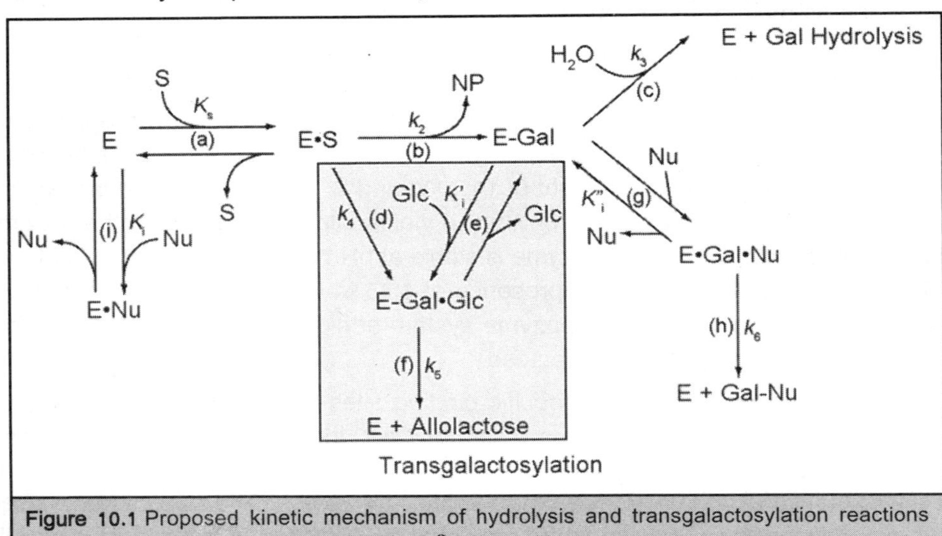

Figure 10.1 Proposed kinetic mechanism of hydrolysis and transgalactosylation reactions catalyzed by the *E. coli lacZ* β-galactosidase.

In the transgalactosylation reaction, the apparent rate of the 2-nitrophenol release is increased, decreased, or unchanged with increasing acceptor concentrations, depending on the nature of the acceptor and the relative magnitude of k_2, k_3, and k_6. These observations have been kinetically analyzed to estimate the K_i'' and k_6 values of a variety of acceptors, which are the measures of how well they bind at the glucose site and how readily they react to form a galactosyl adduct, respectively.

Reactions with Lactose

The reaction with lactose, a natural substrate, proceeds with a mechanism that is somewhat different from that of the ONPG reaction. Steps d, e, and f, which are shown with shadowing in Figure 10.1, must be considered instead of step b, because glucose, a product of lactose hydrolysis, has an affinity (K_i'') to the glucose site, in contrast to 2-nitrophenolate, which has no affinity to the enzyme; released glucose can reversibly bind to form a ternary complex, and then reacts with galactose to yield allolactose.

During the lactose hydrolysis, allolactose is only transiently formed because it is also a substrate of the enzyme. The reported rate constants (at 25°C, pH 7.0) are $k_4 = 60s^{-1}$, $k_5 = 380\ s^{-1}$, and $k_3 = 1{,}200\ s^{-1}$. When high concentrations of lactose are reacted, lactose also serves as a galactosyl acceptor (Nu), and the transfer product is galactosyllactose. The galactosyllactose can, in turn, serve as a galactosyl acceptor to produce tetrasaccharides. In this manner, a mixture of oligosaccharides consisting of di-, tri-, and higher saccharides is produced from lactose.

Active Site

Extensive chemical modification, affinity labeling, and site-directed mutagenesis studies have been carried out to find functional amino acid residues at the active site of the *Escherichia coli lacZ β*-galactosidase; these have shown that Glu461, Glu537, and Tyr503 are important for enzyme activity. All these catalytically important residues are located in the third domain of the enzyme subunit and, indeed, are shown to be located at the active site by X-ray crystallographic studies. However, some argument still exists on the details of the role(s) of each active site residue. Here we briefly describe the currently proposed roles of these amino acid residues in *β*-galactosidase catalysis. The glycosidic oxygen atom of the enzyme-bound substrate undergoes protonation, which facilitates aglycon departure. Intensive kinetic studies with site-directed mutants suggest that the Glu461 serves as a Bronsted general acid–base catalyst at this step, although some evidence also argues for the role of Tyr503 being the general acid–base catalyst.

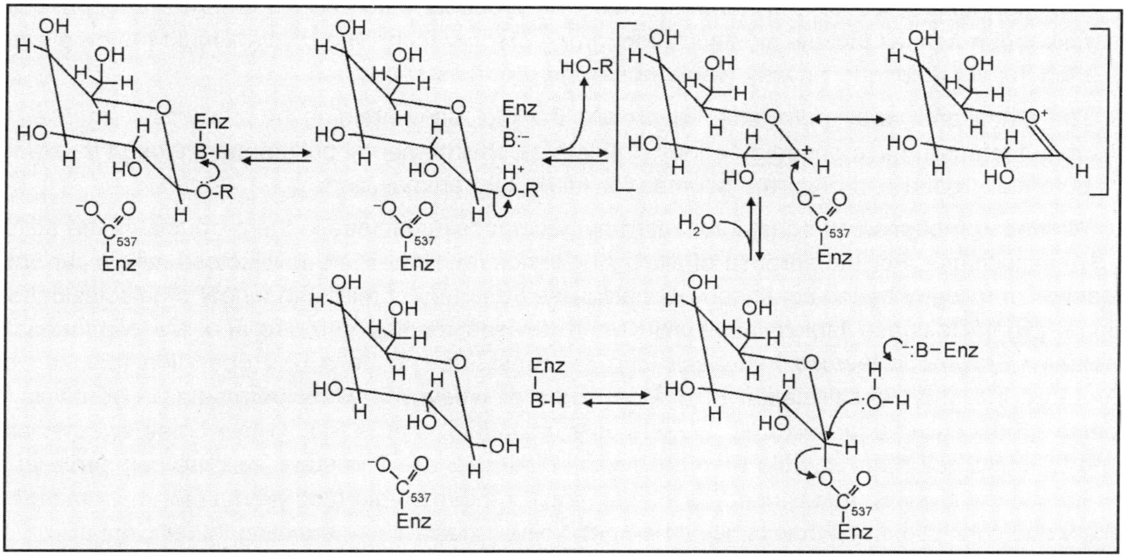

Figure 10.2 Catalytic mechanism of the *E. coli lacZ* β-galactosidase.

The elimination of the aglycon probably produces an oxocarbonium ion intermediate that is planar at its anomeric end. This reaction is assumed because the planar and positively charged analogs of galactose serve as strong inhibitors of the *E. coli lacZ β*-galactosidase. Affinity-labeling studies provided evidence suggesting that Glu537 is a nucleophile that attacks the anomeric carbon of the cationic intermediate, yielding a galactosyl enzyme in which the galactose moiety is covalently bound to the enzyme through an ester linkage.

The water (or acceptor) then reacts with the covalent intermediate to produce galactose (or galactosyl adduct) and free enzyme. Although the enzyme requires monovalent and divalent ions for its activity, the role of the metal activators is not yet known. The bound Mg^{2+} (1 atom/active site; K_d, 0.65 μM) has been suggested to act as an electrophilic catalyst that accelerates aglycone departure, whereas there is alternative evidence suggesting that it is only important in maintaining the proper active site structure. The Glu416, His418, and Glu461 are ligands to the bound Mg^{2+} ion.

SIGNIFICANCE OF β-GALACTOSIDASE-CATALYZED REACTIONS

Effect of Water

Cow milk generally contains 4.5–5.0% (w/v) lactose, whose low digestibility, low solubility, and low sweetness pose the following problems, the so-called lactose problems, in the milk and dairy industries.

1. Lactose intolerance is widespread in the world's adult population, with 70% being unable to metabolize large quantities of lactose in milk because of lack of intestinal lactase.
2. The high lactose content in nonfermented milk products, such as ice cream and condensed milk, often causes lactose crystallization during preservation. The crystals must not be greater than 10 μm in length, or they would make the products sandy.
3. The whey and whey permeate, which are the lactose-containing by-products from cheese manufacturing, have posed a serious problem as environmental pollutants, although these are potentially good carbohydrate sources for foods and animal feeds.

All these problems can be circumvented by decomposing lactose in the products by the action of β-galactosidase. For the purpose of alleviating lactose maldigestion, low-lactose milk is currently produced in industry by the continuous or batchwise treatment of milk with fungal β-galactosidases. Fungal β-galactosidase tablets have also been devised for this purpose and are commercially available. Enzymatic lactose hydrolysis can be considered to be a unit operation that can be integrated into the transformation of a wide range of whey and whey permeates. Hydrolysis of lactose produces a sugar mixture, glucose plus galactose, that is four to five times sweeter and three to four times more soluble than the parent disaccharide. The sugar mixture can serve as a condensed sugar syrup that is used as a sweetener or intermediate food product for manufacturing bakery and confectionary products, ice cream, and other specialities, permitting the efficient utilization of wheys and whey permeates.

Industrial Lactose Hydrolysis

It is important to achieve industrial lactose hydrolysis with minimum cost. Immobilization of β-galactosidases is a means to reduce the cost of industrial lactose hydrolysis. Methods of immobilization and the properties of the immobilized β-galactosidases have been summarized. Baret pointed out the following practical aspects in attaining the best performance of the immobilized enzyme system for the processing of whey and its permeates: (i) the nature of the substrate wheys and permeates, (ii) the design of the reactor system, and (iii) the operating strategy. The key objective of these considerations is to maximize the amount of hydrolyzed lactose per unit weight of catalyst, in other words, to maximize the half- life of the system. Semiindustrial-scale investigations were carried out to examine these aspects using immobilized *Aspergillus niger* and *A. oryzae* β-galactosidases that are covalently bound on a porous silica carrier:

1. Omission of demineralization of permeates had no adverse effect on the performance and stability of the systems. The need for demineralization is related to the application of the end product. However, controlling the levels of suspended solids and colloids in wheys was specifically important during operations in a fixed-bed reactor.
2. Among the fixed-bed, fluidized-bed, and stirred-tank reactors, the fixed-bed reactor gave the best kinetic results, in which the minimum reactor volume was necessary.
3. When no particular care is taken, the efficiency of the immobilized enzyme system immediately decreases because of deposition of material within the bed, the formation of a coating around the particles, microbial contamination, channeling, plugging, and other reasons. Operating strategy must therefore be considered to minimize these problems. The strategy required includes the removal of solids or colloidal substances in substrates, sanitation of substrates, introduction of the cycling of cleaning/sanitation procedures of the reactor during the operation, and programming the operating temperatures. These strategies greatly enhanced the activity and stability of the reactors.

It should be emphasized that the prevention against microbial contamination during operation is of specific importance for the processing of wheys and their permeates, which are nutritionally rich and excellent growth media for microorganisms. Methods for sanitizing the substrates, reactor, and other possible sources of microbial contamination were evaluated. Dilute acetic acid, which has been commonly used in the laboratory for disinfection of the immobilized enzyme matrices, served as only a poor sanitizer under industrial conditions. The substituted diethylenetriamines were effective for the disinfection of the immobilized enzyme matrices without causing loss of enzyme activity and have been used in large-scale operations. The technology obtained from these large-scale examinations was transferred to fully industrial-scale operations; a plant processing 20,000 L/day of sweet whey has thus been operated, in which the hydrolysis reactor can process 1,000 L/h of nondemineralized acidified sweet whey using 40 kg of the immobilized β-galactosidase matrices. The processed products have been used as ingredients in the food industries.

Lactose-Related Oligosaccharides

It was shown in the 1950s that oligosaccharides are formed during the hydrolysis of lactose in milk with the fungal β-galactosidases. These oligosaccharides, the so-called galactooligosaccharides (GO), which are the products of the intermolecular transgalactosylation of lactose, have long been regarded as unwanted by-products in lactose hydrolysis. The *Escherichia coli lacZ* and *ebg* β-galactosidases were also shown to have transgalactosylation activity; their physiological significance has attracted the attention of biochemists because one of the transgalactosylated products, allolactose, serves as an inducer for the lactose operon of the bacterium. During the past decade, GO has been shown to have some beneficial effects in maintaining human health (see later) and thus has been developed as a food material with health-giving functions.

For the efficient production of GO from lactose, many attempts have been made to find microorganisms producing the GO or the β-galactosidase with high transgalactosylation activity. To date, a large number of such microorganisms have been found, among which *Bacillus circulans, Streptococcus thermophilus, Saccharopolyspora rectivirgula, Aspergillus oryzae, Cryptococcus laurentii,* and *Sterigmatomyces elviae* produce enzymes with particularly high transgalactosylation activities. Also, the cells of yeasts, such a *Lipomyces* sp., and those of the lactic acid bacteria, such as *Bifidobacterium bifidum,* can produce GO abundantly from lactose. The structures of GO depend on the enzymes or microorganisms used as catalysts.

The *B. circulans, S. rectivirgula, C. laurentii, S. elviae,* and *A. oryzae* enzymes as well as *B. bifidum* cells produce from lactose the oligosaccharides with a general structure of Gal-$(Gal)_n$-Glc* as main products, where Gal and Glc* indicate the galactose and reducing terminal glucose residues and n the integer ($n \geq 1$), whereas the *S. thermophilus* and *E. coli lacZ* enzymes mainly produce disaccharides (Gal-Glc*). Currently, GO is industrially produced by means of the *A. oryzae* and *S. thermophilus* β-galactosidases. In the reported system, lactose (80%, w/v) is first reacted with the *A. oryzae* enzyme at 65°C for 4 h. The content of GO in the reaction mixture reaches 31% of the total sugar.

The remaining lactose is further converted to the other disaccharidal transfer products by means of the *S. thermophilus* β-galactosidase. Industrial production of 4′-galactosyllactose is also achieved with *C. laurentii* cells entrapped in calcium alginate gels, which is used batch-wise, yielding 4′-galactosyllactose whose content reaches 35% of total sugar. The immobilized cells can be used repeatedly, more than 30 times. The lactose-related oligosaccharides that can be produced by the action of β-galactosidase also include the lactitol oligosaccharides, which are produced by the *A. oryzae* β-galactosidase-catalyzed transgalactosylation to lactitol [β-galactopyranosyl-(1→4)-D-glucitol], a reduced form of lactose.

Additional Properties

GO is of low sweetness with excellent taste quality and is stable under acidic conditions during food processing. Thus, the oligosaccharides serve as food materials potentially applicable to a wide variety of food manufacturing and cooking uses. GO also has beneficial physiological effects on human health, as described in the following subsections.

GO is a functional sweetener

Ingested GO passes through the human small intestine without being digested; GO is therefore a low-calorie sweetener. *Streptococcus mutans,* a causative agent for dental caries, cannot metabolize GO or produce from GO the insoluble glucan that is responsible for dental plaque formation. Thus, GO is also a sweetener beneficial for prevention of dental caries.

GO can improve intestinal and fecal conditions

The observation that GO can improve intestinal and fecal conditions has been explained in terms of the selective promotion by GO of bifidobacterial growth in human colons. The most dominant indigenous bacteria in the neonatal gut are the bifidobacteria, whose predominance in the gut has been suggested to cause beneficial effects in maintaining human health; these bacteria, as well as their metabolites, provide protection from infection and facilitate the normal functions of the gut. However, the bifidobacterial population in the human intestine decreases with aging processes, and the etiological relevance of this fact to the increased incidence of intestinal diseases in the middle- and advanced-aged generations has been implicated. Many in vitro experiments showed that GO specifically promotes the growth of the intestinal bifidobacteria and lactobacilli. Ito et al. further examined the in vivo effects of the ingestion of transgalactosylated disaccharides (TD), which were derived from lactose by the action of *Streptococcus thermophilus* β-galactosidase, on fecal microflora and their metabolism with 12 middle-aged male volunteers.

The results showed that TD ingestion resulted in a specific increase in the fecal populations of bifidobacteria and lactobacilli, as well as a decrease in those of the genera *Bacteroides* and *Candida*. It also lowered fecal pH and caused a decrease in fecal concentrations of the products of amino acid catabolism, such as isobutyric and isovaleric acids, ammonia, *p*-cresol, and indole. These results suggest that the ingested TD reached the colon and selectively promoted the growth of bifidobacteria and lactobacilli to depress the microbial decomposition of amino acids. The GO containing di-, tri-, tetra-, or higher oligosaccharides showed a similar growth promotion of the fecal bifidobacteria and affected the metabolism of fecal microflora. It should be noted that the lactitol oligosaccharides also showed a selective promotion of bifidobacterial growth and improved the metabolism of fecal microflora.

Effect on intestinal absorption of Ca^{2+} and lipid metabolism

Osteoporosis, a loss of bone mass causing an increased risk of bone fracture, is associated with the aging processes and in particular with menopause in women; estrogen deficiency resulting from menopause results in more rapid bone resorption than formation, resulting in a net loss of bone mineral density. Estrogen deficiency also causes hyperlipemia, which produces a higher risk of coronary heart disease. The effects of GO ingestion on calcium absorption, prevention of bone loss, and total serum cholesterol content were examined.

Ovariectomized Wistar rats fed on a diet containing GO (5% in place of sucrose in the control diet) absorbed calcium more efficiently than those on the control diet, and the bone ash weight and tibia calcium content of the GO-ingesting rats were significantly higher than those of the controls. GO also lowered the total concentration of serum cholesterol. It is suggested that the ingested GO is metabolized in the large intestine by intestinal bacteria to enhance the production of volatile acids, such as propionate, which facilitates intestinal calcium absorption and lowers the total serum cholesterol. These observations argue for possible prevention by GO against osteoporosis and hyperlipemia caused by postmenopausal estrogen deficiency.

Modification by Transgalactosylation

Glycosylation is a powerful means for structural and functional modification of medicines and many other bioactive compounds; it enhances their solubilities, physicochemical stabilities, biological half-lives, membrane permeabilities, and intestinal absorption and also improves their taste qualities. In the biosynthesis of glycosides and oligosaccharides, sugar nucleotides, such as UDP-glucose, serve as substrates for enzymes that catalyze glycosidic linkage formation. However, the use of these systems in biotransformations has been limited by the restricted availability of the corresponding enzymes and substrates. An alternative is to use transglycosylation catalyzed by glycosidases; there exists a variety of glycosidases with different specificities, and their substrates are inexpensively available. Because the transfer product is always a substrate for the glycosidase, the success of the procedure as a preparative method in part depends on the rate of hydrolysis of the product being slower than that of the glycosyl donor.

In practice, these conditions can be attained rather readily. Thus, there has been a steadily growing interest in the glycosidases for glycoside and oligosaccharide synthesis. In this context, β-galactosidases with high transgalactosylation activities have also been used for obtaining the galactosyl adducts of sucrose, cyclodextrins, oligosaccharides, glycosides, amino acids, peptides, alcohols, vitamins, and other bioactive compounds. We here describe two interesting examples of synthesis of galactosides using β-galactosidases in further detail.

Synthesis of β-galactosylthiamin

Thiamine (vitamin B_1) occurs in its phosphorylated forms in nature and plays important roles in sugar metabolism as a cofactor. 5′-*O*-(β-Galactopyranosyl)-thiamine has been produced by the *Aspergillus oryzae* β-galactosidase-catalyzed transgalactosylation using ONPG as a galactosyl donor. In contrast to thiamine having a specific odor and strong tongue-pricking taste, the β-galactosylthiamine obtained was odorless and mildly sweet with no such stimulative taste, suggesting its usefulness as a food additive and medicine.

One-pot synthesis of sialyl oligosaccharides using a multienzyme system

Sialyl oligosaccharides are important recognition elements on the surface of cells and serve as ligands of some cell adhesion molecules. Herrmann et al. described a new efficient synthesis of α(2,6)sialyl *N*-acetyllactosamine and analogs by combined use of the *Bacillus circulans* β-galactosidase and pig liver α(2,6)sialyltransferase. In this system, the enzymes work sequentially in one pot to form a sialyl *N*-acetyllactosamine. CMP-sialic acid, a substrate for the sialyltransferase, is generated in situ with a coupled recycling enzyme system.

The β-galactosidease-catalyzed transgalactosylation proceeds irreversibly because the galactosyl adduct is irreversibly sialylated with the sialyltransferase to form a sialyl *N*-acetyllactosamine, which is no longer a substrate for the β-galactosidase. To explore more efficient applications of β-galactosidase in bioprocess technology, efforts are still being made to discover a novel microbial β-galactosidase in nature. In addition, attempts are being made through genetic and protein engineering approaches to improve the specificity and enhance the stability of known β-galactosidases. These attempts, along with basic mechanistic and structural studies, will further enhance our understanding of the molecular mechanism and the structure–function relationship of β-galactosidase.

11

PRODUCTION METHODS OF DEXTRAN

Over the past 25 years a new class of microbial products, the microbial exopolysaccharides, has grown in industrial importance. These products can be used as alternatives to other synthetic and natural water-soluble polymers that are used as emulsifiers, stabilizers, binders, coagulants, lubricants, film formers, and thickening, gelling, and suspending agents. Despite their costly collection and extraction, plant and algal polysaccharides such as gum guar, locust bean gum, starch, alginate, carageenan, and agar are widely used in the food and pharmaceutical industries. These polymers, although valuable, have certain drawbacks such as rheological properties that may not match exactly with a specific requirement, production that is vulnerable to crop failure, climatic conditions and marine pollution that cause the quality, supply, and cost to fluctuate. The production of microbial exopolysaccharides provides a valid alternative to plant and algal polysaccharides.

The exopolysaccharides, produced by a wide variety of microorganisms, show considerable diversity in their composition, structure, and unique physical properties. Their industrial production has been stimulated by the opportunity to provide a guaranteed supply, constant quality, superior properties, and stable price. Among all of the exopolysaccharides, dextran exemplifies attainment of these objectives. Dextran produced from a bacterial strain of *Leuconostoc mesenteroides* NRRL-B512F has a unique value in medical applications as a blood plasma substitute and a blood flow improver, properties that have provided the impetus for initiating much work on its production and further applications.

Dextran is a microbial exopolysaccharide made up of α-D-glucose units linked mainly by α, $1 \rightarrow 6$ bonds. Traditionally, dextran is produced by fermenting dextran-producing bacteria in sucrose-rich media. The prime bacterial product of fermentation in the presence of substrate sucrose is not dextran itself but the enzyme dextransucrase. The enzyme is secreted by growing cells into the medium, where it converts the sucrose to dextran according to the following equation:

$$n \text{ Sucrose} \xrightarrow{\text{Dextransucrase}} (\text{Glucose})_n \text{ or Dextran} + n \text{ Fructose} \quad \ldots(1)$$

Therefore, dextran can be produced from sucrose in cell-free environments by using the enzyme dextransucrase. The main dextransucrase-producing bacteria belongs to the genera *Leuconostoc* and *Streptococcus*. The enzyme dextransucrase uses sucrose as a high-energy glucopyranosyl donor for dextran synthesis, and the properties of the enzyme and the dextran produced are specific to the microorganisms used.

In addition to sucrose, α-D-glucopyransosyl, α-L-sorbofuranoside, lactulosucrose, α-D-glucopyranosyl fluoride, and *P*-nirophenyl α-D-glucopyranoside also serve as glucopyranosyl donors, with sucrose having the lower Michaelis constant. These substrates have glucosidic bond energies of 5.0-5.5 kcal/mol. Hehre and Hamilton and Barker et al. reported that *Acetobactor capsulatum* and *A. viscosum* can transform amylodextrins to dextran. Dextran produced from microbial fermentation is called native dextran and typically has a molecular weight of 5×10^7 to 5×10^9. Native dextran is usually separated from the producing cells and precipitated by alcohol. This native dextran may then be subjected to controlled acid hydrolysis to obtain the desired molecular weight for specific applications.

PREVIOUS STUDIES

Dextran first came under close examination more than 100 years ago. Dextrans develop naturally in sucrose-containing solutions that have become inoculated with dextran-producing bacteria from air, plants, or soil. The resulting transformation of sucrose solution to syrupy, viscous, or ropey fluids or even gelled masses doubtlessly has plagued humankind since the inception of accumulating and storing sucrose-containing food and beverages and has caused much trouble in sugar beet industries. In 1861 Louis Pasteur identified the viscous polymer as a polysaccharide. In 1874 Scheubler determined its empirical formula to be $C_6H_{10}O_5$ and named it dextran because it was closely related to starch and dextrin.

Van Tieghem named the causative bacteria *Leuconostoc mesenterides*. Cell-free enzymatic synthesis of dextran was first demonstrated by Hehre and Sugg. The enzyme involved in this synthesis was given the name dextransucrase by Hestrin and Shapiro-Avineri, and it was soon found to be a member of the general class of transglycosylases. In 1937 Flower and coworkers made the first systematic investigation of the chemical structure of dextran. A large body of work carried out between 1950 and 1970 on dextran production and its structure, properties, and uses is reviewed by Neely, Jeanes, Murphy and Whistlkr, and Sidebotham.

STATUS OF DEXTRANS

The term *dextran* collectively describes a large class of bacterial extracellular hydrocolloid homopolysaccharides consisting of α-D-glucopyranosyl units polymerized predominantly in an α, $1 \rightarrow 6$ linkage. The diversity within the class of dextran molecules derives from the existence of non–α, $1 \rightarrow 6$ linkages that lead to branching from the main polymer chain. Branching mainly occurs by α, $1 \rightarrow 3$, α, $1 \rightarrow 4$, and α, $1 \rightarrow 2$ linkages. The type, amount, length, and arrangements of branches define the differences in dextran structure and influence the water solubility and other rheological properties of dextran. Hucker and Pederson classified the dextran-producing microorganisms into family, tribe, genus, and species. In 1954 Jeanes et al. reported the classification and characterization of dextrans by examining 96 strains of bacteria. First they classified dextrans into three groups: water-soluble dextrans, water-insoluble dextrans, and heterogeneous dextrans. Water-soluble dextrans are further divided into less-soluble dextrans (can be precipitated by 34-37% of alcohol) and more-soluble dextrans (can be precipitated by 40-44% of alcohol).

Different kinds of dextrans produced by the same strain of bacteria are called heterogeneous dextrans, and these mixtures of heterogeneous dextrans can be separated by differential alcohol precipitation. Further, Jeanes et al. classified the dextrans into class A, class B, and class C dextrans on the basis of the percentage of their α, 1 $\rightarrow$ 3 linkages. They characterized the dextrans by optical relation, viscosity, periodate oxidation, and physical appearance after alcohol precipitation. Dextran structures have two common features:

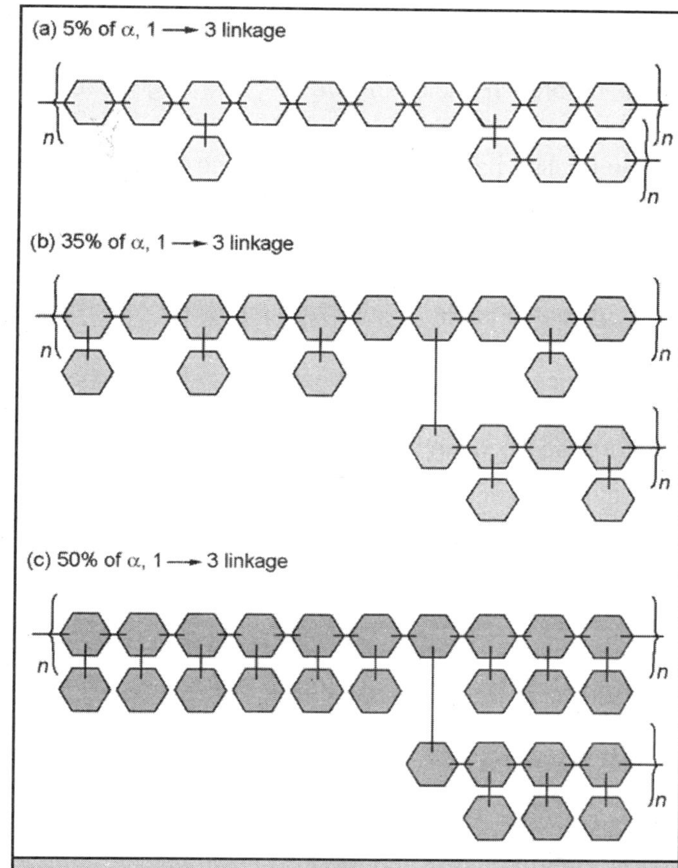

Figure 11.1 Structure of dextrans: Hexagonal units represent α-D-glucose units, horizontal lines joining glucose units are α, 1$\rightarrow$6 bonds, vertical lines joining glucose units are α, 1$\rightarrow$3 bonds: (a) clinical dextran, (b) alternating bifurcated comb structure, and (c) bifurcated comb structure.

1. Main chain and branching chain are made by α-D-glucose units linked together by α, 1 $\rightarrow$ 6 bonds.
2. Branching chains (or branching units) are connected to the main chain by α, 1 $\rightarrow$ 3, α, 1 $\rightarrow$ 4, or α, 1 $\rightarrow$ 2 bonds, called branching linkages.

Examples of α, 1 $\rightarrow$ 3 Branch–Linked Dextrans

Commercially produced *L. mesenteroides* B-512F dextran has 95% of the total glucouspyranosyl linkages as α, 1 $\rightarrow$ 6 linkages and 5% as α, 1 $\rightarrow$ 3 linkages. The α, 1 $\rightarrow$ 3 branching linkages occur in two ways:

1. A single α-D-glucose unit attached to the main chain by an α, 1 $\rightarrow$ 3 bond
2. A branching chain attached to the main chain by an α, 1 $\rightarrow$ 3 bond

This dextran appears as a translucent gel. Senti et al. reported that 77% of the α, (1 $\rightarrow$ 3) linkages are single glucose residues and the remaining branch chains have 50 or more glucose residues. Walker and Pulkownik, studying the hydrolysis of B512F dextran by dextran glucosidase, also concluded that it has 85% single glucose branches and 33 or more glucose residues in the remaining branch chains to give the yield of glucose that they observed.

Streptococcus mutans 6715 produces water-soluble and water-insoluble dextrans. The water-soluble fraction has 35% α, 1 $\rightarrow$ 3 linkages and 65% α, 1 $\rightarrow$ 6 linkages. Branching single glucose units are uniformly connected to the main chain and to extended side chains. The resulting structure was named an alternating bifurcated comb polymer by Robyt et al., is resistant to hydrolysis by endodextranase, and appears as a heavy, opaque material.

L. mesenteroides B-742 strain produced more-soluble and less-soluble dextrans. The more-soluble fraction of dextrans has 50% α, 1 $\rightarrow$ 6 linkages and 50% α, 1 $\rightarrow$ 3 branching linkages. Some branching chains are attached to the main chain by α, 1 $\rightarrow$ 3 branching linkages, but single glucose units are attached to all other glucose units in the main chain by α, 1 $\rightarrow$ 3 bonds. This structure results in the most highly branched dextrans. Robyt et al. named this structure a, in which each single α, 1 $\rightarrow$ 3 linked glucose branch is like a tooth of a comb and the α, 1 $\rightarrow$ 6 linked chains are the backbone of the comb, bifurcated comb polymer. This structure is slightly resistant to hydrolysis by endodextranase. This dextran appears as a finely divided material. The α, 1 $\rightarrow$ 4 and α, 1 $\rightarrow$ 2 branched dextrans are less common.

DISTINGUISHED FEATURES OF DEXTRANSUCRASE

The unit of *dextransucrase* (α, 1 $\rightarrow$ 6-D-glucan: D-fructose 2-glucosyltransferase) activity, DSU, is defined as the amount of enzyme that will convert 1 mg of sucrose to dextran in 1 h at 30°C in acetate buffer solution at pH 5.2. Another unit, U, is defined as the amount of enzyme that liberates 1 μmol of reducing sugar from sucrose in 1 min at pH 5.2 and 30°C. The relation between these two units is 1 U/mL = 20.27 DSU/mL. The extensive purification of dextransucrase was first reported by Ebert and Schenk, and the molecular weight of the enzyme was determined to be 280,000 by ultracentrifugation methods. In 1980 Kobayashi and Matsuda separated two components of dextransucrase having different affinities for dextran gel by a column of Sephadex G-100. The major component of the purified enzyme had a molecular weight of 64,000–65,000, an isoelectric point (pI) of 4.1, and was 17% carbohydrates. Most of the carbohydrates in the purified enzyme were found to be endogenous dextran, covalently bound to the enzyme molecule and released from it by mild treatment with endodextranase. They concluded that dextransucrase exists in multiple forms and that endogenous dextran molecules might be responsible for the multiple forms. Miller et al. and Fu and Robyt found that the molecular weight of dextransucrase produced from a mutant strain of *Leuconostoc mesenteroides* B-512 FM initially secreted into the culture medium was 177,000; later, however, the enzyme exhibited a drop in molecular weight to 158,000 without any loss in activity.

Biochemical Properties

The irreversible reaction of dextransucrase with sucrose is simply formulated in equation 1. Dextransucrase is one of a very few synthesizing enzymes that do not require the presence of cofactors or high-energy phosphorylated intermediates. The energy required for the condensation of two glucosyl units is provided by the hydrolysis of sucrose, distinguishing it from other mechanisms of polysaccharide synthesis.

Mechanism of main chain synthesis

When dextransucrase was first described in 1941 by Hehre, Cori and Cori were studying the action of muscle phosphorylase and Hanes was studying potato phosphorylase. These investigators observed that phosphorylase could elongate glycogen and starch chains by the transfer of glucose from α-D-glucopyranosyl phosphate (Glc-1-P) to the non-reducing end (C_6 moiety of glucose) of the polysaccharide chain according to the following equation:

$$\begin{bmatrix}\text{Polysaccharide}\\ \text{chain (primer)}\end{bmatrix}+\begin{bmatrix}\alpha-\text{D-glucopyranosyl}\\ \text{phosphate (Glc-1-P)}\end{bmatrix}\xrightarrow{\text{Phosphorylase}}\begin{bmatrix}\text{One glucose unit added}\\ \text{to the polysaccharide chain}\end{bmatrix}\quad ...(2)$$

It was believed that the enzyme phosphorylase that catalyzed the preceding reaction requires a preformed polysaccharide chain, called a primer, and a high-energy monosaccharide donor, Glc-1-P. Later it was proven that the enzyme phosphorylase was a degradative enzyme that catalyzed the reaction of inorganic phosphate with the non-reducing end of the polysaccharide chain to give Glc-1-P according to the following equation:

$$\begin{bmatrix}\text{Polysaccharide}\\ \text{chain}\end{bmatrix}+\begin{bmatrix}\text{Inorganic}\\ \text{phosphate}\end{bmatrix}\xrightarrow{\text{Phosphorylase}}[\text{Glc-1-P}]+\begin{bmatrix}\text{One glucose unit reduced in the}\\ \text{polysaccharide chain (primer)}\end{bmatrix}\quad ...(3)$$

The reverse of the preceding reaction, which requires a primer (a polysaccharide chain), produces a polysaccharide chain by adding one glucose monomer to the primer at the nonreducing end (C_6 moiety). Thus in the 1940s and 1950s the primer mechanism was assumed for dextran synthesis.

Evidence for the primer-dependent mechanism for dextran synthesis was strengthened by Germaine and coworkers in the 1970s. They found that addition of dextran (as a primer) to dextransucrase increased the rate of dextran synthesis and that the rate reached a maximum at an average size of the added dextran of 30 glucose residues. Kobayashi and Matsuda also reported that purified dextransucrase was stimulated by dextran (as a primer), although dextransucrase could synthesize dextran without the addition of primer. Based on these results, Kobayashi et al. proposed a primer-dependent mechanism for dextran synthesis. They proposed that the active site has two types of substrate binding sites, one for sucrose and the other for primer dextran.

The elongation of the primer was hypothesized to take place when an enzyme nucleophile (presumably a carboxylate anion) attacks C_1 of the glucose moiety to give the glucosyl–enzyme covalent intermediate. This glucose is then transferred to the 6-hydroxyl group of the nonreducing end of the primer chain attached to the primer binding site of the enzyme, forming the next α, $1 \rightarrow 6$ linkage. For the synthesis to continue, the primer, with newly added glucose residue, must dissociate from the primer binding site and a new primer molecule must bind in the primer binding site with its nonreducing glucose residue in position to accept the next glucose unit. The synthesis in this scheme is thus a discontinuous process. However, Robyt and Corrigan found that a dextran (as a primer) modified by a blocking group (triisoproprylbenzenesulfonyl, tripsyl) on the C_6 position of the nonreducing end of the glucose residue was as effective in increasing the rate of dextran synthesis by dextransucrase as the unmodified dextran primer.

The modified dextran could not act as a primer because its sixth position was blocked, proving that the added modified dextran stimulated synthesis by some other mechanism. Robyt et al. have shown that the activation of dextransucrase by addition of dextran is by an allosteric mechanism rather than by an extending primer mechanism as distinguished from the chain elongation of glycogen and starch by phosphorylase. In 1962 Ebert and Patat proposed another mechanism for dextran synthesis, insertion type polymerization, in which the addition of glucose units takes place at the reducing end (C_1 of glucose moiety) of the growing dextran polymer. In their mechanism they assumed that the growing chain and the enzyme remain in contact in the form of a complex throughout the chain elongation cycle. This enzyme–polymer complex (EP_n) reacts in the first step with substrate to form a substrate complex $(ESP_n)_2$, which is transformed in a second step in such a way that the polymer chain in the reaction complex is extended by one unit. This can only happen if the new monomer unit is inserted into the covalent bond between the enzyme and the polymer chain in the EP_n complex.

$$EP_n + S \underset{k_{-1}}{\overset{k_1}{\rightleftharpoons}} ESP_n \qquad \text{...(4)}$$

$$ESP_n \text{ (F : fructose)} \xrightarrow{k_2} EP_{n+1} + F \qquad \text{...(5)}$$

The bond between the enzyme and polymer chain does not dissociate during the reaction but must be flexible enough so that the new unit can be inserted into it. From their kinetic studies, Ebert and Schenk assumed that the substrate undergoes a second reaction with the EP_n complex in which it reacts as an acceptor.

$$ESP_n + S^* \underset{k_{-3}}{\overset{k_3}{\rightleftharpoons}} S^*ESP_n \rightarrow ES + S^*P_n \qquad \text{...(6)}$$

In the preceding reaction, the acceptor (sucrose molecule acting as an acceptor identified by the asterisk) does not compete with the substrate for the enzyme but reacts with the intermediate ESP_n complex, terminating polymer extension. This would explain why the molecular weights of the products decrease with increasing substrate concentration. From the preceding mechanism Ebert and Patat proposed the following reaction rate expressions:

$$\nu = \frac{V_m[S]}{K_m + [S]} \cdot \frac{K_a}{K_a + [S]} \qquad \text{...(7)}$$

where V_m is the maximum reaction rate (k_2E_0), K_m is the Michaelis constant of the chain extension reaction, and K_a is the corresponding constant of acceptor reaction. Ebert and Schenk conclude that there is strong evidence for their proposed mechanism because the results of their kinetic studies on dextran formation match the proposed rate in equation 7 quite well. They further conclude that primers are not necessary to initiate polymer growth, and sucrose alone can initiate polymerization. The early theories of the mechanisms of dextran synthesis were described by Neely.

Ebert and Schenk propose the sequential addition of monomer units to the reducing end by insertion between a carrier (the enzyme) and growing polysaccharide chain. Sidebotham continued the discussion in 1974, and a series of papers by Robyt and coworkers appears to have conclusively resolved the mechanism. In 1974 Robyt and coworkers proposed a two-site insertion mechanism for dextran synthesis in which the glucose unit is added to the growing polymer chain at the reducing end (C_1 moiety of glucose). They used pulse and chase techniques with ^{14}C-labeled sucrose and dextransucrase immobilized in Bio-Gel-P2. In their two-site insertion mechanism, there are two sucrose binding sites (A and B) that attack two sucrose molecules to give two covalently attached glucosyl–enzyme intermediates (C).

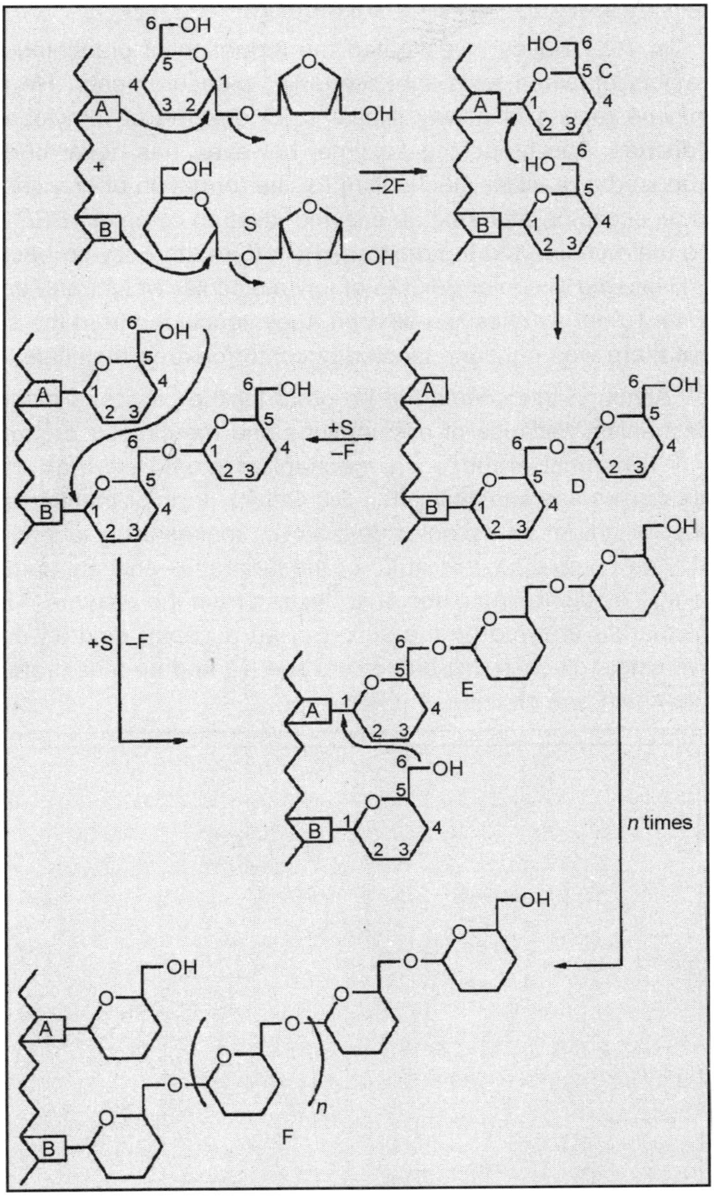

The C_6 hydroxyl of one of these glucosyl–enzyme intermediates makes a nucleophilic attack on C_1 of the other glucosyl–enzyme intermediate with the formation of an α, $1 \rightarrow 6$ glucosidic linkage and results in an isomaltosyl–enzyme intermediate (D). The vacant sucrose binding site (A) attacks another sucrose molecule to form a new glucosyl–enzyme intermediate to give a second α, $1 \rightarrow 6$ linkage and the formation of the isomaltotriosyl–enzyme intermediate (E). This process continues in a similar fashion by inserting a glucose residue between the enzyme and growing chain, which results in a dextranyl–enzyme complex (F). In this mechanism the growing dextran chain is transferred from one site to another during the synthesis. Robyt and Eklund considered the stereochemistry of the reaction and concluded that the linkages of glucosyl and dextranyl units to the enzyme must be β to retain the configuration of the glucose residue in growing from sucrose to dextran.

Figure 11.2 Two-site insertion mechanism: A and B, enzyme active sites; S, sucrose; F, fructose; C, glucosyl-enzyme complex; D, isomaltosyl-enzyme complex; E, isomaltotriosyl-enzyme complex; F, dextranyl-enzyme complex.

Mechanism of branch linkage formation

In 1959 Bovey investigated the formation of branching in dextrans using data on molecular weights obtained from light-scattering measurements. He proposed that a separate branching enzyme (acting in a way similar to Q enzyme on amylose) is responsible for the formation of branches. The branching enzyme, however, has never been found. In 1967 Ebert and Brosche proposed a reaction mechanism for the formation of branches in which a dextran chain itself acts as an acceptor, attacking an enzyme–dextran complex (ESP_n). The acceptor dextran then becomes the main chain, and the dextran chain from the enzyme becomes the side-branched chain. Using a ^{3}H-labeled acceptor dextran of low molecular weight and assuming an average molecular weight of 4×10^5 for synthesized dextran, they concluded from the specific activity of synthesized product that there was only one labeled acceptor dextran molecule in each synthesized dextran.

Although this seemed to be proof for the proposed mechanism, some doubt was cast on the mechanism because of assumptions and the circular arguments that were made. In 1976 Robyt and Taniguchi proposed a mechanism based on their study using ^{14}C-labeled sucrose and dextransucrase immobilized in Bio-Gel-P2. In their mechanism, a C_3 hydroxyl group of an interior glucose unit of an acceptor dextran (D) makes a nucleophilic attack at C_1 of either the glucosyl–enzyme complex (C) or at C_1 of the dextranyl–enzyme complex (E) to form an α, $1 \rightarrow 3$ branch linkage by displacing glucose or dextran from the enzyme. Thus, branching can take place without a separate enzyme by the action of an acceptor dextran on a glucosyl–enzyme complex (C) to give single glucose unit branch linkage (F) and on a dextranyl–enzyme complex (E) to give a long side chain branch linkage (G).

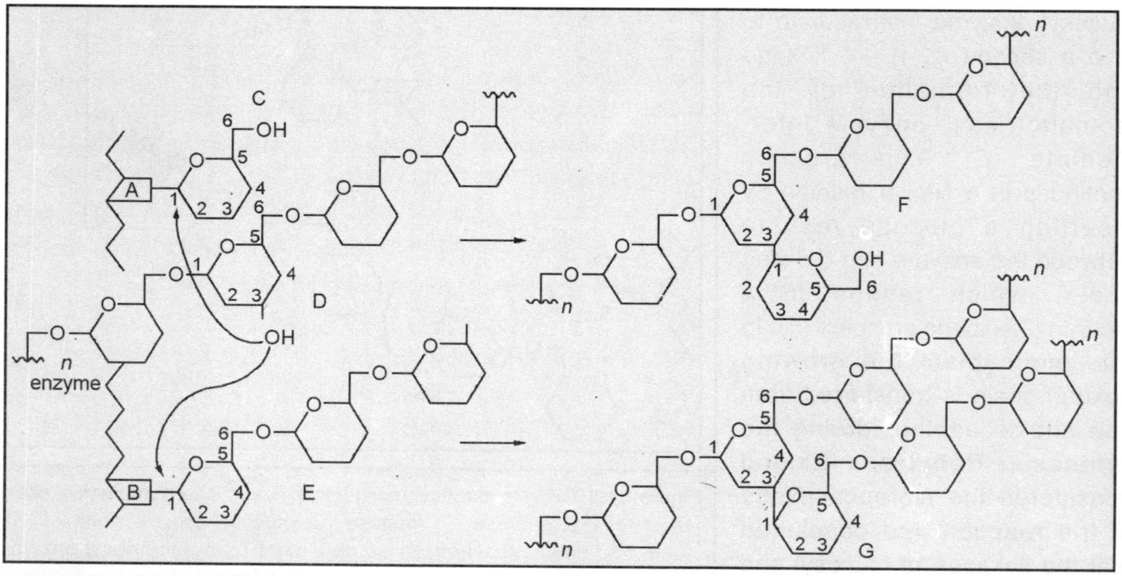

Figure 11.3 Mechanism of α, 1→3 branch linkage synthesis: D, acceptor dextran main chain; C, glucosyl-enzyme complex; E, dextranyl-enzyme complex; F, resulting α, 1→3 branch-linked product when C and D are reacted together; G, resulting α, 1→3 branch-linked product when E and D are reacted together.

Acceptor Reactions and Acceptor Products

In addition to catalyzing the synthesis of dextran from sucrose, dextransucrase also catalyzes the transfer ofthe glucose moiety from sucrose to other carbohydrates that are present or added to the reaction mixture. These other carbohydrates are acceptors. Actually, there are two classes of acceptors—those that give a homologous series of oligosaccharides, each differing one from the other by one glucose residue, and those that form only a single acceptor product that contains one glucose residue more than the acceptor. Koepsell et al. and Tsuchiya et al. first observed that the presence of low molecular weight carbohydrates (acceptors) shifted the course of dextransucrase reaction from synthesizing a high molecular weight dextran to synthesizing of a low molecular weight dextran. These acceptors not only terminate the dextran synthesis but also produce a novel series of oligosaccharides. Su and Robyt systematically studied maltose acceptor reactions with dextransucrase. In this maltose acceptor reaction the D-glucose moiety of sucrose is diverted from synthesis of dextran but transferred to the nonreducing end of maltose (a disaccharide) to form panose (a trisaccharide). This first acceptor product, panose, can itself act as an acceptor to give a tetrasaccharide. This tetrasaccharide can again act as an acceptor to give a pentasaccharide and so forth. This series of homologous acceptor products is produced by acceptor reactions.

Mechanism of acceptor reaction

As described earlier, dextransucrase catalyzes the synthesis of dextran by initially forming two covalent glucosyl–enzyme complexes in which the C_6 hydroxyl moiety of one glucose attacks the C_1 of the other to form an α, 1 $\rightarrow$ 6 glucosidic bond to produce an isomaltosyl–enzyme complex. This is the beginning of the dextran main chain growth; the dextran chain is covalently attached to

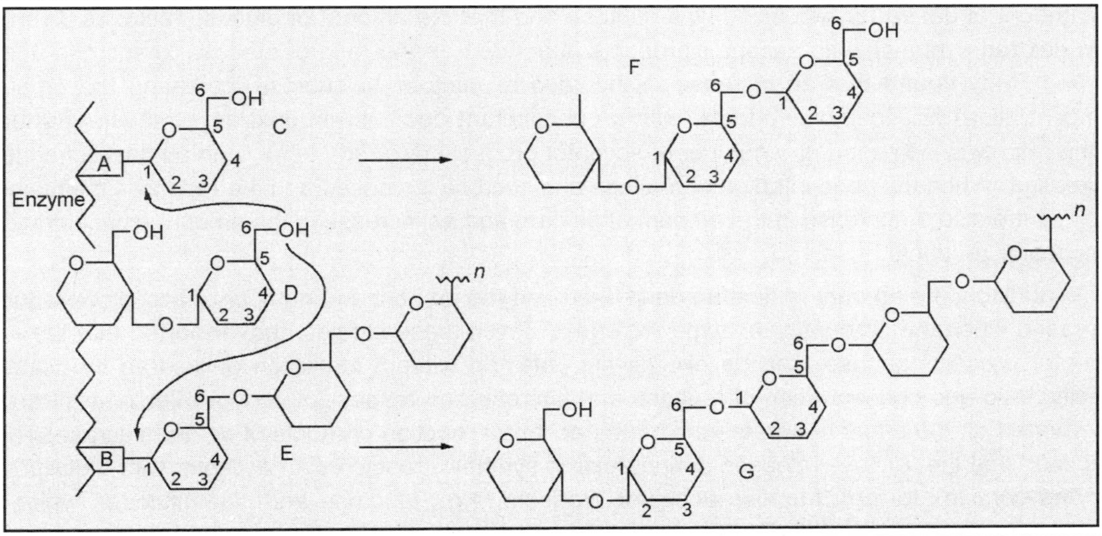

Figure 11.4 Mechanism of the acceptor reactions: D, maltose molecule bound at acceptor binding site; C, glucosyl-enzyme complex; E, dextranyl-enzyme complex; F, resulting acceptor product when C and D are reacted together; G, resulting acceptor product when E and D are reacted together.

the enzyme active site and elongated by the insertion mechanism. In the presence of an acceptor, such as maltose, the transfer of glucose into a growing dextran chain is interrupted as the maltose binds to an acceptor binding site. Robyt and Walseth proposed a mechanism for the acceptor reaction. In this mechanism, the acceptor bound at an acceptor binding site, and if the maltose molecule is an acceptor, its C_6 hydroxyl group at the non-reducing end attacks C_1 of the glucosyl–enzyme (C) or dextranyl–enzyme (E) complex to give an oligosaccharide (F) or a dextran acceptor product (G), respectively. Su and Robyt, using maltose as an acceptor in an equilibrium dialysis experiment, showed that there was one acceptor binding site and two sucrose binding sites (A and B).

Effect of acceptor on dextran production

The acceptors affect dextran production in two ways:

1. They terminate the growth of the dextran chain by reacting with the dextranyl–enzyme complex, which results in the low molecular weight dextran (product G).
2. They interfere with the transfer of the glucose moiety of sucrose to the growing dextran chain by reacting with the glucosyl–enzyme complex, which results in production of a homologous series of acceptor products and reduction in the amount of dextran produced (product F).

Paul et al. demonstrated the effect of the sucrose-to-acceptor ratio in the synthesis of controlled molecular weight dextrans. Robyt and Eklund showed that the amount of dextran synthesized decreased as the molecular ratio of maltose to sucrose increased. Not all of the acceptors reacted with equal efficiency. In a series of reactions with different acceptors at a 1:1 ratio of acceptor concentration to sucrose concentration, they found that the most effective acceptor for decreasing the amount of dextran synthesized was maltose and that the amount of glucose residues diverted from dextran synthesis into acceptor products depended on the ratio of maltose to sucrose used. Su and Robyt found that an increase in the ratio of maltose to sucrose increased the amount (concentration) of acceptor products, with a concomitant decrease in dextran synthesis. Further, as the ratio was increased, the number of acceptor products (e.g., tri-, tetra-, and pentasaccharides) decreased. When the concentration of maltose and sucrose increased and the ratio was maintained at 1:1, there also a decrease in the amount of dextran and an increase in the amount (concentration) of acceptor products observed.

In addition, the amount of dextran decreased and the amount and number of acceptor products increased when the amount of enzyme increased. From these studies, they reported that the first acceptor product (panose) can be exclusively obtained without formation of dextran by using a specific ratio and concentration of maltose and sucrose and a specific amount of enzyme. Part et al. reported on the effect of borate ions on the acceptor reaction products of dextransucrase. They explained that the fructose released during dextran synthesis reacts as an acceptor with a glucosyl–enzyme complex to form fructose acceptor products (e.g., leucrose and isomaltulose), which in turn reduce the dextran yield. This fructose can be complexed with borate ions, and its acceptor effects can be minimized to increase dextran yield. They concluded that with 0.115 M sodium tetraborate in the reaction mixture, leucrose and isomaltulose formation can be minimized up to 75% with consequent increase in dextran yield. Therefore, the acceptor reaction between fructose and the dextranyl–enzyme complex, which leads to diminished low molecular weight dextran, can be overcome by adding borate ions to the reaction medium.

Table 11.1 Acceptors and their efficiency on dextransucrase

Acceptors	*Efficiency (%)*
Maltose	100
Isomaltose	89
Nigerose	58
Methyl α-D-glucopyranoside	52
D-Glucose	17
Lactose	11
Cellobiose	9
D-Fructose	6.4
Maltooctaose	6.2

Lee et al. used the idea of producing low molecular weight dextran by adding an acceptor to the reaction medium to directly produce clinical dextran (MW 70,000). They studied the effect of dextran T-10, dextran T-40, and maltose as acceptors in the batch and continuous reactors. They reported that increasing the concentration of dextran T-10 reduces the production of high molecular weight dextran and increases the production of clinical dextran, whereas increasing the dextran T-40 concentration decreases the production of clinical dextran. Maltose reduces high molecular weight dextran production but increases the production of acceptor products, thereby reducing clinical dextran production.

Regulation of Dextransucrase

Germaine et al. observed that addition of dextran to the dextransucrase (produced from the *Streptococcus mutans* reaction mixture) increased the rate of dextran synthesis. They found that the rate depended on the size of the dextran chain and reached a maximum when the average size of the added dextran was 30 glucose residues. Kobayashi and Matsuda also observed that purified dextransucrase was stimulated by addition of dextran, although dextransucrase could synthesize dextran without addition of dextran to the reaction mixture. The rate of dextran synthesis in dextran-free reaction mixtures was accompanied by a lag period that could be eliminated by addition of exogenous dextran. Robyt et al. reported that the addition of dextran to dextransucrase reaction mixtures affects the initial velocity of dextran synthesis in a sigmoidal manner characteristic of an allosteric binding of dextran to the enzyme. Further, they proposed that the allosteric activation of dextransucrase occurs when the dextran binds to the dextran binding site, where it acts as a matrix to join or induce the subunits to associate into an active form.

Kobayashi and Matsuda reported that purified B-512 F dextransucrase contains 17% carbohydrates, mostly as endogenous dextran covalently bound to the enzyme, which can be eliminated by the mild action of endodextranase. They suggested that exogenously added dextran could stabilize the enzyme not only during storage but also during long-term incubation. Miller and Robyt, from their rate kinetics of dextran synthesis, showed that concentrations of Ca^{2+} ion below 1 mM activate the enzyme by increasing V_m and decreasing the K_m for the sucrose. Above 1 mM Ca^{2+}, the ions act as weak competitive inhibitors. Although Ca^{2+} is an activator at low concentrations, it is not required for dextran synthesis. Miller and Robyt proposed a model for Ca^{2+} ion activation and inhibition behavior in which the enzyme has two Ca^2 binding sites. One Ca^2 site is an activating site and the other is an inhibitory site at which Ca^{2+} ions prevent the binding of sucrose to the sucrose binding site.

Kobayashi and Matsuda reported that the purified enzyme can be activated by Ca^{2+} ions and strongly inhibited by EDTA. The inhibitory effect of EDTA can be reduced by adding Ca^{2+} ions. However, they further observed that the purified enzyme was strongly inhibited by Cu^{2+}, Fe^{2+}, and Mn^{2+} ions in this order. Contrary to the crude enzyme, which contains endogenous dextran, the purified enzyme shows optimum activity at pH 6.0. Miller and Robyt found that the dextransucrase could be stabilized against the activity losses caused by heating or by dilution through the addition of low concentrations nonionic polymers (e.g., dextran, PEG 20000, and methyl cellulose) or of nonionic detergents at or slightly below their critical micelle concentrations.

Enzyme Concentration and Dextran Synthesis

In 1962 the effect of *enzyme concentration* on dextran synthesis was reported by Braswell et al. They observed that at higher concentrations of enzyme the molecular weight of dextran synthesized was decreased and the intrinsic viscosity and branching ratio (α, 1 $\rightarrow$ 6 linkages to α, 1 $\rightarrow$ 3 linkages) was increased. Tsuchiya et al. made a related observation that increasing the enzyme concentration decreased the amount of high molecular weight dextran synthesis. At the time, neither Braswell and coworkers nor Tsuchiya and coworkers put forward an explanation. However, in 1993 Su and Robyt explained in their report that the results are due to the increased acceptor reactions at higher enzyme concentrations. Another important point is that improving the yield of enzyme from a given fermentation does not necessarily produce an improved yield of dextran because the conditions under which the enzyme is allowed to act on sucrose solution determine the dextran yield.

PRODUCTION OF DEXTRANSUCRASE AND DEXTRAN

The dextran fermentation process involves three phases: (i) the cell growth phase, (ii) the dextransucrase production phase, and (iii) the enzyme action phase. The enzyme action phase that produces dextran is the third phase (i.e., it occurs after cell and enzyme production). Thus, the rate of dextran production under given conditions is a function of the enzyme production rate, which in turn is a function of the cell production rate. Moreover, dextransucrase stability and its activity under given conditions influence dextran synthesis. To optimize the dextran fermentation process, one must consider the following rates of reactions: (i) cell growth rate, (ii) enzyme production rate, (iii) enzyme deactivation rate (stability), and (iv) enzyme action rate.

The interdependency of these reaction rates under specific fermentation conditions is a key factor for in situ dextran production. Because each rate of reaction has its own optimum level, a set of conditions given for a fermentation may not be optimal for all four rates of reaction. For example, a temperature of 30°C is optimal for cell production and enzyme action but not for enzyme production and its stability. The three phases of dextran fermentation can be separated and the optimum level of each phase can be used one by one for maximum dextran yield. Therefore, each phase and its reaction rates must be understood separately and in great detail to employ process control techniques, to find optimal conditions, and to design a high- performance reactor.

Factors Affecting Fermentation of Dextran

Diversity of bacteria

The genera of bacteria recognized to produce the enzyme dextransucrase, which is capable of synthesizing dextran from sucrose, are primarily *Leuconostoc* and *Streptococcus*. These two genera are gram- positive, facultative anaerobic cocci that are closely related. The notable difference between these two genera is that *Leuconostoc* species require sucrose in the culture medium to induce the formation of dextransucrase, whereas the *Streptococcus* species do not require sucrose in the culture medium to induce enzyme formation. Thus, the *Leuconostoc* species are normally inducible for the formation of dextransucrase and the *Streptococcus* species are normally constitutive for their formation. However, Kim and Robyt have obtained mutant strains of *Leuconostoc mesenteroides* B-512 FM, B-742, B-1299, and B-1355 that are constitutive for dextransucrase formation.

Sucrose concentration

Sucrose is an essential substrate for dextransucrase and dextran synthesis. Although *L. mesenteroides* can grow on various carbohydrates such as glucose, fructose, and maltose, it prefers sucrose over other carbohydrates as a carbon and energy source. The carbohydrates are fermented through the hexose monophosphate pathway, giving lactic acid, carbon dioxide, and ethanol under anaerobic conditions. An aerobic oxidative metabolism that yields carbon dioxide, lactic acid, and acetic acids also exists. Tsuchiya et al. found in batch culture that increasing sucrose concentration from 5 to 50 g/L resulted in an increase in enzyme productivity from 6 to 120 DSU/mL. However, they recommended 20 g/L sucrose for optimum dextransucrase production because higher sucrose concentrations increase the viscosity of the broth due to dextran synthesis, which makes cell removal difficult.

The enzyme activity that they achieved at 20 g/L sucrose is 86 DSU/ mL. It is essential to maintain the lower level of sucrose concentration (5–10 g/L) throughout fermentation to produce a higher activity of dextransucrase. This is possible in semicontinuous or continuous operational mode. Lopaz and Monsan achieved the enzyme activity of 180 DSU/mL by addition of sucrose at the rate of 20 g/L/h over the cell growth phase. Lawford reported that the enzyme can be synthesized only after a critical concentration of sucrose, the inducer, has been reached or surpassed. The highest activities of dextransucrase reported in the literature are by Schneider et al. and Barker and Ajongwen. The former obtained 400 DSU/mL in laboratory scale and 289 DSU/mL at pilot plant scale. Barker and Ajongwen obtained activity of 450 DSU/mL under nonaerated fed-batch fermentation and 350 DSU/mL under aerated fed-batch fermentation. At higher sucrose concentrations, the enzyme is inhibited and the conversion of sucrose to dextran decreases. Moreover, the molecular weight of synthesized dextran decreases at higher sucrose concentrations, where sucrose itself acts as an acceptor to terminate polymer growth.

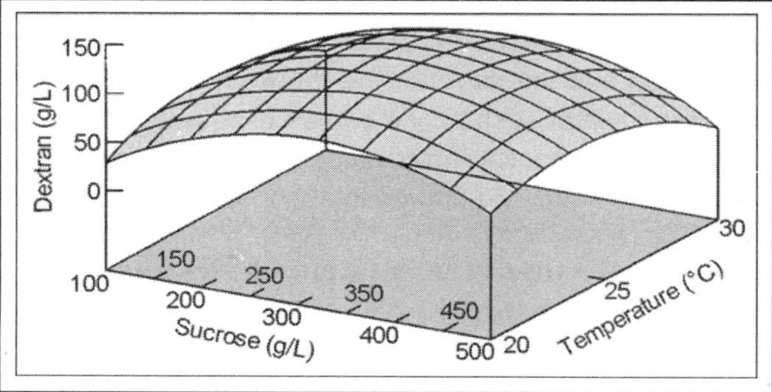

Figure 11.5 A calculated response surface plot of dextran yield (g/L) as a function of sucrose and temperature with a constant level of yeast extract (10 g/L), (30 g/L) K_2HPO_4, and initial pH 8.3.

Braswell et al. observed that increasing the sucrose concentration from 50 to 300 g/L increases the intrinsic viscosity and the branching ratio of dextran produced while decreasing the molecular weight of dextran and the rate of dextran formation. Alsop reported that higher dextran yields have been obtained at lower initial sucrose concentrations, reducing costs as a function of sucrose alone. However, the net amount (and so the concentration) of dextran produced was lower, reducing plant productivity and substantially increasing ethanol and labor costs in downstream processing. Alsop calculated the optimum value on the basis of ethanol precipitation costs (which depend on the molecular weight of the dextran synthesized; high molecular weight dextrans can be precipitated with lower concentrations of alcohol) and reported that the optimum value is 170 g/L of initial sucrose concentration.

However, the same author reported another optimal value, 125 g/L of sucrose, based on the operating costs of the plant. Martinez-Espindola and Lopez-Munguina reported that in cell-free enzymatic synthesis of dextran, an initial sucrose concentration of 150 g/L is optimum for 99% conversion of sucrose to dextran. The optimum value was based on reaction kinetics of dextransucrase and a specific design for a batch reactor. In our work, we found the optimum value (based on yield) of sucrose to be 300 g/L for batch dextran production in whole cell culture based on response surface methodology. We observed a yield of 135 g/L dextran at this optimum level. Further, we observed substrate inhibition kinetics for cell growth at more that 100 g/L sucrose and that at more than 300 g/L sucrose the dextran yield decreases considerably.

Nitrogen source

The nitrogen sources used to grow *L. mesenteroides* and produce enzyme are corn steep liquor, yeast extract, soya meal preparations, and peptone. These complex nutrients also provide essential growth factors and trace elements. Ammonium ions used as the nitrogen source appear to have an adverse effect on cell growth. Tsuchiya et al. reported that an increase in the concentration of corn steep liquor as a nitrogen source from 2.5 to 20 g/L increases the enzyme yield but only when higher amounts of phosphate are present in the medium. Barker and Ajongwen, who obtained the highest enzyme activity, used 40 g/L yeast extract as a nitrogen source in fed-batch culture. However, they observed variable performance with different yeast extract types, because different yeast extracts contain varying levels of vital nutrient components. In our work, we optimized yeast extract concentration to 10 g/L for conversion of 300 g/L sucrose to dextran and further observed that below 10 g/L yeast extract the growth of cells and enzyme production were affected considerably.

Concentration of phosphate

Phosphates such as K_2HPO_4 and KH_2PO_4 act as buffers in the fermentation medium in addition to being nutrients. Tsuchiya et al. reported that 35 g/L KH_2PO_4 is optimum for enzyme production. We observed that 30 g/L K_2HPO_4 is optimum for batch dextran production to convert 300 g/L sucrose to dextran. Lawford et al. reported that larger amounts of phosphate can remove the Ca^{2+} ions in the medium by forming a complex with Ca^{2+} ions. Calcium is an essential element for dextransucrase stability. They observed loss of activity of dextransucrase during batch culture at 10 g/L potassium phosphate, whereas at 1 g/L phosphate the produced activity remained stable for a long time. London and Webb optimized the concentration of Na_2HPO_4 to 20 g/L, and Barker and Ajongwen used K_2HPO_4 at 20 g/L in their fed-batch fermentation for higher enzyme production.

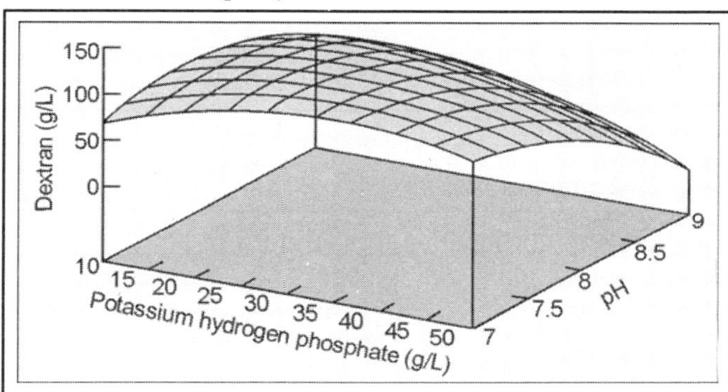

Figure 11.6 A calculated response surface plot of dextran yield (g/L) as function of initial pH and K_2HPO_4 with a constant level of sucrose (300 g/L) 10 g/L yeast extract, and 23°C.

In batch fermentation without controlled pH the pH of the medium starts to fall as the cell starts to grow. However, the rate of fall can be controlled by using higher phosphate concentrations. Cell growth, enzyme production, and dextran production are faster at pH 7.0, 6.7, and 5.2, respectively. However, in pH-controlled fermentations, the higher amounts of phosphates are not required and the adverse effect of Ca^{2+} ion removal can be avoided. pH control is recommended over a higher phosphate concentration in the culture.

Growth factors and trace elements

Carlson and Carlson reported that the growth factors nicotinic acid, thiamine, pantothenic acid, and biotin and the amino acids valine and glutamic acid are required for *L. mesenteroides* growth. However, the complex nitrogen sources (e.g., yeast extract, corn steep liquor, and peptone) are expected to supply those vital nutrients and necessary trace elements. In 1952 a salt solution called R salts that contained 4.0% $MgSO_4 \cdot 7H_2O$, 0.2% NaCl, 0.2% $FeSO_4 \cdot 7H_2O$, and 0.2% $MnSO_4 \cdot H_2O$, was first introduced by Koepsell and Tsuchiya. London reported that the trace inorganic salts generally used were not found to be necessary, suggesting that nutrient requirements are culture dependent. The effect of Ca^{2+} ions on enzyme stability was discussed earlier.

pH and temperature

The optimum levels of pH and temperature for cell, enzyme, and dextran production are given in Table 11.2, and the enzyme dextransucrase is stable at pH 5.2 and low temperature. The effects of pH and temperature on enzyme production and its action and stability are interrelated. Although pH 6.7 is optimum for enzyme production in this example, this value is applicable only at the optimum temperature of 23°C. At higher temperatures, pH 6.7 inactivates the enzyme rapidly.

Table 11.2 Optimum pH and temperature for cell growth, enzyme production, and dextran production by *Leuconostoc mesenteroides*

Parameter	*Cell growth*	*Enzyme production*	*Enzyme action*
pH	7.0	6.7	5.2
Temperature (°C)	30	23	30

Similarly, the optimal temperature of 30°C for cell growth and enzyme activity is too high for enzyme production and stability. The enzyme activity in culture filtrate is retained at pH 5.2 for at least 24 h at 25°C, but 35% of the activity is lost at pH 6.7 within 1 h at 25°C. Moreover, at higher temperatures the deleterious effect of pH is accentuated markedly. Kobali and Reilly reported that the half- lives of enzyme at pH 4.9 are 13.75, 7.02, and 3.28 min at 30, 32, and 35°C, respectively. London and Webb observed that the maximum enzyme production rate was achievable at pH 6.1, although the maximum activity was observed at pH 6.7.

They further stated that at pH 6.0 and below, the enzyme was relatively stable, retaining 100% of activity for 5 h at pH 5.4 and 26°C. Above pH 6.0 the enzyme was far less stable, retaining only 10% of its original activity at pH 6.6 and 26°C. The effect of temperature as the velocity of the reaction corresponds to an Arrhenius activation energy of 10.6 kcal/mol. However, Kobali and Reilly reported a lesser value of 8.6 kcal/mol of activation energy. Dextran synthesis temperature also plays an important role in the properties of the dextran produced. Braswell et al. reported that decreasing the reaction temperature from 30 to 0°C lowers the molecular weight of dextran and increases the intrinsic viscosity. One would normally expect lesser intrinsic viscosity for lower molecular weight polymers, but a more linear polymer is produced at lower temperatures, increasing the intrinsic viscosity observed. Sabatie and Choplin also reported that dextran produced at 5°C was more linear than dextran produced at 30°C. They observed 5% α, 1 $\rightarrow$ 3 branch linkages at 5°C of synthesis temperature compared with 4.8% at 30°C.

Aeration and agitation

The maximum productivity and product concentration achievable in industrial microbial processes for production of biomass as product are typically limited by the rate of oxygen transfer. *L. mesenteroides* is a facultative anaerobe. However, some researchers used the aerobic fermentation for dextransucrase production. Schneider and coworkers observed a higher enzyme activity of 400 DSU/mL under highly aerated conditions in which the oxygen level was maintained at 40–80% of the initial saturated value throughout the course of fermentation. However, Barker and Ajongwen observed the higher activity of 450 DSU/mL under nonaerated fed-batch fermentation but achieved only 350 DSU/mL of activity under aerated fermentation. Foster also found that aeration gives a lower yield of enzyme. Koepsell and Tsuchiya and Tsuchiya et al. noticed that enzyme yields were higher in shaken than in still flasks, and Johnson and McCleskey also observed similar results. This may be due to the mixing requirements for limiting nutrients. London and Webb compared the stirred and unstirred cultures for enzyme production. They observed a lower cell growth rate and higher enzyme production in unstirred culture than in stirred culture.

This is due to the limiting substrate concentration around the cells being lower under poor mixing conditions (i.e., the transfer from the liquid to the cells is insufficient to meet their demand, which results in poor cell growth rate). But as the cells produce the enzyme and acid, the enzyme and acid are expected to collect in pockets around the cells in nonagitated conditions. This lowers the pH around the enzyme to its stability optimum. As the dextran concentration increases in the fermenter during its synthesis, the viscosity of the broth also increases, which requires proper mixing conditions for better conversion of substrate to polysaccharide and scale-up of polysaccharide fermentation. Margaritis and Bokkel studied the rheological behavior of dextrans (low and high molecular weight) in different concentrations. Their data fit well the following power law equation, which represents the pseudoplastic behavior of fluid:

$$\log\tau = \log K + n\log\gamma \qquad \text{...(8)}$$

where τ is the shear stress applied on the fluid (N/m^2), K is the consistency index, n is the flow behavior index, and γ is the shear rate (s^{-1}).

From Margaritis and Bokkel's observations, the flow index n of the high molecular weight dextran solution was much less than the flow index for the low molecular weight solution. The consistency index K of high molecular weight dextran solutions was found to be higher than the K values of low molecular weight solutions for the same concentrations. These results show that high molecular weight dextran shows a higher degree of pseudoplastic behavior than low molecular weight dextran.

Batch Culture

Traditionally, dextran is produced by anaerobic, unstirred batch culture using dextransucrase-producing bacteria in sucrose-rich media. Sugar solutions are fortified with a nitrogen source such as yeast extract or corn steep liquor, a phosphate source such as K_2HPO_4 or KH_2PO_4, and trace metal salts. Little further process control is employed. The pH, which was neutral at the start of the fermentation process, falls as cells grow. Dextransucrase is not produced until near the beginning of the stationary phase. At this point the enzyme is secreted rapidly into the culture medium and reaches a maximum after a few hours. Dextransucrase kinetics have generally been described by the Michaelis–Menten model and applied for sucrose levels below 100–150 g/L, where the saturation of the enzyme is achieved. Martinez-Espindola and Lopez-Munguina proposed a simple substrate inhibition model for enzyme action at higher sucrose levels as follows:

$$V_i = [V_{max} + S]/[K_m + S + (S^2/K_i)] \qquad \text{...(9)}$$

For enzymatic synthesis of dextran in a batch reactor, Martinez-Espindola and Lopez-Munguina derived the following design equation:

$$V_{max}t = S_0X - K_m \ln(1-X) + (K_iS_0^2)(2X - X^2)/2 \quad ...(10)$$

where S_0 is the initial sucrose concentration, X is the conversion, and K_i is the inhibitor constant. They predicted that 150 g/L of sucrose concentration is optimum for 99% conversion of sucrose to dextran in terms of productivity in a batch reactor.

London and Webb optimized both the medium and inoculum for batch production of dextransucrase in an unaerated and unstirred culture. They achieved a final enzyme activity of 142 DSU/mL. Although it is lower than the fed-batch achievement, it was the best reported value for batch operation. However, London and Webb's scaled-up process to 1,000 L yielded only 45 DSU/mL of activity.

Table 11.3 Optimal conditions for production of dextransucrase used by various authors

Variables	*El-Sayed et al.*	*London and Webb*	*Barker and Ajongwen*	*Karthikeyan et al.*
Sucrose	20 g/L	40 g/L	10 g/L	300 g/L
Yeast extract	20 g/L	20 g/L	40 g/L	10 g/L
Tween 80	2 g/L	—	—	—
KH_2PO_4	2 g/L	—	—	—
K_2HPO_4	—	—	20 g/L	30 g/L
Na_2HPO_4	—	16 g/L	—	—
$CaCO_3$	10 g/L	—	—	—
R salts	5 mL	—	0.5% v/v	—
pH	6.7	7.1	6.7	8.3
Temperature	26°C	26°C	23°C	23°C
Inoculum	—	1% v/v	—	10% v/v
Mode	Fed-batch	Batch	Fed-batch	Batch
Sucrose feed	10 g/L/h	—	Connected with alkali addition	—
Product	Dextransucrase	Dextransucrase	Dextransucrase	Dextran

We reported the optimum conditions for batch dextran production by using response surface methodology (RSM) and multistage Monte Carlo optimization (MSMCO) optimization techniques with a predicted yield of 154 g/L dextran. The optimum values are given in Table 11.3; under these conditions we observed 135 g/L dextran yield. The optimum initial pH of 8.3, which was set before inoculation and sterilization, dropped to 7.5 at the start of fermentation after adding 10% v/v acidic inoculum. Batch operational mode is a suitable method to produce dextran through either whole cell fermentation or cell-free enzymatic synthesis due to the possibility of a high final dextran concentration (hence the productivity), which minimizes the alcohol precipitation costs in recovery of the product. However, if our aim is to produce maximum enzyme activity, fed-batch operation that keeps lower levels of sucrose concentrations in the fermenter is the best operational mode.

Discontinuous Culture

Because dextran fermentation shows substrate inhibition kinetics, the substrate level can be maintained at the desired level by using fed-batch operational mode. This is shown by the various authors who produced higher enzyme activity than in batch cultures. Lopez and Monsan and Monsan et al. achieved 180 DSU/mL enzyme activity by fed-batch fermentation. They fed sucrose solution at a rate of 20 g/L/h and achieved maximum cell density and enzyme activity. Barker and Ajongwen achieved the highest enzyme activity (450 DSU/mL) under low agitated, nonaerated fed-batch fermentation. El-Sayed and coworkers observed higher enzyme activity in fed-batch operation than batch mode in both immobilized-cell and free-cell cultures. They observed that when the sucrose feed rate is increased from 5 to 10 g/L/h, enzyme activity increases from 73 to 118 DSU/mL in free-cell culture and 110 to 223 DSU/mL in immobilized-cell culture. Although fed-batch culture (used to maintain a low substrate level) is suitable for enzyme production, it may not be suitable for dextran production because of lower plant productivity and increased alcohol precipitation costs in downstream processing.

Regular Culture

Ringpfeil and Selenia theoretically calculated from batch culture experiments the maximum yield of dextransucrase that would be obtained from sucrose-limited continuous culture at low dilution rate. The maximum enzyme activity achieved at a dilution rate of 0.53 h^{-1} was 16 DSU/mL. However, Lawford et al. reported that the enzyme can be synthesize only after a critical concentration of sucrose has been reached because they could not synthesize dextransucrase at dilution rates less than 0.2 h^{-1}. They calculated the theoretical dilution rate for maximum cell growth to be 0.67 h^{-1}. However, they achieved maximum cell growth rate at 0.35 h^{-1}, concluding that either some essential metabolite other than the energy source limited the growth or some components of the growth medium inhibited the growth. Lee et al. produced low molecular weight clinical dextran in a two-stage continuous reactor using acceptors such as maltose, dextran T-40, and dextran T-10. Cells cultivated in the first reactor were transferred to the second reactor, where sucrose and acceptor were added for clinical dextran production.

Use of Immobilized Cells and Enzymes

Use of immobilized dextransucrase, if it were of sufficient stability, would allow the production of dextran without the requirement for repetitive fermentation. Earlier reports of dextransucrase immobilization were issued from several groups. The enzyme has been immobilized ionically to DEAE sephadex A-50, covalently to Bio-Gel-P2, and covalently to polyacrylamide gel, cellulose acetate membranes, and polysulfone hollow fibers. In 1980 Kobali and Reilly examined 13 carriers (e.g., Enacryls, DEAE cellulose, DEAE sephadex, SP sephadex, and porous silica) for dextransucrase immobilization. Among the 13 carriers surveyed, only alkaline amine porous silica retained more than 10% of activity on immobilization (10 DSU/g carrier). The optimum pH (5.2) and temperature (30°C) of enzyme activity are not changed on immobilization, but the activation energy below 30°C was 1.7 kcal/mol for immobilized enzyme, which is lower than for soluble enzyme (8.6 kcal/mol). Monsan and Lopez immobilized dextransucrase on an aminoporous silica support activated with 2% glutaraldehyde by covalent coupling. They examined the specific surface area of the carrier silica for immobilization efficiency and observed that (i) lowering the specific surface area increased the coupling efficiency for dextransucrase and that (ii) adding maltose to the dextransucrase preparation during enzyme coupling greatly increased the coupling efficiency. The production of high molecular weight dextran with immobilized dextransucrase results in extremely viscous solutions, which makes recovery of the support difficult at the end of the synthesis. They also observed from their kinetic study diffusional limitation in the immobilized catalyst because of high molecular weight dextran synthesis.

Chang et al. successfully immobilized the enzyme on porous phenoxy acetyl cellulose beads by hydrophobic adsorption with high efficiency. The enzyme was immobilized at more than 100 times the activity attained on porous silica beads by Kobali and Reilly and at nearly double the activity found by Monsan and Lopez. El-Sayed and coworkers published a series of papers in which they studied the effect of different carriers (i.e., calcium alginate, three types of celites, celites coated with calcium alginate, stainless steel, and stainless steel coated with calcium alginate) for immobilization, mode of operation (batch and fed-batch), and the performance of different reactors (bubble column and airlift bioreactors). They further compared all the results with whole cell fermentation under the same conditions. El-Sayed and coworkers used the immobilized cells in calcium alginate beads for sequential production of dextransucrase followed by dextran production in three cycles. Each cycle consisted of fed-batch enzyme production for 24 h followed by batch dextran production for 24 h. The maximum dextran yields reached 40%, 44%, and 46% based on the added sucrose and were achieved 15, 7, and 3.5 h from the beginning of the first, second, and third dextran production periods, respectively. Increasing the sucrose feed rate from 5 to 10 g/L/h led to an increase in the total enzyme activity of about 88%.

El-Sayed et al. compared the results with free-cell fermentation where only one cycle was possible. They observed that immobilized cells produced more than three times as much dextran as free cells during one cycle. When comparing different carriers they observed that the immobilized cell cultures were significantly affected by pore and particle size. Enzyme production with cells immobilized in celite R 648 resulted in the highest total enzyme activity followed by celite R 633, alginate-coated celite R 630, celite R 630, and the calcium alginate beads. A layer of calcium alginate on porous celite R 630 particles improved their mechanical stability, increased the amount of soluble dextransucrase activity, and decreased the cell leakage from the highly porous support. Immobilizing *L. mesenteroides* on a porous stainless steel support also increased dextransucrase and dextran production. Coating calcium alginate layers on stainless steel supports enhanced the cell loading capacity of the support and increased dextransucrase and dextran production.

Other Strategies

Dextran fermentations consist of three phases—cell, enzyme, and dextran production—and each phase shows different optima for pH, temperature, and sucrose concentration. A major advantage in dextran fermentation is that these individual phases can be performed separately and sequentially in a reactor system that allows flexibility in operational strategy. Their separation could be achieved by either of the following simple techniques:

1. Single reactor system with multistage operation: a stepwise operation of the individual phases with their individual optimum condition.
2. Multistage reactor systems in series: simultaneous operation of individual phases (i.e., each individual phase performed with unique optimum levels in separate reactors in series).

In Figure 11.7, reactors a, b, and c represent single reactor–multistage operational systems with batch, fed-batch, and continuous mode of operation, respectively. In these systems, for example, if we divide the total fermentation time into three stages so as to perform cell, enzyme, and dextran production in a stepwise fashion, we can control the pH at 7 and temperature at 30°C for the first stage to maximize cell growth. In the second stage, after reaching a high biomass concentration, we set the pH and temperature to 6.7 and 23°C, respectively, for enzyme production. Similarly, in the third stage we can set the pH at 5.2 and temperature at 30°C for dextran production.

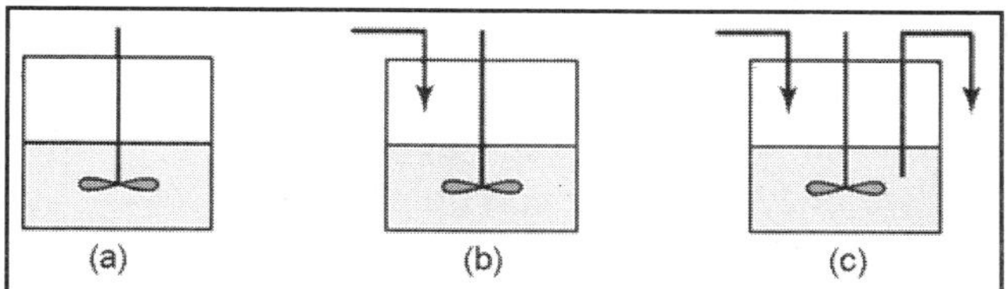

(a) (b) (c)

Figure 11.7 Reactor designs for stepwise process: (a) batch reactor, (b) fed-batch reactor, (c) continuous reactor.

Further, we can control sucrose concentration in fed-batch and continuous cultures for these three stages. El-Sayed and coworkers attempted the sequential production of dextransucrase followed by dextran production using immobilized cells in an airlift bioreactor. If we divide the fermentation time for individual phases in the stepwise process the questions to be answered are as follows:

1. What is the required amount of biomass and enzyme activity to be produced to get a maximum conversion of given sucrose to dextran in the shortest time?
2. How many stages are required? Are three stages necessary, because dextran fermentation consists of three phases?
3. What is the duration required for each stage?

The term *required amount* in question 1 is important because the aim of fermentation is to produce maximum dextran, not cells or enzyme. If we produce more cells and enzymes than the required amount, it will result in low yield from the fermentation medium. Therefore, in the first and second stages we need not produce maximum cells and enzyme but a required amount to optimally convert sucrose to dextran in the overall process. Optimization can be solved by employing mathematical techniques such as dynamic programming and the continuous maximum principle. These techniques require the mathematical models for cell growth, enzyme formation and deactivation, and dextran synthesis. The model parameters should explain the effect of pH and temperature. In the stepwise process, we need not necessarily divide the total time into only the three stages suggested earlier. In fact, more than three stages will prove advantageous. The number of stages required for maximum production can also be obtained through the previously mentioned mathematical techniques. Multistage reactors in series allow simultaneous operations. In all simultaneous reactor systems, the symbol F_a represents additional feed of sucrose to the respective reactors. This feed can be utilized to control sucrose concentration in the respective reactor.

The optimal operational conditions for simultaneous operation can also be obtained by the previously mentioned mathematical techniques. In reactor system f, three reactors labeled A, B, and C are connected in series. Reactor A is kept at optimum conditions for cell growth and, similarly, B and C are kept at optimum conditions for enzyme and dextran production, respectively. Another reactor system has only two reactors, labeled A and B. Here, reactor A supports both cell growth and enzyme production and reactor B performs the dextran synthesis. Separate sucrose feed (F_a) can also be employed in reactor B. Lee et al. reported the performance of a two-stage continuous reactor for dextran synthesis. In the d reactor system, reactor A can also be utilized with immobilized cell systems to produce enzyme, which can be supplied to reactor B for dextran production. The reactor system e has a cell recycle unit R that separates the cells in the medium before it goes to reactor B and maximizes the cell concentration in reactor A. Another interesting system g consists of a packed-bed reactor (A) connected to a stirred tank reactor (B). Here immobilized cells are packed and are controlled at the optimum conditions for enzyme production. Using the enzyme from reactor A, reactor B supports dextran synthesis under optimum conditions. The stirred tank reactors can be replaced by other reactors such as airlift, bubble column, and so forth. A major advantage in reactor systems e and g, is that dilution of the final viscous broth for cell removal by centrifugation, is not needed, thereby saving on alcohol precipitation costs. The additional investment in terms of equipment and control systems and the operational difficulties are major disadvantages in systems with reactors in series. The overall economics of such a process should be worked out after taking into consideration the product value for such a process.

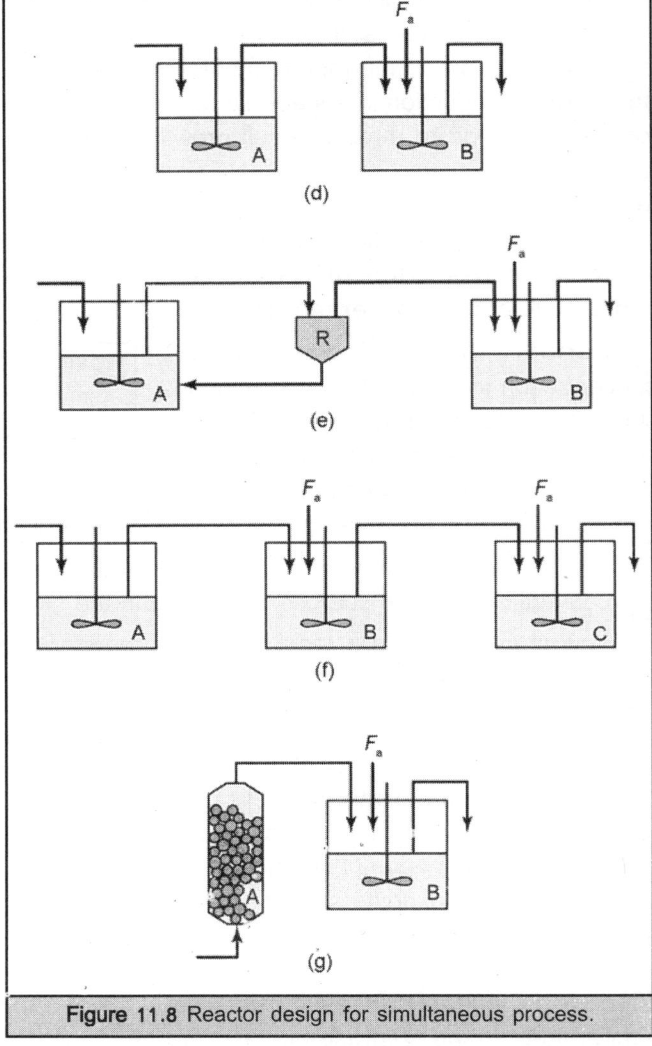

Figure 11.8 Reactor design for simultaneous process.

SIGNIFICANCE OF DEXTRANS

Depending on the microbial species employed and the fermentation conditions, a wide variety of dextrans can be produced. For example, dextrans commercially produced from *L. mesenteroides* NRRL-B512F consist of 95% α, 1 $\rightarrow$ 6 linkages and 5% α, 1 $\rightarrow$ 3 linkages. Many types of applications of this particular dextran and its derivatives have been developed. The important applications of dextran and its derivatives are described in the following sections.

Clinical Significance

Dextran fractions of average molecular weight 70,000 can be used as blood plasma substitutes (or blood volume expanders). This single most important application initiated extensive research in dextran production. Clinical dextran (6% w/v) in isotonic saline provides viscous, colloidal, and osmotic properties equivalent to those of plasma proteins. Clinical dextran solutions can be administered intravenously for treatment of shock and burns, especially when blood is not available, to restore blood pressure and circulatory volume.

The advantages of clinical dextran solution are its stability to sterilization by heat and in storage, independence of the blood types of recipients, low incidence of adverse reactions, and freedom from transmissible diseases. Clinical dextran (10–12% w/v) in aqueous or dextrose solutions administered intravenously provides high colloidal and osmotic activity and is used for treatment of hypoproteinemia and nephrotic and cerebral edema.

Rheomacrodex or Low-Viscosity Dextran

Dextran fractions of average molecular weight 40,000 can be used as blood flow improvers. Dextran (10% w/v) in sodium chloride solution (0.9% w/v) shows hematorheological properties that counteract capillary stagnation and improve the suspension stability of blood. When intravenously administered, it increases the fluidity of erythrocytes.

Cross-Linked Dextran Gel

Dextran is used in the manufacture of molecular sieves. It can be cross-linked with reagents such as epichlorohydrin in the presence of sodium hydroxide. The degree of cross-linking of dextran determines the porosity of the gel structure, water regain value, and molecular sieveing properties. These cross-linked gels, having three-dimensional structure, revolutionized the purification and separation, of proteins, nucleic acids, and polysaccharides based on differences in molecular size. These gels have well-defined exclusion limits in the molecular weight range of 5,000–200,000. Commercial cross-linked dextran was introduced in the 1960s by Pharmacia Fine Chemical Limited under the tradename Sephadex.

Hydroxyl propyl derivatives of Sephadex have been made to separate materials according to molecular size in organic solvents and have had use in separation of lipids, fatty acids, steroids, and so forth. Diethylaminoethyl and carboxymethyl derivatives of cross-linked dextran gels provide adsorbants with different porosities and fractionation ranges used in the fractionation of proteins, nucleic acids, and polysaccharides on the basis of different adsorption characteristics. Cross-linked dextrans and their derivatives have found wide application in the separation and purification of biological molecules based on charge, molecular size, and so forth on both laboratory and preparative scales. Proteins are not usually denatured by hydrophilic polymer networks, and nonspecific adsorption is usually very low.

Dextran Sulfate

Dextran sulfate can be prepared by esterification of polysaccharide of molecular weight 5×10^5 with chlorosulfonic acid. This product, a polyelectrolyte containing 17% sulfur (an average 2.3 sulfate groups on each glucosyl residue), has various applications. It shows reversible interaction with plasma proteins, and under some conditions selective precipitation, and thus permits isolation of substances such as β-lipoproteins and human antihemophilic globulin.

Dextran of sufficiently low molecular weight (5,000–10,000), substituted with sulfate, has anticoagulant activity. It mimics the action of heparin but with 15% of the activity of heparin. Dextran with a molecular weight of 7,000–8,000, substituted with sulfate, is used for selective stabilization of γ-globulin and fibrinogen to heat. The wide spectrum of biological properties of dextran sulfate has produced studies involving enzyme inhibition, activation, and release; adjuvant effects on immune responses; cellular interactions; and decreased virus infectivity and multiplication. It has shown some efficacy as a drug in the treatment of AIDS and tumors.

Dextran and Polyethylene Glycol in Aqueous Two-Phase Systems

The need for an extracting phase that does not harm labile enzymes, proteins, cellular organelles, and macromolecules has motivated consideration of liquid–liquid two- phase extraction systems in which each phase is primarily aqueous. The production of separate aqueous phases is achieved by dissolution of two incompatible polymers such as dextran and polyethylene glycol. Dextran may be replaced by its derivatives, including sodium dextran sulfate, sodium carboxymethyl dextran, or diethylaminoethyl dextran, in this two-phase system. These systems offer a new approach for extraction that is different from conventional techniques. Pullulanase and phosphorylase are extracted from *Klebsiella pneumoniae* using the dextran–polyethylene glycol (PEG) system.

Activated Dextrans

Several attributes can be achieved by coupling ligands to dextran, including the following:

1. Stability and lifetime of ligands can be increased.
2. Because of the general compatibility of the dextran fraction in vivo, drugs that require a longer circulatory lifetime can be covalently coupled with dextran. Insulin has been coupled with dextran, and the hypoglycemic action of the complex was reported to be superior to native insulin. The complex also had a longer lifetime in the bloodstream than free insulin.
3. The antigenicity of protein ligands can be decreased.
4. Enzymes can be immobilized for enzyme reactors to produce various products.
5. Covalent linking of ligands for affinity chromatography can be used in purification processes.

Because dextran, in a well-defined molecular size range, is an established blood plasma substitute, the coupling with hemaglobin may provide a blood substitute. The preparation of hemaglobin covalently coupled with dextran was studied. These preparations were cleared slowly by the kidney, and it was found that the oxygen molecule bound more tightly than to free hemaglobin.

Industrial Applications

Dilute aqueous solutions of dextrans with molecular weights 18,000, 74,000, 228,000, and 7,400,000 are used as superior media for suspension polymerization of unsaturated monomers. Dextran (5–50% w/w; MW 10,000) incorporated into the total solids in gelatin-silver halide emulsions has significant use in the photography industry. It improves film surface coverage, maximum density contrast, and speed of gelatin-silver halide emulsions for radiologic and lithographic films.

Native Dextrans

Linear-type, water-soluble native dextrans are effective for a number of other potential uses. Dextran is superior to many polymers tested in viscous water flooding for secondary recovery of petroleum. The dextran used for this purpose has the requisite rheological properties—thermal, chemical, and solution stability and innertness to ionic effects; it does not plug earth formations and is not adsorbed on rock surfaces.

Dextran is superior to carboxymethyl cellulose and other common colloidal agents when used in high-calcium drilling muds even at elevated temperatures. Muds containing dextran or aldehyde-modified dextran show superior stability and performance at high pH and in saturated brines. Dextran shows relatively high resistance to deterioration in soil aggregates. It also makes effective protective coatings for seeds and deflocculants for production of fine papers. Certain film-forming dextrans have value for protecting selected areas from deposition of metals in gas plating processes while remaining water soluble for ease of subsequent removal.

Other Uses

An iron–dextran complex has achieved significant commercial success in the treatment of iron-deficiency anemia. This complex is prepared by mixing a hot alkaline solution of dextran with $FeCl_3$ solution followed by neutralization. This complex containing 5% iron and 20% dextran is used for intramuscular or intravenous injections for treating iron deficiency. It is currently being used to treat anemia in newborn piglets. Dextran is used as a stabilizer in freeze-drying sensitive materials such as vaccines and bacteria.

As recently reported, it is one of the best preservative and stabilizing agents for transplant organs. Compatibility of linear-type dextrans with living tissues and their ability to be metabolized permits their use in binding multiple collagen fibers into strands for surgical suture. Dextran's use in foods is based on its ability to stabilize syrups, confections, and the like against crystallization; to stabilize and provide viscosity to liquid sugars; to act as humectants and adhesives; and to show high tolerance for a low pH. Acting alone, dextran is superior for stabilizing texture in ice creams and sherbets.

12

MAMMALIAN OPTIMIZING EXPRESSION

Gene expression systems offer a way to produce a heterologous protein encoded by a specific gene of interest through its introduction into a specific host cell. Heterologous gene expression of recombinant proteins has many applications, for example, in the biochemical and structure and function analysis of complex proteins, promoter analysis of genes of interest, and development of cell/membrane-based assays. One of the most crucial applications of mammalian cell expression systems is the production of therapeutic proteins such as growth factors and recombinant monoclonal antibodies (mAbs). Mammalian gene expression systems are also being used in gene-therapy-related experiments for the treatment of human diseases. Moreover, various expression systems utilizing mammalian cells can be used to determine functions for novel genes that are being identified through genomics.

The efficiency of expression of heterologous genes in mammalian cells is determined by the choice of host cells, method of gene transfer, sites at which the heterologous genes are integrated into the host genome (in the case of stable expression), and *cis*-acting DNA elements (e.g., enhancers) present in the vector carrying the gene of interest. In addition, efficiency of transcription, mRNA transport and stability, efficiency of translation, and secretion of the expressed protein into the media may also play a role in the successful expression of recombinant proteins. Steps such as posttranslational modification, glycosylation, phosphorylation, and proteolytic processing must also be considered because they can affect the quality of the protein made in or secreted by recombinant host cells.

GENERAL CONSIDERATIONS

Types of Host Cells

Mammalian cells such as chinese hamster ovary (CHO) cells, myeloma cells, human embryonic kidney (HEK)293 cells, baby hamster kidney (BHK) cells, mouse Ltk$^-$ cells, and NIH3T3 cells have been frequently used for stable expression of heterologous genes. In contrast, cell lines such as Cos and HEK293 are routinely used for transient expression of recombinant proteins. CHO-K1 and DHFR⁻ CHO cells DG44 and DUK-B11 have been extensively used for high-level protein production because these cells enable the amplification of genes of interest via selection with the amplifiable marker DHFR.

A wide variety of proteins have been expressed in CHO cells, including enzymes, recombinant mAbs, blood coagulation factors, 7TM-receptors, and other cell surface markers, chemokines, adhesion molecules, growth factors, and their receptors. Myeloma cells such as Sp2/0, YB2/0, NSO, and P3X63.Ag8.653 have also been commonly employed for high-level production of mAbs. HEK293 cells, which were derived after transformation of primary human kidney cells by adenovirus (Ad virus) early-region-encoded oncoproteins, have been widely used to stably express many 7TM receptors that can functionally couple to G-proteins. It has been shown that the human cytomegalovirus (hCMV) enhancer/promoter in an expression plasmid is highly active in these cells because Ad virus E1A protein expressed in HEK293 cells transactivates the hCMV enhancer/promoter. Examples of 7TM receptor expression in these cells are the *Drosophila* dopamine D1 receptor, the human PTH receptor, the human muscarinic cholinergic receptor Hm1, the human CGRP type 1 receptor, and the rat bradykinin B2 receptor.

In addition to these receptors, a number of chemokine receptors such as the human CC CKR5 receptor, the mouse mCCR2 receptor, the mouse eotaxin receptor, the human MCP3 receptor, and the human MCP-1 receptor have been successfully expressed in HEK293 cells. High-level expression on the cell surface and functional coupling of most the 7TM receptors expressed in HEK293 cells have been achieved. HEK293 cells have also been used for expression of various enzymes, as exemplified by the bovine constitutive nitric oxide synthase (eNOS), and the human prostacyclin synthase. HEK293 cells have been used for heterologous expression of a wide variety of proteins, for example, interleukin 1 receptor–IgG fusion protein (immunoadhesion), the human insulin growth factor binding protein 3, platelet integrin $\alpha_{IIb}\beta_3$ (an integral membrane protein that mediates adhesion of cells to the extracellular matrix), and the human tissue factor (a procoagulant factor that serves as a cofactor for the blood coagulation factor VII).

Cos cells have been used to transiently express a large number of heterologous proteins. Cos cells constitutively express simian virus 40 (SV40) T-antigen and therefore support replication of expression plasmids containing the SV40 origin of replication. This results in high copy number (~40,000 copies cell^{-1}) of the transcribed coding sequences. Cos cells are primarily used for three reasons: (i) to determine if the expression plasmid produces the protein of interest, (ii) for rapid expression of recombinant proteins to perform initial functional studies, and (iii) expression cloning. Hundreds (if not thousands) of proteins have been expressed transiently in Cos cells. For example, Patel and Grundy used a Cos cell expression system to study biogenesis of apolipoprotein B–containing lipoproteins.

Functional expression of 7TM-receptors, for example, the guniea pig α2A, α2B, and α2C-adrenoreceptors, and the rabbit bradykinin B1 receptor, have been successfully achieved in Cos cells. Cos cells have been used commonly to express small amounts of functional enzymes, such as the rat liver microsomal glutathione transferase, the human 11β-hydroxylase and aldosterone synthase. Functionally active recombinant mAbs have also been expressed transiently in Cos cells. It should be remembered however, that the Cos expression system is a transient expression system, and therefore, multiple repeat transfection experiments, which are time-consuming and labor intensive, are needed to produce milligram quantities of proteins for initial studies. More importantly, data obtained from a Cos expression system cannot necessarily be extrapolated to the level of expression expected from stable expression systems.

Engineered hosts

Some of the host cell lines just described have been engineered to produce milligram quantities of recombinant proteins in a rapid manner. In most cases this has been achieved by expressing (constitutively) a transcriptional trans-activator that stimulates transcription from the promoter linked to the gene of interest. Following are descriptions of some of these modified hosts.

CHOE1a

To quickly express a heterologous protein, Cockett et al. have established a CHO cell line constitutively expressing the Ad virus E1A, which transactivates the CMV promoter in the expression plasmid. Transfection of a human procollagenase gene into this cell line produced a 13-fold higher level in stable expression compared with that in CHO-K1 cells. This system has been used to express mAbs and single-chain Fvs.

CHO/Pit-1

CHO cells were engineered to express the rat Pit-1 transactivator (a pituitary specific transcription factor). Stable transfection of a reporter gene (tPA) under the control of the rat prolactin promoter showed an enhancement of expression levels (~20- to 60-fold) compared with the same promoter in parallel in CHO cells.

BHK/VP16

Hippenmeyer and Highkin have established a BHK cell line that stably expresses the herpes simplex (HSV) VP 16, a transcriptional transactivator. Using this cell line, tPA and bGH were expressed at high levels (1–20 μg 10^6 cells^{-1} day^{-1}) as compared with that in the parental BHK cell line (<0.2 μg 10^6 cells^{-1} day^{-1}). Since then, a wide variety of proteins have been stably expressed in the BHK/VP16 cell line.

293/EBNA

HEK293 cell lines have been established after stably transfecting the Epstein–Barr virus (EBV) nuclear antigen 1 (EBNA-1) protein, which is required for the maintenance of a plasmid carrying the EBV origin of replication extra-chromosomally. Such cell lines have been used for heterologous protein expression. For example, using a secreted form of human placental alkaline phosphatase (SEAP) as a reporter gene, Cachianes et al. obtained 10-fold higher expression in 293 cells expressing EBNA-1 as compared with that in wild-type 293 cells. A number of other proteins (e.g., tPA, vascular endothelial growth factor, hepatocyte growth factor, soluble tumor necrosis factor) have been produced in this cell line at levels ranging from 0.2 to 10 μg ml^{-1}.

T7 RNA polymerase

Mouse Ltk⁻ cells, CHO cells, and mouse myeloma Sp2/0 cells expressing T7 RNA polymerase have been established. In these cells, T7 RNA polymerase has been modified by fusion of a nuclear localization signal to the N-terminus of the polymerase so that it can function in the cell nuclei. Transfection of a heterologous gene under the control of a T7 promoter into these modified host cell lines enables high-level expression of the gene of interest. For example, using this system the human growth hormone has been stably expressed at levels of 20–30 μg 10^6 $cells^{-1}$ day^{-1}.

Mammalian Expression Vector

A typical mammalian expression vector contains a cDNA copy of the gene of interest placed downstream of a promoter sequence to ensure efficient transcription; placed upstream is a polyA addition signal, which is required for efficient transcription termination. Enhancer sequences, which stimulate transcription, are usually placed upstream of the linked promoter, whereas a selectable marker (e.g., amp^r gene) and a plasmid origin of replication are included for propagation of the expression construct in *E. coli*. In some instances, a drug selectable/amplifiable marker expression cassette for use in eukaryotic cells is linked to the heterologous gene expression unit. As mentioned, for transient expression in Cos cells, the SV40 origin of replication is included in the expression plasmid. A wide variety of promoters and enhancers and a combination thereof have been used for heterologous expression in mammalian cells and are included in vectors that are commercially available. Among the more popular promoters are those from (i) animal viral genomes such as SV40, human CMV, Ad virus, Rous sarcoma virus (RSV), murine leukemia virus (MuLV), HSV, and so forth; and (ii) cellular genes such as β-globin, immunoglobulin, heat shock, and metallothionein genes.

SV40 early and late promoters, hCMV IE promoter, RSV long- terminal repeat (LTR), and SRα promoter (a composite promoter consisting of the SV40 early promoter and a part of the LTR of human T-cell leukemia virus) have all been successfully utilized in CHO-K1 and $DHFR^-$ CHO cells for expression of a variety of proteins. However, the choice of promoter is not usually critical when high-level expression is desired in $DHFR^-$ CHO cells where gene amplification methodology will be employed. Plasmids containing the CMV promoter have been shown to produce high-level expression in HEK293 cells. Most of the expression vectors used for mAb expression in myeloma cells contain both H- and L-chain cDNAs under the transcription control of IgG promoter, although the hCMV promoter has been used successfully for mAb expression in NSO myeloma cells.

A polyA addition signal (AA/ TTAAA) and downstream sequences that influence transcription termination can be derived from many sources including the SV40 genome, the bovine growth hormone gene, and Ig genes. The expression level of the gene of interest can sometimes be increased several fold through the use of an intronic sequence, for example, β-globin or chimeric J/Ck intron. It is advisable to include the intron at the 5′ end of the coding sequence because, in some examples, aberrant processing of transcript was found when the intron was placed at the 3′ end of the cDNA. However, it should be pointed out that the effect of introns in enhancing expression levels is not a general phenomenon and in most cases does not significantly affect expression levels. .

Kozak Sequence

In most vertebrate mRNAs examined to date, a sequence related to GCCACC*AUG*G, called a *Kozak sequence* (the underlined AUG is the initiator methionine), has been identified at the 5′ end, which can promote efficient translation of a linked open reading frame (ORF). Therefore, many investigators include an optimal *Kozak sequence* in their expression vector. For example, Ho et al. studied translation initiation and assembly of rat peripherin, an intermediate filament protein in transfected cells. A full-length peripherin cDNA (designated as p61-11), when expressed in an adrenocarcinoma cell line (SW13 c1.2 vim⁻) without any detectable cytoplasmic intermediate filament, did not yield a functional filamentous network because of low-level expression of p61-11. When the first ATG was placed in the context of a better *Kozak consensus sequence,* the translational efficiency of p61-11 was increased significantly, and the resulting 57 kDa protein was able to form a functional filamentous network in the vim cells.

Enhancers of Translation

Regulatory sequences that stimulate translation, hence called *translational enhancers,* have been identified at the 5′ end of many eukaryotic viral RNAs and some cellular mRNAs. These sequences bind to translation initiation factors with higher affinity than their cellular counterparts. The Ad virus tripartite leader sequence is such a translational enhancer and has been used in mammalian expression vectors. For example, Sheay et al. compared expression of a reporter CAT gene from expression vectors containing the Ad virus tripartite leader sequence placed downstream of various promoters such as CMV, RSV, and MuLV. In transient transfection assays, expression vectors containing the Ad virus tripartite leader sequence gave 5- to 20-fold higher CAT activity when compared with that from vectors without the leader sequence. A translational enhancer has also been identified in the 5′ UTR of a cellular mRNA (e.g., the human β-kinesin mRNA encoding the first exon).

High levels of reporter gene expression have been achieved from transfection of plasmids carrying the kinesin enhancer sequences when compared with that obtained without the enhancer. For example, in Cos transient transfections, the kinesin enhancer placed downstream of the HSV tK promoter gave a four-fold increase in reporter gene (CAT) activity when compared with that from a plasmid without the enhancer. A 34-bp-long translational enhancer has been described in the *Xenopus* β-globin mRNA. When placed upstream of a reporter gene, up to 300-fold increases in translational efficiency have been demonstrated in in vitro systems. Translational enhancers have also been identified in many plant viral genomes, with the tobacco mosaic virus (TMV) enhancer receiving the most attention. However, the use of this and other plant translation enhancers for improving heterologous mammalian expression systems has not yet been explored.

Dominant Selectable Markers

A number of dominant selectable markers such as neomycin resistance (Neo^r), hygromycin resistance (Hyg^r), *E. Coli.* xanthine-guanylyl phosphotransferase (Ecogpt), and glutamine synthetase (GS) have been used for initial selection of clones obtained after transfection of the expression plasmids into appropriate host cells. A mouse DHFR cDNA has been used for both initial selection of DHFR resistant CHO cells, as well as gene amplification. The selectable/amplifiable marker expression unit (under the control of a weak promoter and polyA addition signal) can either be linked to the heterologous gene expression unit or can be on a separate plasmid that is usually cotransfected into the host cells.

With the use of an expression vector containing a linked DHFR gene, the gene copy number in the initially selected $DHFR^+$ clones can be amplified by selecting the clones further in the presence of stepwise increasing concentrations of methotrexate (MTX), a competitive inhibitor of DHFR. Glutamine synthetase can be used as a dominant selectable/amplifiable marker when cells are grown in the presence of methionine sulfoxamine (MSX), an inhibitor of GS. Other gene amplification markers that have been used to a lesser extent are encoded by adenosine deaminase (ADA) and multiple drug resistance (MDR) genes.

Technique for Transfection

A variety of methods can be used to introduce expression plasmids into mammalian host cells. These include $CaPO_4$, DEAE-dextran, electroporation, and lipofection. Receptor-mediated gene transfer has also been described in the literature. Interestingly, a single, standard, optimized method has yet to be described in the literature for establishing mammalian cells expressing heterologous gene products. In a typical stable expression experiment, the plasmid carrying the gene of interest and a drug selectable/amplifiable marker is transfected into host cells using any of the gene transfer methods just mentioned. The transfected cells are selected in the presence of a drug until the drug-resistant colonies are obtained (usually within 2–3 weeks). The resultant colonies are then screened using a variety of methods to determine which are producing the desired heterologous gene product.

The detection method is dependent on the nature of the expressed protein, availability of antibodies, and so forth. For secreted proteins, conditioned media can easily be screened by ELISA, western blotting, and so forth. On average, one-third to one-half of the drug-selected colonies produce the protein of interest at a detectable level. The other clones are either expressing the desired gene product at very low levels or not at all, presumably due to inactivation or rearrangements of the heterologous gene.

Process Development

Growth media and *process development* play an important role in the successful expression of a heterologous gene product in mammalian cells. Serum-free media has been developed for large-scale growth of CHO and myeloma cells. Serum-free or protein-free media offers the following advantages: (i) they are less expensive compared with media containing serum; (ii) downstream purification of heterologous gene products becomes easier, especially for proteins that are secreted into the media; and (iii) contaminating agents present in serum are avoided. Increases in expression levels of heterologous gene products can normally be achieved through media manipulation and process development. For example, Bebbington et al. have reported a three- to fourfold increase in mAb expression in NSO cells amplified with the GS system when cells were grown in a defined serum-free medium.

When an amino acid feed was applied to these cells, an additional increase in expression level up to 560 mg L^{-1} was achieved. Using NSO myeloma cells and the GS system, expression level of a number of mAbs in the range of 0.5–2 g L^{-1} have been reported through medium development, nutrient supplementation, and other process development considerations. Similar results have also been obtained with mAbs produced in $DHFR^-$ CHO cells. Glycosylation patterns of recombinant proteins can vary widely when expressed in different mammalian host cell types. For example, Parekh et al. reported different glycosylation patterns for recombinant human tPA expressed in four different cell types: CHO cells, murine epithelial (C127) cells, the Bowes melanoma cells, and human colon fibroblast (CCD-18Co) cells.

Cell culture conditions also determine glycosylation pattern of recombinant proteins. Borys et al. found that culture pH affects expression rate and glycosylation of recombinant mouse placental lactogen proteins produced in CHO cells. The optimum expression rate was found to be in pH range of 7.6 to 8.0. The extent of glycosylation was decreased at low pH (below 6.9) and high pH (above 8.2). N-linked oligosaccarides were found to be different in composition when recombinant human tissue kallikrein was produced by CHO cells grown in serum-free suspension culture or on microcarrier beads.

Control

Inducible expression systems have been proven to be valuable for elucidating the function of a number of genes in *E. coli*, yeast, and *Drosophila*. In mammalian cells a tightly regulated expression system may also offer an advantage for expression of those proteins that are potentially cytotoxic. One way to achieve regulated expression of a heterologous gene is through the use of an inducible promoter. A number of inducible promoters have been described in the literature, such as IFN-γ, hsp70, metallothionein (Mtn), and MMTV LTR. Following are some popular methods for achieving inducible expression of heterologous genes in mammalian cells.

Lacswitch system

Bacterial transcription control systems such as the *lac* operator/repressor have been adapted for regulated expression of heterologous genes in mammalian cells. The *lac* repressor, made from a *lacI* expression vector, blocks transcription of a reporter gene placed downstream of *lac* operator sequences. Addition of the inducer IPTG results in derepression of reporter gene transcription by up to 10- to 60-fold. Labow et al. modified the *lac* repressor/operator system by converting the *lac* repressor into a transactivator by fusing it to the HSV VP16 transactivation domain. In the absence of IPTG, this *lac* transactivator stimulates expression from a minimal promoter containing *lac* operator sequences up to 1,000-fold.

The tet system

Gossen and Bujard described a regulated expression system based on the control elements of the *E. coli* tetracycline (Tc) resistance operon. In this system the *tet* repressor has been fused to the HSV VP16 transactivator domain, and the resultant transactivator, called tTA, stimulates expression from a hCMV minimal promoter linked to *tet* operator sequences. In the presence of

Tc (1 μg ml^{-1}), the tTA cannot bind to the operator sequence, and transcription from the minimal promoter shuts off, whereas removal of Tc leads to activation of transcription. In stable HeLa clones expression of a reporter gene has been shown to be modulated up to 10^5-fold. Derivatives of Tc, such as anhydrotetracycline and doxycyclin, have also been used to regulate expression using this system.

An expression vector with the *tet*-responsive promoter, tTA expression unit, origin of replication of EBV, EBNA-1 expression unit, and a selectable marker has been described for the use in transient and stable expression of heterologous genes. Using this vector, luciferase activity in transiently transfected cells can be regulated by several orders of magnitude by varying both the Tc concentration and incubation time in the presence of Tc. The *tet*-responsive system has been shown to produce different responses in different cell lines. Howe et al. used the *tet* system to regulate (i) expression of dopamine D3 receptor in stable GH3 cells and (ii) GluR6 (a glutamate receptor subunit) in stable HEK293 cells. In GH3 cells, a high-level induction of dopamine D3 receptor expression was obtained in the absence of Tc, whereas only a modest level induction of GluR6 was seen in HEK293 cells. Recently, a mutant form of the *tet* repressor has been used to create a transactivator after fusion with the HSV VP16. This transactivator, called rtTA, has reversed DNA binding activity as compared with that of wild-type *tet* repressor or tTA. In the presence doxycycline or anhydrotetracycline, more than 1,000-fold induction has been achieved in stable HeLa clones.

Ecdysone-inducible system

The *Drosophila* ecdysone (Ec) receptor, a member of the steroid hormone receptor superfamily, can function as a ligand-dependent transcription factor in animal cells. Functional EcR is a heterodimer of EcR and the *Drosophila* ultraspiracle gene product (USP). The activity of the EcR is regulated by certain ecdysteroids such as ecdysone (Ec) and not by mammalian hormones such as dexamethasone and estradiol. EcR by itself is incapable of high-affinity DNA binding and transcriptional activation. In the presence of Ec, physical association between EcR and USP is greatly enhanced. Structural analysis shows that steroid hormone receptors contain a DNA-binding domain, ligand-binding domain, and transactivation domain. Mammalian retinoic acid receptor (R-X-R) can replace USP to form a functional heterodimer. Recently, No et al. described an ecdysone-inducible gene expression system for mammalian cells.

Mammalian cells cotransfected with EcR and USP gave only a threefold induction upon treatment with a micromolar concentration of muristerone (MurA), a synthetic analog of Ec. Modifications of EcR were made such that the N-terminal trans- activation domain in EcR was replaced by the corresponding domain of the glucocorticoid receptor (GR). This hybrid receptor gave a 34-fold induction in a transient reporter gene assay. A more potent heterodimer was made by combining either R-X-R and VpEcR (an N-terminus of EcR was truncated and fused to the VP16 transactivation domain) or R-X-R and VgEcR (which recognizes a hybrid Ec/GRE). Stable lines generated with these regulatory elements and induced with 100 nM MurA gave 20,000-fold induction of reporter gene (β-galactosidase) expression.

All three of the inducible systems are commercially available. For system it is important to determine the level of leaky expression in the absence of inducer (for the *lac* and Ec system) and the presence of Tc (for the *tet* system). The optimal concentration of the inducer and the kinetics of induction need to be determined. More importantly, one needs to determine if the high-fold induction obtained through any inducible system correlates with the desired high-level expression of a recombinant protein.

Heterologous Gene Expression

In prokaryotic systems, many mRNAs encode two or more gene products in polycistronic messages. Expression of the distal coding sequences are often regulated by the translation of the upstream coding sequence in these messages. In contrast, eukaryotic mRNAs are not generally polycistronic. However, translation of polycistronic message in eukaryotic systems has now been achieved through the use of an internal ribosome entry site (IRES). IRESs have been identified in the single-stranded RNA genome of picornaviruses such as encephalomyocarditis (EMC) virus, foot and mouth disease virus, and polio virus. Ribosomes bind to a 450- to 600-nt-long secondary structure (IRES) present in a 600- to 1,200-nt-long 5′ UTR followed by an ORF in the viral genome. Kaufman first reported the use of an IRES placed in between the gene of interest (a cDNA-encoding eukaryotic initiation factor 2α [eIF-2α]) and a selectable marker (DHFR).

The resultant expression vector is called a dicistronic vector. When transfected into DHFR$^-$ CHO cells, efficient internal translation initiation of DHFR yielded increased transfection efficiency. High-level expression of eIF-2α was observed when the DHFR$^+$ clones were grown in presence of increasing concentration of MTX. A number of dicistronic vectors have been described in the literature for heterologous gene expression. In order to increase the expression level of a gene of interest (5′ cistron), vectors have been designed such that the translation initiation of the selectable marker gene (3′ cistron) is reduced. The resultant high-level expression of the heterologous gene in the marker-selected (drug-selected) clones can be due to the integration of the expression plasmid in a transcriptionally active region of the host genome. For example, Kobayashi et al. constructed a di-cistronic vector in which transcription is under the control of the CMV enhancer/chicken β-actin promoter.

The vector contains an IRES of EMC virus downstream of a reporter β-galactosidase gene, followed by a neomycin resistance gene. The translation of *neo*^{r} gene was weakened by placing spacers of various lengths in between AUG-11 (located at the 3′ end of IRES) and the AUG of *neo*^{r}. The number of Neo^r 3T3 clones obtained after transfection of the dicistronic vector with the weakened *neo*^{r} was 4- to 7-fold less compared with that from a conventional dicistronic vector, but these clones gave about 10-fold higher levels of β-*gal* activity. A similar strategy of weakening the selectable marker gene expression from a dicistronic vector was described by Rees et al. Expression levels of serotonin receptors in 10 independent stable HEK293 clones varied from 1.5 to 11.9 pmol mg^{-1} cellular protein, which was significantly higher than the reported expression levels (98 fmol mg^{-1} in HeLa to 4 pmol mg^{-1} in Lmtk$^-$ cells). Lucas et al. reported a dicistronic intron vector (or DI vector) in which the mouse DHFR cDNA is placed into the intron of the pRK vector and is under the transcriptional control of the CMV enhancer/promoter.

A cDNA encoding a gene of interest (e.g., tPA or TPO) was placed downstream of the intron. When transfected into DHFR-CHO cells, full-length mRNA produced by DI vectors was translated into DHFR, while spliced mRNA produced the desired recombinant protein. DI-vector-derived CHO clones grown for 5 days gave 11-fold more tPA (~1 mg L^{-1}) and 13-fold more TPO than were produced by conventional vector-derived cells. The DI vector system has also been used to produce antibodies at levels ~4–5 mg L^{-1} 10 day^{-1}. The weakening of DHFR expression and/or its linkage to product expression gave fewer clones, but they had more productivity.

SIGNIFICANCE OF EXPRESSION TECHNOLOGY

Examples of CHO Cell Expression

Promoting growth

The following examples describe the use of $DHFR^-$ CHO cells for expression of various growth factors. The expression vectors were either cotransfected with a DHFR expression plasmid to select the $DHFR^+$ clones or contained their own DHFR expression cassette. Expression levels of the growth factors were increased via amplification in the presence of increasing concentrations of MTX, as already described. Madisen et al. achieved high-level expression of TGF-β2 and a precursor TGF-β2 in CHO cells using an expression vector similar to that described by Gentry et al. for TGF-β1. A pSV2 expression vector containing the genes encoding TGF-β2 and DHFR was utilized to obtain $DHFR^+$ clones. These clones were subsequently adapted to 0.5, 2.5, 10.0, and 50 μM MTX. Two clones grown at 10 and 50 μM MTX were analyzed for expression of recombinant mature TGF-β2 (12 kDa under reducing conditions).

In addition to the mature protein, proteins of 85, 105, and 130 kDa, consisting of both mature and proregion sequences were also produced, as determined by immunoblotting. The pro-TGF-β2 was found to be both glycosylated and phosphorylated. Bourdrel et al. expressed recombinant human TGF-β1 after subcloning the cDNA into the pDSVα2 expression vector, a derivative of the pCD vector in which the SV40 polyA addition signal was replaced by the polyA addition signal of the a-subunit of bovine follicular stimulating hormone and a mouse DHFR minigene (downstream of the expression cassette). CHO cells were transfected using the $CaPO_4$ method, and a single $DHFR^+$ clone expressing human TGF-β1 at 7 mg L^{-1} in the conditioned media was identified.

Ferrara et al. used CHO cells to produce the full-length vascular endothelial growth factor (VEGF) after subcloning the cDNA encoding the full-length human VEGF cDNA into a derivative of the vector pCISh, which carries a DHFR gene. The transcription of VEGF cDNA and that of the DHFR are both under the control of the SV40 promoter. The initial selected clones, which released the highest mitogenic activity, were amplified in the presence of increasing concentration of MTX, resulting in clones that were resistant to 1 μM MTX and produced 0.5 μg ml^{-1} day^{-1} of VEGF. The human neurotrophin 3 (NT-3) gene product was purified and characterized after expression in CHO cells. A construct that contained a chimera encoding the human nerve growth factor pre-pro region and NT-3 mature region under the control of MuLV-LTR was transfected into CHO cells.

$DHFR^+$ clones were serially subcultured in media containing 0.1, 1.0, and 10 μM MTX, yielding clones expressing as much as 1 mg recombinant NT-3 L^{-1}. The purified recombinant NT-3 was biologically active, as shown through the induction of neurite outgrowth in neurons prepared from an 8-day-old chicks. Burgess et al. used CHO cells to study biosynthetic processing of the rat *neu* differentiation factors (NDFs). pDSRα2-derived vectors containing NDF-β2c and NDF-β4a were transfected using the $CaPO_4$ method, and the selected clones were screened by western blotting. High-expressing clones for NDF-α2c and NDF-β4a were identified. Expression of NDF-β4a was gradually amplified with MTX (70 nM). The precursor of NDF undergoes typical glycosylation, and a portion of the molecule is proteolytically cleaved, leading to secretion of soluble mature NDF. A portion of the newly synthesized precursors escape processing and are retained on the cell surface.

Immunology

In recent years DHFR$^-$ CHO cells have been successfully used to produce a number of recombinant mAbs at a high level. For example, Page and Sydenham used DUK-B11 CHO cells to obtain expression of the humanized mAb Campath-1 at levels of 100 μg 10^6 cells^{-1} day^{-1}. Using the same host cells, Fouser et al. reported expression of a chimeric ganglioside GD2 antibody at 80–110 μg 10^6 cells^{-1} day^{-1}. Recently, Reff et al. expressed a chimeric antibody against CD20 in DG44 CHO cells at levels of 30 μg 10^6 cells^{-1} day^{-1}, using only one round of gene amplification with MTX. A variety of promoters have been used to obtain expression of H- and L-chains in these CHO cells, for example, the β-actin promoter, the human CMV MIE promoter, the Ad virus major late promoter (MLP), the RSV promoter, and a murine leukemia virus LTR.

A number of vectors for mAb expression are described in the literature in which the two Ig chains are carried by two different plasmids with an independent selectable/amplifiable marker. For example, Fouser et al. linked the Ig H-chain with the ADA gene, and the L-chain with the DHFR gene. Using these vectors, an expression level of 80–110 μg of mAb 10^6 cells^{-1} day^{-1} was achieved. Trill et al. have used vectors containing H-chain linked to a DHFR marker, and an L-chain expression cassette with the Neo^r marker and vice versa to obtain ~180 mg of a humanized mAb L^{-1} 7 day^{-1} in spinner flasks. The methods used for initial selection and subsequent amplification vary from laboratory to laboratory.

In general, high-level mAb expression can be obtained using the following steps: initial selection and subsequent amplification of candidate clones, coselection (e.g., in cases where both H-chain and L-chain expression vectors carry DHFR expression unit) and amplification, coamplification using different amplifiable markers, and initial selection and amplification in mass culture, followed by dilution cloning to identify individual high-expressing clones. Because integration sites may influence the efficiency of H-chain and L-chain expression and overall mAb expression, single vectors have been created in which the two Ig-chain expression units are placed in tandem. These vectors also carry a dominant selectable marker such as Neo^r and the DHFR expression cassette. Using the single-vector system and two rounds of amplification, Newman et al. achieved high-level expression (>200 μg ml^{-1}) of a primatized anti-CD4 mAb. Similar high-level expression of two different humanized mAbs has recently been reported by Trill et al.

Biological Catalysts

Recently, Laubach et al. used the GS system to achieve high-level expression of the human inducible nitric oxide synthase (iNOS) in CHO-K1 cells. The iNOS expression plasmid, a derivative of pEE 14 vector was transfected into CHO cells using Lipofectamine and selected with 25 μM MSX in the presence of ethylisothiourea (EITU), an inhibitor of iNOS (to reduce toxicity due to NO production). Further amplification of initially selected clones in the presence of 400 μM MSX gave stable clones with activity > 3 nmole min^{-1} mg^{-1}. Sodium butyrate treatment of cells increased the level of expression by another 1.5-fold.

Sensory cells

A coding sequence for the soluble form of CD4 was constructed through the introduction of a translation termination codon at the extracellular boundary of the transmembrane domain, and subcloning the cDNA downstream of the SV40 early promoter and upstream of the bovine growth hormone poly A addition signal. This expression cassette was linked to a mouse DHFR expression cassette, and the resultant plasmid was transfected into $DHFR^-$ CHO DXB-11 cells. The initially selected $DHFR^+$ clones were exposed to increasing concentration of MTX, and the expression level in several clones was >3pg $cell^{-1}$ day^{-1}. This specific productivity translated into an expression level of ~40 mg of sCD4 L^{-1} under high-density suspension culture conditions. Newman-Tancredi et al. reported high-level expression of recombinant 5-HT1A receptors in CHO cells using a derivative of pSVL. DHFR-CHO cells were cotransfected with pSV2dhfr using the $CaPO_4$ method.

Out of 35 $DHFR^+$ clones tested for expression, 33 had receptor expression at 20–3,000 fmol mg^{-1}, as determined by ^{3}H-8OH-DPAT binding assays. The highest-expressing clone (expressing at 2.8 pmol receptor mg $protein^{-1}$) was identified, and functional coupling was demonstrated through the action of a series of 5-hydroxytryptamine agonists that inhibited forskoline-stimulated adenyl cyclase (AC) activity. The receptor density and receptor regulation are maintained in suspension grown cells, but the inhibition of AC by the receptor is lost gradually (probably due to a change in G-protein–AC interaction and not due to a change in receptor–G-protein coupling). The GS system has also been utilized to express a 7TM receptor for human thyrotropin (TSH) in CHO cells. Cell lines expressing up to 78,000 receptors, $cell^{-1}$, which efficiently couple to AC, have been identified.

E-selection

The soluble form of human E-selectin (sE-selectin) has been expressed at a high level in $DHFR^-$ CHO cells. The sE-selectin cDNA was subcloned into the RLDN expression vector (in which expression of sE-selectin was under the control of RSV promoter) and transfected into CHO cells. The $G418^r$ clones expressing sE-selectin were further selected stepwise in media containing 10 nM and 100 nM MTX. A 100 nM MTX^r clone was identified. In spinner flasks it produced 100 mg L^{-1} of sE-selectin over a 5-day period.

Promoter

Liao et al. expressed recombinant human Mig, a CXC chemokine, after subcloning the Mig cDNA into the vector pMSXND. Expression of Mig was under the control of the Mtn I promoter. The expression vector contained the SV40 small t-antigen intron and poly A addition signal, a DHFR expression cassette under the control of the SV40 early promoter, and a *neo*^{r} under the control of the tk promoter. The construct was transfected into CHO cells ($proline^-$), and the *neo*^{r} clones were subjected to amplification using 0.2 μM MTX. The MTX^r clones were induced with $CdSO_4$, and the conditioned media screened by SDS-PAGE showed a collection of immunoreactive polypeptides (due to protein processing of the high molecular weight Mig), which were similar to that secreted from IFN-treated peripheral blood monocytes and THP-1 cells.

Other examples

Cockett et al. developed the GS system for high-level expression of heterologous genes in CHO cells. Transfection of an expression vector containing a tissue inhibitor of metalloproteinases (TIMP) cDNA (under the transcriptional control of the hCMV promoter) and a GS mini gene (under the control of the SV40 late promoter) into CHO-K1 cells (followed by selection with 20 μM MSX) yielded clones expressing up to 9 μg TIMP 10^6 $cells^{-1}$ day^{-1}. Further selection with higher concentration of MSX (500 μM) produced a clone expressing TIMP at 110 μg 10^6 $cells^{-1}$ day^{-1}. Yonemura et al. have developed a high-level expression system for blood coagulation factor VIII (consisting of a 90-kDa H-chain and an 80-kDa L-chain). The expression plasmids contained a chicken β-actin/rabbit β-globin hybrid promoter and a DHFR transcription unit under the control of the SV40 early promoter. $DHFR^-$ CHO cells (DG44) were cotransfected with H- and L-chain expression vectors. After selecting for $DHFR^+$ clones (using nucleoside-free media), cells that tested positive in a precoagulation assay were cultured in increasing MTX concentrations (upto 1,000 nM) for gene amplification. A clone resistant to 1,000 nM MTX was established that produced 15 IU 10^6 $cells^{-1}$ day^{-1} of factor VIII activity.

Examples of Myeloma Cell Expression

Immunity

High-level production of mAbs have been reported by several groups. For example, Gillies et al. expressed two chimeric mAbs, Ch 14.18 and ChB72.3, in Sp2/0 cells. H-chain V-region cDNA and a hybrid human–mouse L-chain V-region cDNA were fused to the their respective genomic constant regions. The H-chain expression unit (under the control of the mouse H-chain enhancer/ metallothionein MT I promoter) was linked to a DHFR expression cassette. The L-chain expression unit (under the control of a synthetic murine K-promoter) was subcloned into the H-chain expression plasmid to generate a single vector. The expression vector was transfected into Sp2/0 cells, and the MTX^r clones secreting antibody were adapted to increasing concentration of MTX. Subclones of 5 μM MTX^r clones had expression levels of 120–180 μg ml^{-1}.

Shitara et al. constructed a “tandem vector” in which both H- and L-chain cDNA expression units (both under the control of Moloney virus LTR), DHFR, and Neo^r expression units were linked. Expression from this vector was compared in three host cells: mouse Sp2/0, P3X63Ag8U.1, and rat YB2/0 cells. YB2/0 cells gave higher numbers of $G418^r$ clones and higher levels of mAb expression. Subsequent amplifications with 50–200 nM MTX yielded several clones secreting 70–100 μg of mAb 10^6 $cells^{-1}$ day^{-1}. The GS system has been widely utilized for antibody expression in myeloma cells. Bebbington et al. reported high-level expression of a chimeric antibody in NSO myeloma cells.

An expression vector was created after placing in tandem a GS expression cassette, the H-chain transcription unit, and the L-chain transcription unit. The expression of the GS gene was under the control of the SV40 early promoter, while the expression of both H-chain and L-chain was under the transcriptional control of hCMV promoter. Transfection of this expression plasmid into NSO myeloma cells, followed by selection in glutamine free medium, gave a pool of clones expressing up to ~4 pg antibody $cell^{-1}$ day^{-1}. Further selection with 20–100 μM MSX gave rise to the highest-expressing clone, with a specific productivity of about 10 pg $cell^{-1}$ day^{-1}.

Southern blot analysis showed that this cell line had approximately four copies of the vector per cell. As described in this review, a large number of expression systems are available for heterologous expression of a wide variety of proteins in mammalian cells. The choice of a particular host/vector system depends primarily on the desired expression level, nature of expression (transient vs. stable), intracellular versus extracellular expression, constitutive versus inducibile expression, and so forth. Using the expression systems described, it is now possible to produce therapeutic proteins such as mAbs at very high levels (~1–2 gm L^{-1}). However, the challenge still remains to produce such therapeutic proteins in an even more economical manner. Therefore, alternate expression systems still need to be developed to achieve even higher levels of expression and reduce cycle time (the time interval between construction of an expression vector to production of the protein of interest in large scale). Further understanding of cellular processes such as transcription, translation, and secretion will be essential to achieve such a goal.

13

BIOLOGICAL SYNTHESIS OF DILTIAZEM

Diltiazem, (2S,3*S*)-3-acetoxy-5-[2-(dim ethylam ino)ethyl]-2,3-dihydro-2-(4-methoxyphenyl)-1,5-benzothiazepin-4(5*H*)-one hydrochloride, is a calcium channel blocker and has been clinically used since 1974 as an antianginal and antihypertensive agent in more than 100 countries. Because diltiazem has two stereo centers at positions 2 and 3 in the structure shown in Figure 13.1, four isomers are possible. Among them, only the (2*S*,3*S*)-isomer exhibits the desired coronary vasodilating activity. The conventional production has been a nine-step process starting with *p*-anisaldehyde and methyl chloroacetate. The fourth step is a chemical resolution of the advanced intermediate 6. Resolution at an earlier step should save raw materials and reduce the amount of waste. (±)-MPGM is the first intermediate containing stereo centers. Two of the four isomers are excluded through the chemical synthesis of (±)-MPGM, which yields only *trans* type MPGM. Furthermore, no racemization occurs on the subsequent steps from (–)-MPGM to diltiazem. Therefore, the asymmetric syntheses of (–)-MPGM and resolutions of (±)-MPGM have been investigated for the synthesis of diltiazem. We focus on the resolution of (±)-MPGM as the more effective method of (–)-MPGM production. Previous workers resolved (±)-MPGM by first hydrolyzing to the carboxylic acid, resolving the carboxylic acid with an optically active amine, and then esterifying. This method involves many steps and yields oily MPGM with low purity.

To solve these problems, we resolved (±)-MPGM using an enzyme. Enzymes are now widely recognized as practical catalysts for asymmetric synthesis. Previous examples of enzyme-catalyzed enantioselective reactions of glycidate compounds include en- antioselective hydrolysis of one ester group in *meso-cis*-oxirane dicarboxylic acid diester and of one enantiomer of racemic glycidyl esters. Although these glycidate compounds are stable in aqueous solution, our substrate, (±)-MPGM, is not. The 4-methoxypheny! group increases the reactivity of the oxirane so that (±)-MPGM decomposes easily in aqueous solution under neutral to acidic conditions. (±)-MPGM is more stable in an organic–aqueous biphasic system. In addition, the two-phase mixtures also simplify separation of reactants and products. Enzyme recovery is also very important for cost reduction in industrial enzyme reactions, but effective recovery is difficult from two-phase mixtures due to the formation of stable emulsions. We used the Sepracor membrane reactor based on the principle that permits immobilization of the enzyme and contact between organic and aqueous phases for both efficient hydrolysis and simple separation of products.

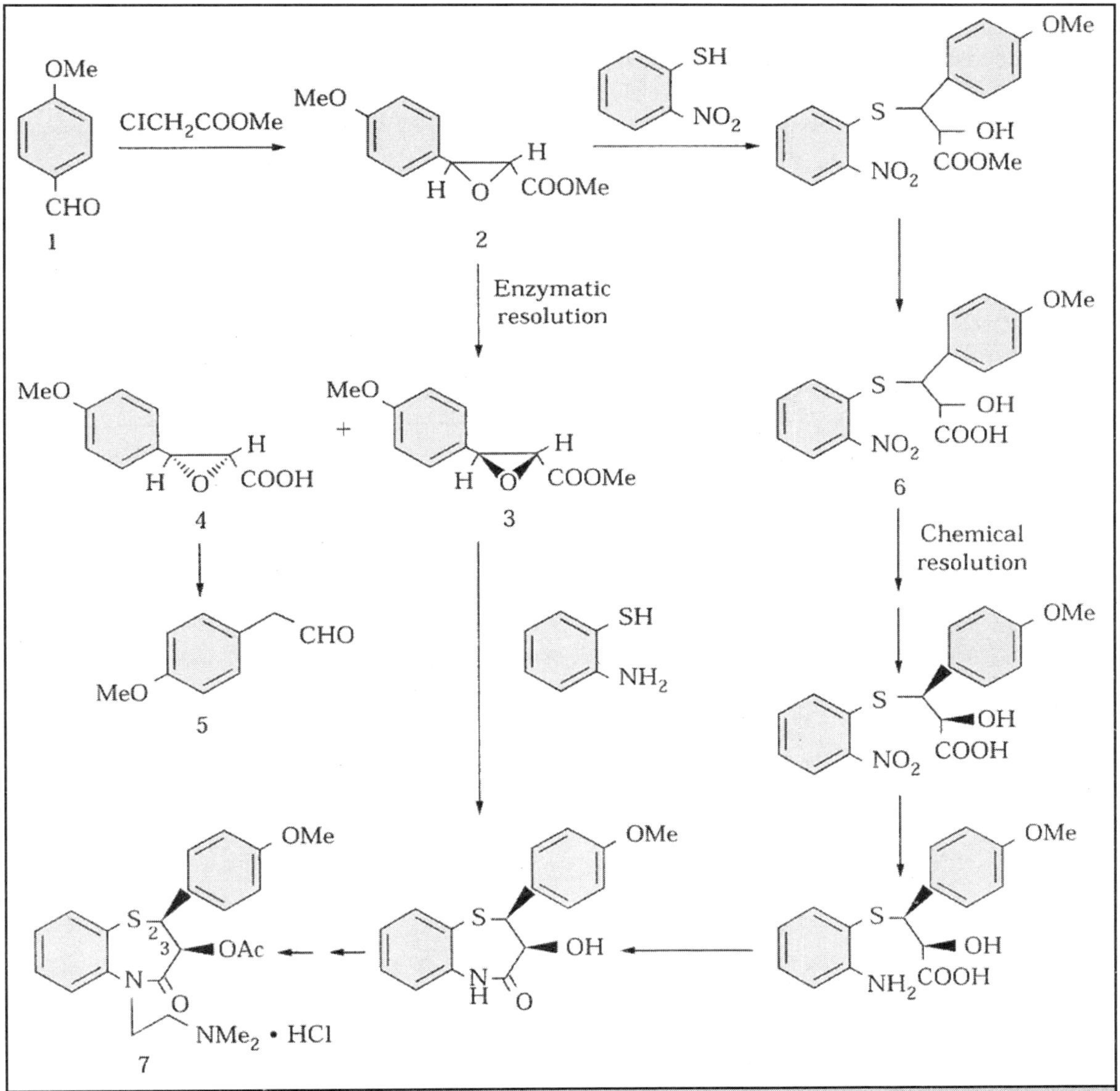

Figure 13.1 The synthetic process for diltiazem hydrochloride.

Furthermore, we overproduced the enzyme using mutation, optimization of culture medium, and recombinant DNA techniques. Our objectives were production of (−)-MPGM without chemical degradation of (±)-MPGM, construction of a lipase-hyperproducing strain, an effective isolation of

(–)-MPGM from an organic–aqueous biphasic system, and a repeated use of enzyme. The screening of hydrolytic enzymes, over-production of the enzyme, optimization of the enzyme reaction, and application of membrane reactor are reported herein.

INDUSTRIAL SCALE PRODUCTION

To successfully apply microbial reduction to industrial scale production it is important to have microorganisms with a high enzymatic activity and a high enantioselectivity. Three production routes for optically active diltiazem intermediates were examined.

Reducing Activities of Microorganisms

Methyl (2*R*,3*S*)-3-(4-methoxyphenyl)glycidate [(–)-MPGM] is an important optically active intermediate for the industrial synthesis of diltiazem. (–)-MPGM can be also obtained selectively from methyl (2*RS*,3*S*)-2-chloro-3-hydroxy-3-(4-m ethoxyphenyl)propionate [(2*RS*,3*S*)-**9**] by treating it with NaOMe in methanol; there is epimerization in the second position of **9**. Microbial reduction of (*RS*)-**8** has been studied for the production of an optically active **9**. The screening of microorganisms was carried out in a buffer adjusted to pH 5, which contained 5% glucose, substrate

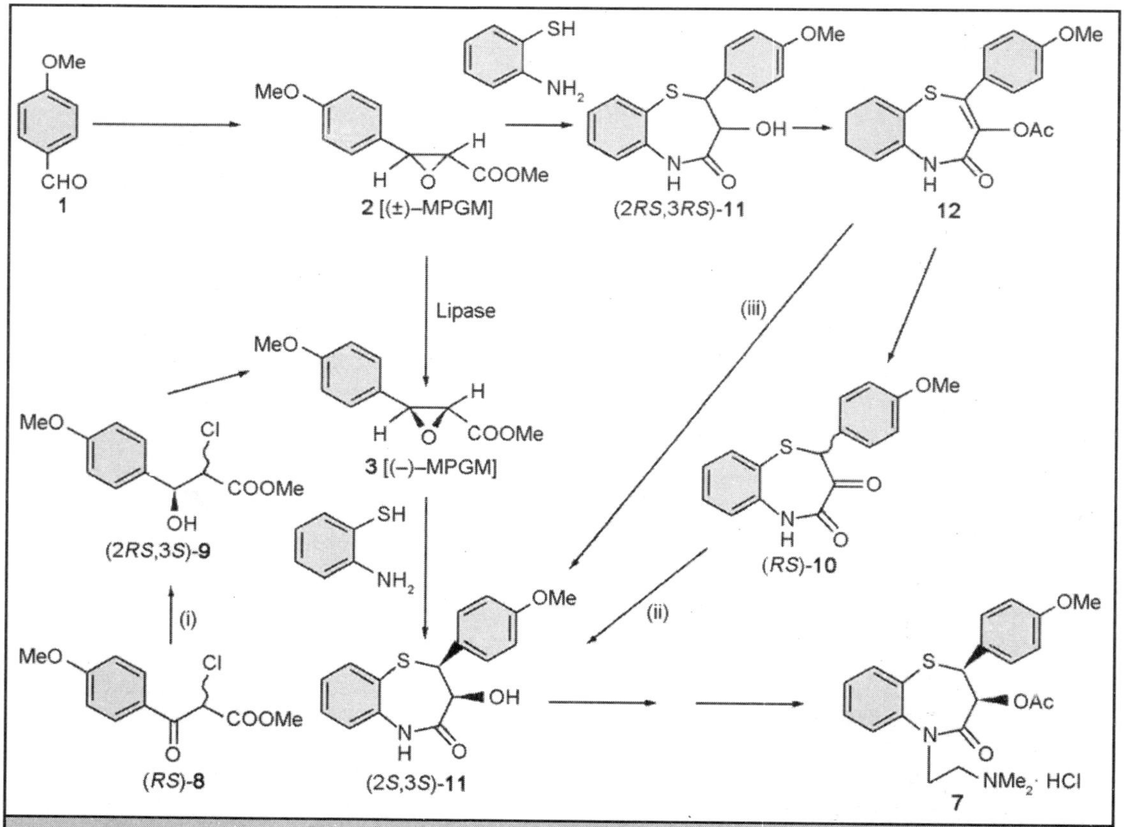

Figure 13.2 Mechanism for diltiazem hydrochloride.

8 and product **9** are unstable above pH 6. The yield and e.e. of **9** were calculated with HPLC, where the yield stands for the total yield of four optically active isomers of **9**, and the e.e. of **9** was determined for each diastereomer of *anti*-**9** and *syn*-**9**.

The terms *syn* and *anti* were defined according to Masamune et al. The microorganisms that produced alcohol **9** with the 3*R* configuration, which can be used for the synthesis of the enantiomer of diltiazem with high enantioselectivity, are in the genera *Absidia, Cryptococcus,* and *Mucor. Mucor ambiguus* IFO 6742 (IFO: Institute for Fermentation) gives only (2*S*,3*R*)-**9** in 58% yield (>99% e.e.) among the four isomers, through keto–enol tautomerization of substrate (*RS*)-**8**. With reaction conditions of pH 4 and 30°C, the yield of (2*S*,3*R*)-**9** increases up to 68% (isolated yield), and the e.e. of (2*S*,3*R*)-**9** is over 99%. Furthermore, screening of microorganisms for the ability to conduct asymmetric reduction of (*RS*)-**8**, revealed that *Trichoderma viride* OUT 4642 gives **9** with the 3*S* configuration, which is a key diltiazem intermediate with high enantioselectivity.

The production ratio of *syn*-**9** to *anti*-**9** was 2:1, and the e.e. of the (2*R*,3*S*)-**9** and (2*S*,3*S*)-**9** is 96% and 87%, respectively. However, the total yield of (2*R*,3*S*)-**9** and (2*S*,3*S*)-**9** is low (7%) due to the formation of the by-products methyl 3-(4- methoxyphenyl)-3-oxopropionate (**13**) and 2-chloro-1-(4-methoxyphenyl)ethanone (**14**). When whole cells are used as a catalyst, there are often unwanted side reactions. In this case, biotransformation by the extracted free enzyme is thought to be more effective than whole-cell biotransformation. However, if microbial reduction is performed with whole cell, an industrial device for cofactor recycling must be introduced. Desiring microbial reduction with high enzymatic activity and high enantioselectivity, two whole-cell reactions were investigated.

Another Reducing Activities

The microbial asymmetric reduction of (*RS*)-2-(4-methoxyphenyl)-1,5-benzothiazepin-3,4(2*H*,5*H*)-dione [(*RS*)-**10**] is a useful method for producing another optically active key intermediate, (2*S*,3*S*)-2,3-dihydro-3-hydroxy-2-(4-methoxyphenyl)-1,5-benzothiazepin-4(5*H*)-one [(2*S*,3*S*)-**11**]. Substrate (*RS*)-**10** is easily obtained from (±)-MPGM through three steps of chemical ring closure with aminothiophenol, chemical oxidation, and chemical hydrolysis. Microorganisms that produce the key intermediate (2*S*,3*S*)-**11** are widely distributed in the genera *Brevibacterium, Candida, Debaryomyces, Pichia, Pseudomonas, Rhodosporidium, Saccharomyces,* and *Streptomyces*. On the other hand, only *Rhizopus* reduces (*RS*)-**10** to (2*R*,3*R*)-**11**. Among these, baker's yeast (*Saccharomyces cerevisiae*) has the highest ability for the formation of (2*S*,3*S*)-**11**, which is useful for the synthesis of diltiazem.

Baker's yeast itself has been widely used as a reagent in organic synthesis and the large-scale preparation of, for example, ethyl (*S*)-3-hydroxybutanoate using baker's yeast has already been reported. In the microbial asymmetric reduction of (*RS*)-**10**, baker's yeast strictly recognizes (*S*)-**10**, and (2*S*,3*S*)-**11** forms selectively through keto–enol tautomerization of substrate. A very small amount of 3-hydroxy-2-(4-methoxyphenyl)-1,5-benzothiazepin-4(5*H*)-one *S*-oxide (**15**) is formed as a by-product of the chemical reaction. In the baker's yeast reduction system, (2*S*,3*S*)-**11** is accumulated with 94% conversion under the following the low substrate concentration conditions: substrate concentration 1 g/L at pH 7 and 42°C for 24 h. Sometimes, it is difficult to separate a product from the enzymatic reaction mixture. In this case, separation is carried out with the ease as follows: At the end of the reaction, acetone was added as a volume of three times the reaction mixture. The cells in acetone were dried to powder form and easily removed by filtration through Celite. Then, acetone in the filtrate was evaporated under reduced pressure.

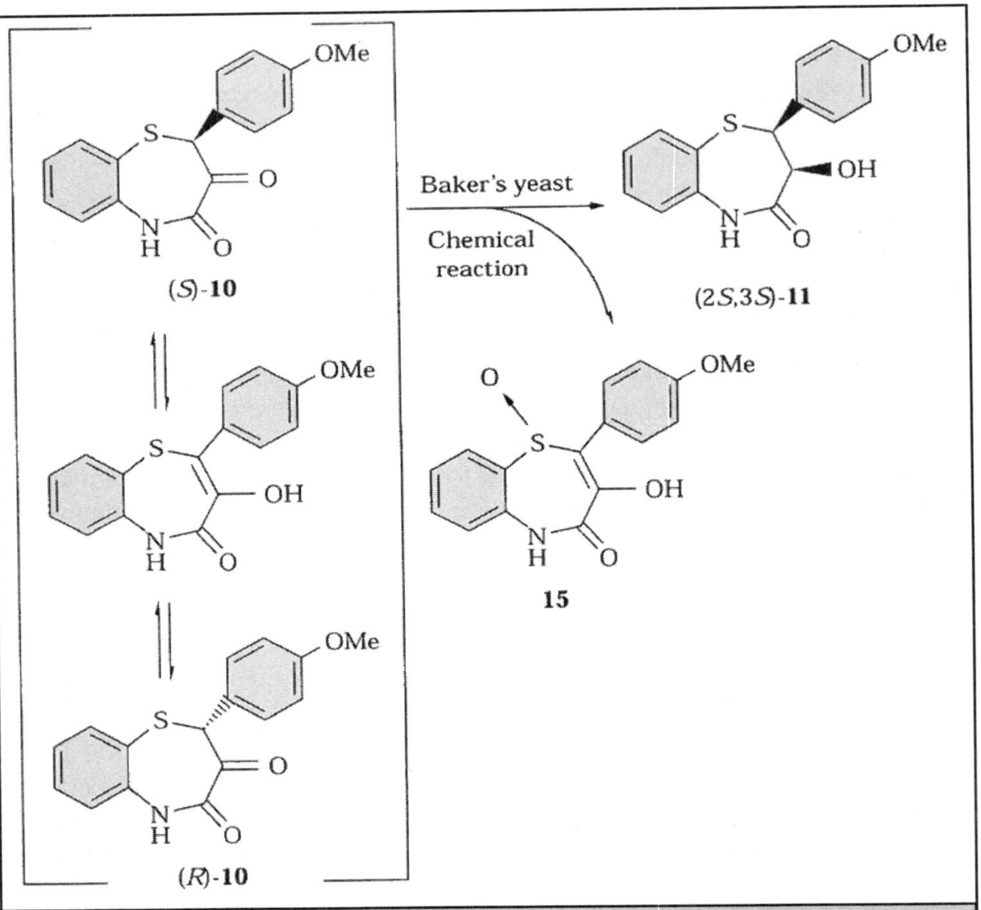

Figure 13.3 Reduction of (RS)-10 by Baker's yeast.

The deposited precipitate in the aqueous layer was recovered by filtration and washed with water in the amount of four times the volume of the reaction mixture. The precipitate was redissolved in acetone, and then the solution was dehydrated by the addition of anhydrous magnesium sulfate. After filtration, crude crystals were obtained by evaporation of the organic layer to dryness under reduced pressure. The residue was recrystallized from methanol (10 mL methanol/1 g residue). The isolated yield was 85% based on (*RS*)-**10**, and the e.e. was over 99%.

Biological Conversion of 3-Acetoxy-2-(4-methoxyphenyl)-1,5-benzothiazepin-4(5*H*)-one

An optically active key intermediate, (2*S*,3*S*)-**11**, can be also obtained by the microbial conversion of **12**. For this type biotransformation, the microbial conversion of enol ester using

Pichiafarinosa has been reported. The microorganisms that have the ability to catalyze the conversion of **12** into (2*S*,3*S*)-**11**, with excellent enantioselectivity. The conversion yield of **12** to **11**, and the e.e. of **11** are calculated on the basis of HPLC analysis, and the isolated yield is based on **12**. No accumulation of **15** was observed in this reaction. By employing this microbial conversion, diltiazem can be synthesized in only six steps with a high yield. Figure 13.4 shows the time courses of the microbial conversion of (*RS*)-**10** and **12** using *Bacillus sphaericus* FERM P-6746. In both cases, the products are optically active (2*S*,3*S*)-**11** (>99% e.e.) and isolated easily by the same procedure already described. With (*RS*)-**10** as the substrate, the reduction proceeds directly from the beginning of the reaction, whereas with **12**, an obvious lag phase is observed at the beginning of the reaction.

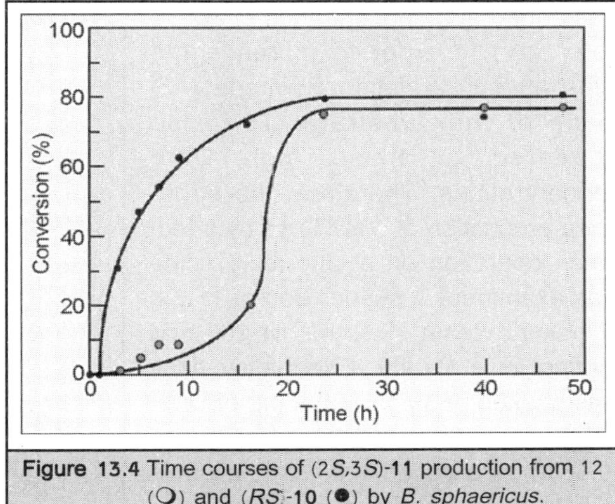

Figure 13.4 Time courses of (2*S*,3*S*)-**11** production from 12 (○) and (*RS*)-**10** (●) by *B. sphaericus*.

Later, however, the yield of (2*S*,3*S*)-**11** from **12** increases to the same level as that from (*RS*)-**10**. Therefore, in the reaction pathway of microbial conversion of **12** to (2*S*,3*S*)-**11**, it is thought that (*RS*)-**10** is an intermediate, that hydrolysis and reduction proceed in turn, and that the hydrolysis of **12** is a rate-determining step. Furthermore, it is also confirmed that the hydrolytic enzyme is induced by **12** and that asymmetry is introduced during the (*RS*)-**10** reduction step. Because all the other microorganisms also reduce (*RS*)-**10** to (2*S*,3*S*)-**11** (79–88% conversion, >99% e.e.), the microbial conversion of **12** into (2*S*,3*S*)-**11** is thought to proceed through a similar pathway in these microorganisms. Another approach to the synthesis of (2*S*,3*S*)-**11** from **12** in the presence of the hydrolytic enzyme from *B. sphaericus* and baker's yeast gives (2*S*,3*S*)-**11**, with a conversion of 65% (99% e.e.).

INDUSTRIAL APPLICATION OF REDUCTION

In general, industrial application of enzymatic reduction is difficult due to the low concentration of the substrate in the reaction medium. Among the three reactions catalyzed by microorganisms mentioned earlier, baker's yeast–mediated reduction of (*RS*)-**10** is the most promising, for industrial application if the substrate concentration is increased, because of the high optical purity of the product, easy product isolation, and a simple reaction system (not a plural enzyme system). Because the substrate (*RS*)-**10**, obtained as crystals by recrystallization from organic solvent, is almost insoluble in an aqueous medium, the reaction was carried out by adding an *N,N*-dimethylformamide (DMF) solution of (*RS*)-**10**, and by using ethanol as an energy source under aerobic conditions.

Although it is widely accepted that sugar is essential as an energy source, the use of ethanol as the energy source has also been considered because it is clean and efficient compared with sugar. In the reduction of (*RS*)-**10**, the reduction activity of baker's yeast decreased when DMF was added at a concentration of above 10% (v/v) in order to increase the substrate concentration. By-product, *S*-oxide of the substrate (**15**), also appeared at these high DMF concentrations. Therefore, the large-scale production of (2*S*,3*S*)-**11** by a fed-batch operation on a laboratory scale was examined. A suspension of 11.2 g of baker's yeast in 50 mL of 200 mM ethanol in a 300-mL Erlenmeyer flask was shaken at 30°C, and 6 mL of DMF solution containing 1.0 g of (RS)-10 (0.56 M) was added in 0.15 mL aliquots every 1 h. Since ethanol was consumed within 12 h, 0.575 mL of ethanol was added three times. After 42 h, the reaction was stopped and the product was isolated and purified. The yield of (2S,3S)-**11** was 0.83 g (83%), and e.e. was over 99%. By fed-batch operation, substrate concentration increased to 20.0 g/L. In order to increase the substrate concentration further, as a yeast-mediated reduction would be inhibited by a high concentration of DMF, the reactions were also carried out in an aqueous solution without DMF. However, in the case of crystalline (*RS*)-**10** (an average grain size of 21.3 μm) in an aqueous solution directly, reduction did not proceed successively. Increasing the surface area of the crystalline (*RS*)-**10** by grinding (an average grain size of 2.3 μm) was not very effective in increasing the reduction rate.

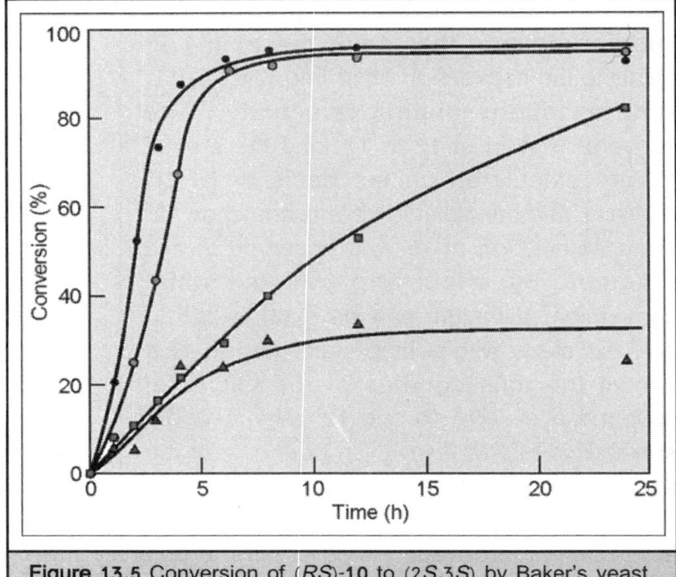

Figure 13.5 Conversion of (*RS*)-**10** to (2*S*,3*S*) by Baker's yeast.

On the other hand, when a DMF solution of (*RS*)-**10** was added to water with stirring, a dispersed colloid of (*RS*)-**10**, an emulsion was formed, and this emulsion gradually changed to a suspension, producing small precipitates. These precipitates of (*RS*)-**10**, called aggregates, were easily collected by centrifugation (an average grain size of 13.0 μm) and were reduced with a rate comparable to that of a DMF solution of (*RS*)-**10** under low concentration of substrate. The aggregate was soluble in an aqueous medium at about 20 mg/L, and DMF did not need to be added to the reaction medium.

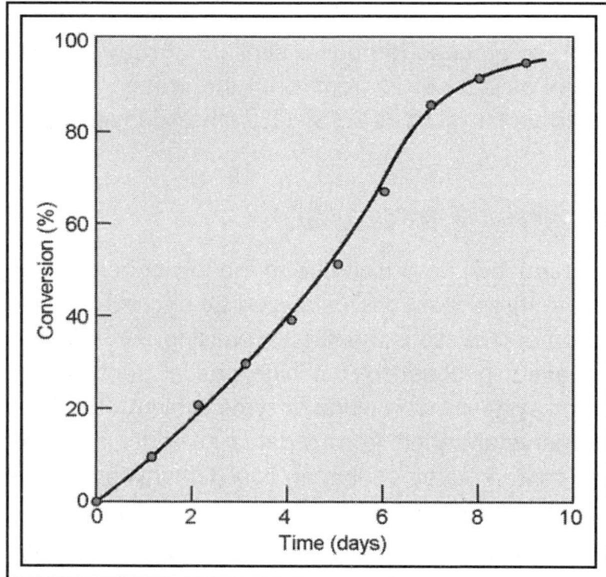

Figure 13.6 Changes in conversion ratio of (*RS*)-**10** to (2*S*,3*S*)-**11** in a high-concentration reaction.

By using the aggregate as the substrate, substrate concentration was increased up to an industrial production level (100 g/L, yield 80%) by means of the following procedure: A suspension of 11.2 g of baker's yeast and the aggregate, prepared from 5.0 g of the crystalline (*RS*)-**10**, in 50 mL of 200 mM ethanol solution was shaken at 30°C for 9 days. Ethanol (0.46 g) was added every 12 h. After processing as described earlier, over 99% e.e. of (2*S*,3*S*)-**11** was obtained with 80% yield. The substrate concentration was improved by applying this procedure. The disadvantages of (i) a relatively low concentration of substrate, (ii) a tedious procedure for isolation and purification of the product, and (iii) a relatively low optical purity of the product in the microbial reduction were resolved. This method of using baker's yeast, which is commercially available and cheap, is expected to be developed for industrial application.

ENZYMATIC SCREENING

Procedure

The instability of (±)-MPGM in an aqueous solution is a serious problem in the enzymatic resolution of (±)-MPGM. However, (±)-MPGM is both more stable and more soluble in a mixture of toluene–water. Chemical hydrolysis of the oxirane ring in (±)-MPGM increases at a pH lower than 8, yielding 2,3-dihydroxy-3-(4-methoxyphenyl)propanoic acid methyl ester. We used the toluene–water biphasic system at pH 8 to screen for hydrolase with high activity and enantioselectivity toward (±)-MPGM. Because the reaction system contains an organic solvent, we focused on lipases that are more stable in organic solvents and can act on insoluble substrates such as oils by adsorbing to the oil droplet. (+)-MPGM hydrolyzing activity is assayed using the emulsion method. One unit of esterase activity is defined as the amount of enzyme that hydrolyzes 1 μmol (+)-MPGM/min.

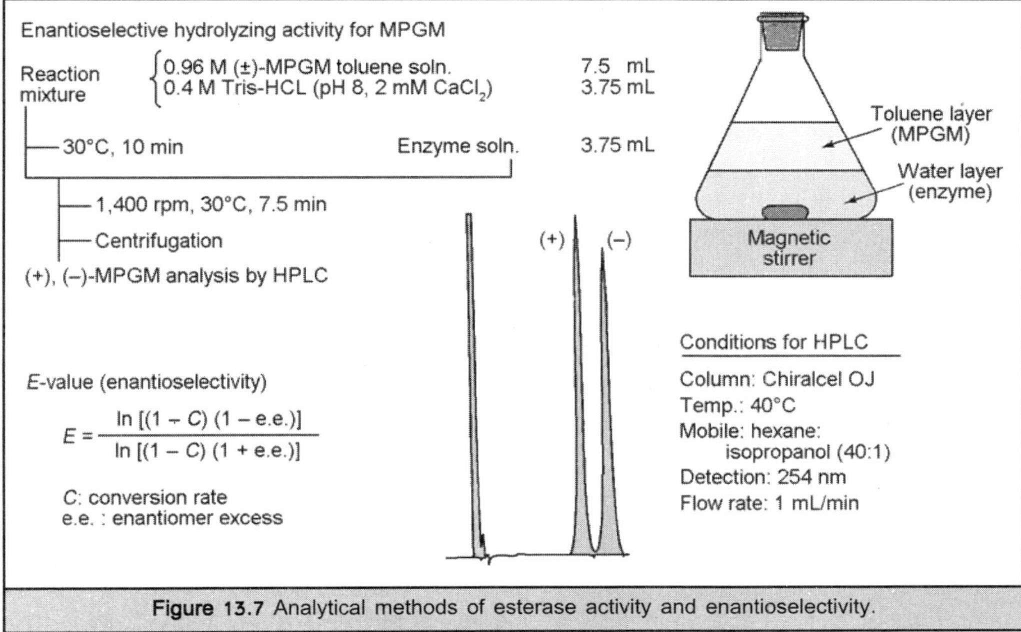

Figure 13.7 Analytical methods of esterase activity and enantioselectivity.

One unit of lipase activity is defined as the amount of enzyme that hydrolyzes 1 μmol of ester group from 2,3-dimercaptopropan-1-ol tributyrate, as measured with the Lipase Kit S. The enzyme activity in "Overproduction of Enzyme" is shown by the lipase activity. *E*-values are calculated from the enantiomeric excess of the recovered (–)-MPGM and extent of conversion and are not corrected for chemical hydrolysis of the methyl ester.

Table 13.1 The solubility and stability of (±)-MPGM in organic solvents

Organic solvent (organic solvent:water = 1:1)	*Solubility (mg/mL)*	*Stability (%)*
Water only (suspension)	<0.1	10
N,N-Dimethylformamide	>20	29
Ethanol	>20	21
Acetone	>20	70
Tetrahydrofuran	>20	85
Methyl isobutyl ketone	219	90
Ethyl acetate	560	90
n-Butyl acetate	267	90
tert-Butyl methyl ether	79	95
Toluene	251	95
Carbon tetrachloride	205	95
p-Xylene	171	95

Biological Hydrolysis of (±)-MPGM

Among the enzymes from 700 type strains and 30 commercially available enzymes, 453 enzymes hydrolyzed the ester of (±)-MPGM. Hydrolytic enzymes for (±)-MPGM were widely distributed in bacteria, yeasts, and molds. Many favored the (+) enantiomer of MPGM. The *E*-values of 12 enzymes were more than 10, and *Serratiam arcescens* lipase has the highest enantioselectivity ($E = 135$) and hydrolyzing activity. The ratio of the esterase activity to the hydrolyzing activity on olive oil is the highest for *S. marcescens* lipase. This indicates that *S. marcescens* lipase is the most effective enzyme for hydrolysis of (+)-MP GM. In addition, since the *E*-value includes chemical degradation of (±)-MPGM, it shows a lower value than the true value of the enzyme. *S. marcescens* lipase does not enzymatically hydrolyze (–)-MPGM; the *E*-value of this lipase itself is considered to be infinity.

Nature of *S. marcescens* Lipase

Three main protein bands exist in the supernatant after centrifugation of the culture broth. The molecular masses of these bands are 62,000, 51,000, and 40,000 Da on SDS-PAGE. The hydrolytic activity toward (+)-MP GM exists in the 62,000-Da band. The esterase can be purified from culture supernatant (about 21-fold) at a yield of 20%, and its (±)-MPGM hydrolyzing activity is 570 units/

mg protein. Esterase corresponds to 5% of the extracellular proteins in cells grown with dextrin as the carbon source and Meast S (beer yeast extract) as the nitrogen source.

Table 13.2 Asymmetric hydrolysis of (±)-MPGM by enzymes from various microorganisms

Enzyme source	*Enzyme used (mg)*	*Esterase activity (units/mg protein)*	*Conversion (%)*	*e.e. (%)*	*E*
Serratia marcescens Sr41 8000	1.4 mL	36	48.2	89.0	135
Pseudomonas putida ATCC 17426	3.75 mL	18	44.2	75.1	85
Corynebacterium primorioxydans ATCC 31015	3.75 mL	16	44.6	75.6	73
Pseudomonas mutabilis ATCC 31014	3.75 mL	13	44.7	74.7	57
Corynebacterium alkanolyticum ATCC 21511	3.75 mL	18	45.8	77.4	53
Candida cylindracea (Lipase OF)	4.0	34	50.0	82.2	26
Mucor javanicus (Lipase M)	150	0.82	48.0	75.0	22
Bacillus licheniformis (Alcalase)	1.9 mL	0.91	41.2	57.0	17
Rhizopus japonicus (Lipase saiken 100)	100	2.3	42.8	58.4	15
Bacillus subtilis (Protease 27)	300	–	45.1	62.6	14
Mucormiehei (PalataseM)	0.25 mL	27	45.3	60.3	12
Pseudomonas sp. (Lipase P)	300	0.20	49.8	68.6	11

The activity of the purified enzyme is 3,100 units/mg protein for olive oil, and 520,000 units/mg protein for 2,3-dimercaptopropan-1-ol tributyrate. Because this esterase selectively hydrolyzes the 1,3-positional ester of triglyceride, the esterase is a lipase. The protein exhibiting a molecular mass of 51,000 on SDS-PAGE is a protease with hydrolytic activity toward casein. The protease does not have (±)-MPGM hydrolyzing activity. (±)-MPGM as well as other esters having different alcohol moieties may be used as a key intermediate in the synthesis of diltiazem. The enantioselectivity of the lipase did not change upon altering the chain length of the alcohol moiety, except for the acid ethyl ester. (±)-MPGM crystallizes more easily from toluene and is less expensive than other esters, so it was selected for preparing enantiomerically pure 3-(4-methoxyphenyl) glycidic acid ester.

INDUSTRIAL PRODUCTION OF ENZYME

Multiplication of *S. marcescens lipA* Gene

S. marcescens produces the lipase at a level of 3×10^3 units/mL when grown in LB medium. The amount of lipase increased 50-fold by optimizing the culture condition, and 2.5-fold as a result of mutation using *N*-methyl-*N*-nitro-*N*-nitrosoguanidin. Extracellular lipase activity reaches 400×10^3 units/mL of culture medium when the lipase-hyperproducing mutant is cultured in a fed-batch culture on proline. To further increase production of the lipase, the *lipA* gene, encoding the lipase, was cloned from *S. marcescens*. The nucleotide sequence of the *lipA* gene showed a major open reading frame encoding a 64.9-kDa protein of 613 amino acid residues; the deduced amino acid sequence contained a lipase consensus sequence, GXSXG. The *S. marcescens* lipase has 66 and 56% homologies with the lipases of *Pseudomonas fluorescens* B52 and *P. fluorescens* SIK W1, respectively, but does not show any overall homology with other lipases. The *Escherichia coli* cells carrying the *lipA* gene do not secrete the lipase into the medium. The *S. marcescens* lipase lacks a conventional N-terminal signal sequence and is not subjected to any processing at either the N-terminal or C-terminal regions. A specific short region, GGXGXDXXX, which is often found in secretory proteins having no N-terminal signal peptide, is observed in the amino acid sequence. The region is characterized as a glycine- and aspartic acid–rich region. Thus, these results reveal that *S. marcescens* lipase is secreted by the N-terminal-signal-peptide-independent system.

Secretion of Lipase by *S. marcescens*

The 20-kb *Sac*I DNA fragment allowing the lipase secretion was cloned from *S. marcescens* by lipase selection phenotype based on the clear halo formation by recombinant *E. coli* cells. The subcloned 6.5-kb *Eco*RV fragment contains three open reading frames, *lipB, lipC,* and *lipD,* which are composed of 588, 433, and 437 amino acid residues and constitute an operon. The *E. coli* cells carrying the *lipA, lipB, lipC,* and *lipD* genes secreted the *S. marcescens* lipase, but the cells carrying only the *lipA* gene did not. Deletion mutation analysis shows that all three genes are essential for the extracellular secretion of the *S. marcescens* lipase. Thus, the secretion of the lipase requires three proteins: LipB, LipC, and LipD. Recent papers reported that the genes encoding the secretion device for the extracellular proteins lacking an N-terminal signal sequence are found in several Gram-negative bacteria and has been categorized as being in the ABC (ATP binding cassette) transporter family: the *hlyB, hlyD,* and *tolC* genes for *E. coli* hemolysin; the $prtD_{EC}$, and $prtE_{EC}$, and $prtF_{EC}$ genes for *Erwinia chrysanthemi* metalloprotease; the *aprD, aprE,* and *aprF* genes for *Pseudomonas aeruginosa* alkaline protease; the *cyaB, cyaD,* and *cyaE* genes for *Bordetella pertussis* cyclolysin; the *lktB* and *lktD* genes for *P. haemolytica* leukotoxin; and the *hasD, hasE,* and *hasF* genes for *S. marcescens* heme-binding protein and metalloprotease.

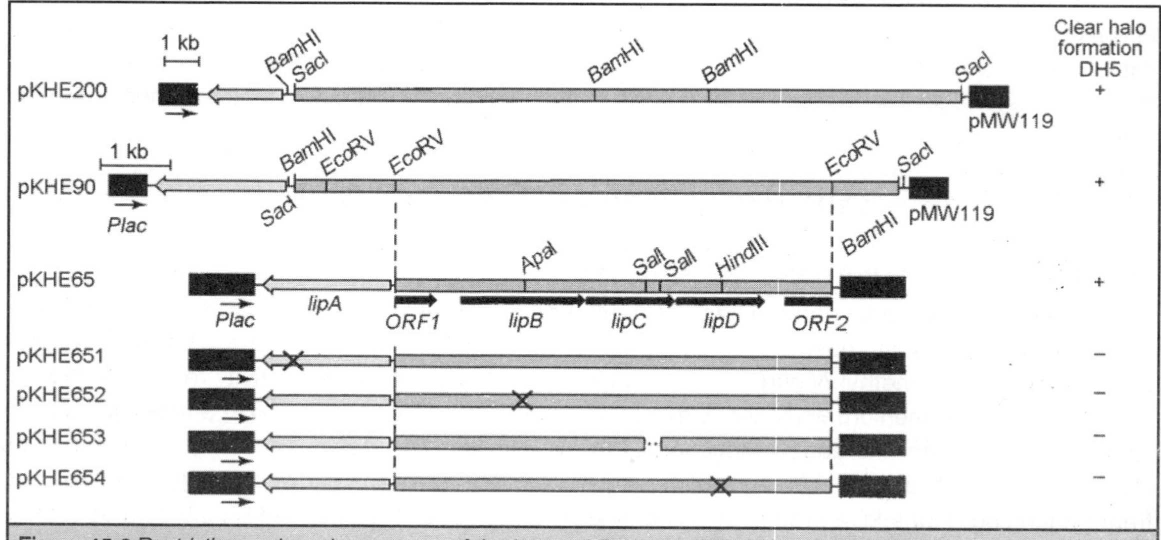

Figure 13.8 Restriction endonuclease maps of the inserted DNA of pKHE200, deletion derivatives, and the ability of the recombinant plasmids to give a lipase phenotype to the *E. coli tolC** strain DH5.

The deduced amino acid sequences of the *lipB* and *lipC* genes showed high homology to those of the *E. chrysanthemi prt*D_{EC} and *prt*E_{EC} genes (56 and 46% identity, respectively), and also the *S. marcescens has*D and *hasE* genes (54 and 45% identity, respectively). The LipB protein belongs to a member of the ABC protein family, and its putative ATP-binding site was found in the central region of the C-terminal half. No N-terminal signal sequence was identified in the LipB and LipC proteins. The LipB and LipC proteins were revealed to be inner membrane components involved in transport across the membrane. The outer membrane component corresponding to the *E. chrysanthemi prt*F_{EC} gene product was encoded by the *lipD* gene in *S. marcescens* (42% identity). The products of the *lipB, lipC,* and *lipD* genes also showed 26, 20, and 22% identity with the *hlyB, hlyD,* and *tolC* gene products of *E. coli,* respectively. These results indicates that LipB-LipC-LipD is included in the ABC transporter family and is the secretion device of the *S. marcescens* lipase.

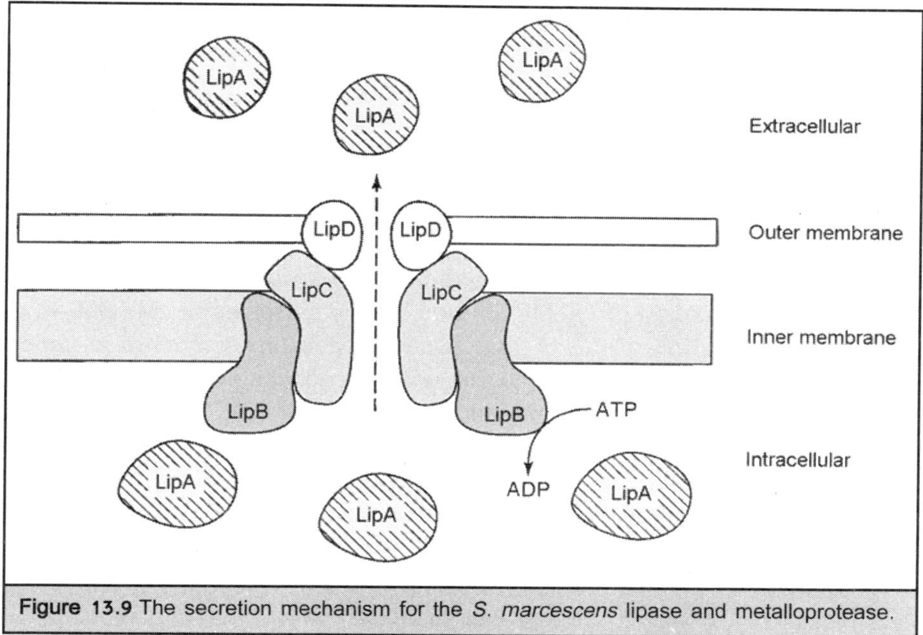

Figure 13.9 The secretion mechanism for the *S. marcescens* lipase and metalloprotease.

Industrial Production of Lipase by *S. marcescens*

The *S. marcescens lipA* gene was cloned into the high-copy-number vector pUC19, resulting in pLIPE121. This plasmid was introduced into the cells of *S. marcescens* 8000, a wild-type strain of Sr41. The extracellular lipase activity of the recombinant strain reached 220 x 10^3 units/mL in the lipase production medium and is about nine fold greater than that of the strain carrying pUC19. However, we expected an even greater increase, based on the high copy number of the plasmid. The *S. marcescens lipB, lipC,* and *lipD* genes were cloned into the low-copy-number vector pMW219, resulting in pMWBCD10. The *S. marcescens lipA* gene was cloned into the high-copy-number vector pBR322, resulting in pBRE121. The lipase productivity of *S. marcescens* cells carrying pMWBCD10 and pLIPE121 (or pBRE121) is higher than that carrying the *lipA* plasmid only. The extracellular lipase activity of the hyperproducing strain carrying pMWBCD10 and pBRE121 reaches 420 x 10^3

units/mL of the culture medium and is 13-fold greater than that of the wild-type strains. The extracellular lipase activity of the hyperproducing strain carrying pMWBCD10 and pLIPE121 reaches 820 x 10^3 units/mL of the culture medium and is 25-fold greater than that of the wild-type strains. We are now investigating large-scale lipase production using this recombinant strain.

Table 13.3 Effect of *lipA* and *lipBCD* expression on lipase production

Plasmid		*Growth*	*Lipase activity*
lipA	*lipBCD*	*(OD_{660})*	*($x10^{-3}$ units/mL)*
–	–	12.2	31.8
pBRE121	–	11.9	258
pBRE121	pMWBCD10	11.2	421
pLIPE121	–	12.2	220
pLIPE121	pMWBCD10	10.9	820

CONDITIONS FOR ENZYMATIC ACTIVITY

Asymmetric Hydrolyzing Activity

Screening for hydrolases, as already described, used the toluene–water biphasic system, which suppresses chemical degradation of (±)-MPGM and readily dissolves (±)-MPGM. In this section, (±)-MPGM hydrolyzing activities of the *S. marcescens* lipase are compared in various biphasic systems. The hydrolyzing activities differ with the kind of organic solvent. The lower the dielectric constant of the solvent, the higher the enzyme activity obtained. Among the biphasic systems tested, the carbon tetrachloride–water biphasic system gives the highest hydrolyzing activity. However, carbon tetrachloride is not suitable for industrial application because of its toxicity. Consequently, the toluene–water biphasic system, which gives the second highest hydrolytic rate, was selected as the reaction solvent.

Table 13.4 Esterase activity of *S. marcescens* lipase in various organic solvent–water (1:1) biphasic systems

Organic solvent (organic solvent:water = 1:1)	*Dielectric constant of solvent (ε)*	*Esterase activity (units/mg protein)*
Carbon tetrachloride	2.2	85
Toluene	2.2	36
p-Xylene	2.2	31
n-Butyl acetate	5.0	26
Ethyl acetate	6.1	19
Methyl isobutyl ketone	13.1	11
Trichloroethane	7.5	3
Ethanol	24.6	0

The lipase shows maximum activity at about pH 8. When the lipase was kept at 30°C for 24 h at different pHs, we found it to be stable between pH 6 and 8. The apparent enantioselectivity of hydrolysis of (±)-MPGM by this lipase in a toluene–water biphasic system decreases at pH values below 8 because the chemical degradation of the oxirane ring in (±)-MPGM increases below pH 8. An Arrhenius plot indicates that the apparent activation energy of the enzymatic hydrolysis is 26 kJ/mol. The thermal inactivation of the enzyme occurs above 50°C. On the other hand, the apparent activation energy of chemical degradation of the oxirane ring in (–)-MPGM or (+)-MPGM is 66 kJ/mol. These findings show that reaction at a lower temperature minimizes the relative amount of chemical degradation of (±)-MPGM. The best conditions were a toluene–water emulsion at pH8 and 30°C. As the toluene to water ratio in the reaction mixture decreases, the rate of enzymatic hydrolysis also decreases. At a toluene to water ratio lower than 1:1, the chemical degradation of (±)-MPGM increases. Therefore, we use the condition of a toluene to water ratio of 1:1 as a typical reaction for the asymmetric hydrolysis of (±)-MPGM.

Factors Affecting Stability of the Enzyme

Attempted reuse of the *S. marcescens* lipase showed much slower reaction rates, indicating deactivation of the lipase. When fresh lipase is incubated with the toluene–water biphase without substrate before the enzymatic reaction is started, the progress of reaction is nearly equal that of untreated enzyme. Therefore, the esterase activity of the reaction decreased only slightly. However, when the first incubations are carried out in the toluene–water biphase containing (±)-MPGM or (–)-MPGM, the progress of reaction is slowly toward that of untreated enzyme. The esterase activity of the second incubation decreases from 36 units/mg protein to 6 and 12 units/mg protein, respectively. Inactivation of the lipase is greatly influenced by MPGM.

When the first incubation is carried out in the toluene–water biphase containing (±)-MPGM, the degree of inactivation of the lipase is higher than that of the lipase treated with the toluene–water biphase containing (–)-MPGM. This is because *p*-methoxyphenyl-acetaldehyde (aldehyde 5) derived from chemical degradation of the hydrolyzed product of (+)-MPGM probably inactivates the lipase. Treatment of the lipase with the toluene–water biphase containing 20 mM and 100 mM aldehyde 5 lowers the esterase activity of the second incubation to 28 and 18 units/mg protein, respectively. Although the lipase is highly stable in a toluene–water biphase, it is unstable under the biphasic system containing (±)-MPGM and aldehyde 5. Hence, the repeated use of the lipase is difficult for this batch process.

Isolation of (–)-MPGM

In the organic–aqueous biphasic systems, the reaction rate is influenced by the outer diffusion and contact area on the biphase, and the highest rate is obtained by the formation of fine emulsion. However, the fine emulsion was unfavorable for industrial purposes because the destruction of the emulsion after the enzymatic reaction and subsequent phase separation are difficult, and the desired product cannot be recovered in high yields. The destruction of the emulsion after the enzymatic reaction and subsequent phase separation is a serious problem for industrial applications of the emulsion bioreactor.

Several means involving an electrical method such as sonication, a physical method such as filtration, centrifugation, or heating, and a chemical method such as addition of surfactant or solvent have already been proposed to overcome this problem. Among them, the chemical method using surfactants would be the most favorable for industrial purposes because of the ease of operation. When a 0.01% (w/v) sodium dodecyl sulfate is added to the reaction mixture, the destruction of emulsion and phase separation are accomplished.

In the hydrolysis of (±)-MPGM with the lipase, aldehyde 5 is accumulated in the toluene phase, with the desired (–)-MPGM. Aldehyde 5 suppresses crystallization of (–)-MPGM. To isolate (–)-MPGM efficiently, it is necessary to remove the aldehyde 5 from the toluene solution. It is well known that aldehyde compounds generally form a water-soluble adduct through the reaction with bisulfite anion. The aldehyde 5 can be removed by transfer to the aqueous phase by the formation of an adduct (RCHOH · SO_3Na) with the addition of sodium bisulfite (pH 6). After the phase separation and removal of the aldehyde, the toluene phase is neutralized with sodium hydrogen carbonate, washed with water, dried over anhydrous sodium sulfate, and the (–)-MPGM is crystallized. The yield of (–)-MPGM isolated from (±)-MPGM is 40–43%, and the optical purity of (–)-MPGM is nearly 100% e.e.

APPLICATION OF MEMBRANE REACTOR

Production of (–)-MPGM

The enzymatic hydrolysis of (±)-MPGM proceeds efficiently in a conventional emulsion reactor with a toluene– aqueous biphasic system. However, destruction of the emulsion and separation of the biphase are required for effective isolation of the product. Enzyme stability and recovery are also required for its repeated use. In particular, the recovery of enzyme is very important for cost reduction in industrial processes involving enzyme reactions. However, efficient recovery is difficult in the case of a stable emulsion. We proposed an immobilized enzyme system to solve these disadvantages of the emulsion reactor. With respect to the immobilization of lipase, various methods have already been proposed. When these lipase-immobilized reactors are utilized, lipase can be efficiently immobilized. However, it is difficult to achieve reaction and product separation simultaneously.

Furthermore, because the hydrolysis of oily fat by lipase is an interfacial reaction involving the adsorption of enzyme to the surface of oil droplets, the activities of most immobilized lipases are low compared with other immobilized enzymes. A membrane reactor permits both physical adsorption of enzyme and contact between the organic and aqueous phases; this configuration enables efficient hydrolysis and simplifies product separation. Since a pioneering work on the membrane reactor with liquid–liquid contact mode by Hoq et al., several membrane bioreactors with industrial promise have been developed.

Structure of the Membrane Reactor

The membrane reactor used in this study is a commercially available Sepracor membrane reactor model MBR 500-tm. The membrane module is a hydrophilic hollow-fiber ultrafiltration membrane with a molecular weight cutoff of 50,000. The membrane module contains approximately 8,500 fibers with an internal diameter of 0.21 mm and a wall thickness of 0.05 mm. The effective membrane area is 0.75 m^2. The saturated adsorption amount of proteins is 3.0 g/m^2. The membrane, composed of polyacrylonitrile, is an asymmetric membrane having a tight skin layer on the lumen side and a loose spongy layer on the shell side. Immobilization of enzyme is carried out by ultrafiltration of the lipase solution from the shell side to the lumen side. The lipase is physically immobilized on the spongy layer of the shell side. The reaction using a membrane reactor is achieved by circulating both toluene solution containing (±)-MPGM in the shell loop and an equimolar solution (pH 8.5) of sodium sulfite in the lumen loop. Flow rates in both loops are maintained at 750 mL/min and 500 mL/min, respectively; these are sufficient to keep mass transfer resistances at a minimum. The pressure in the shell loop is maintained at overpressure at 0.4–0.5 kg/cm^2 over the pressure in the lumen loop, to avoid the filtration of aqueous-phase solution into the shell loop through the hydrophilic membrane.

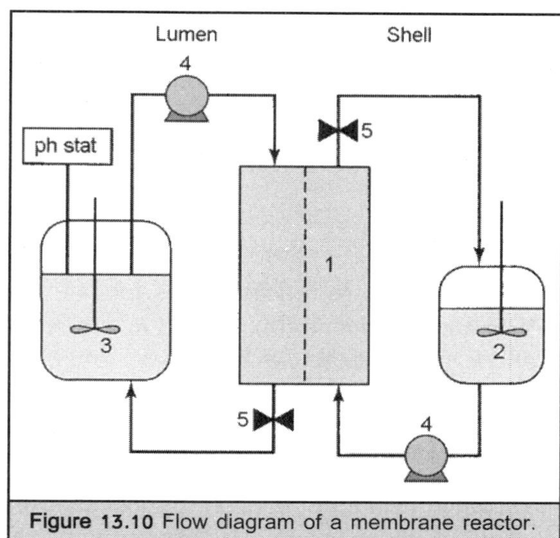

Figure 13.10 Flow diagram of a membrane reactor.

The enzyme reaction is carried out at 20°C and pH 8.5 by controlling the pH of the aqueous phase. As discussed earlier, 30°C was selected for the optimum temperature of reaction. As for the lipase-catalyzed asymmetric hydrolysis on a membrane reactor, a long period of reaction was required, and 20°C was selected in order to suppress chemical degradation of (±)-MPGM. If sodium bisulfite was not added to the aqueous phase in the lumen loop, the aldehyde 5 accumulated in the toluene phase, and deactivated the lipase. The stability of the lipase is not influenced when the sodium hydrogen sulfite concentration is below 0.1 M at pH 8.5 and 22°C. From these results, 0.0935 M sodium hydrogen sulfite solution (pH 8.5) was applied to the aqueous phase to prevent the lowering of enzyme activity. Another by-product, methanol, is transferred from the toluene phase to the aqueous phase. Because the concentration of methanol in the aqueous phase was below 0.2% (w/v), the hydrolysis of (±)-MPGM was not influenced.

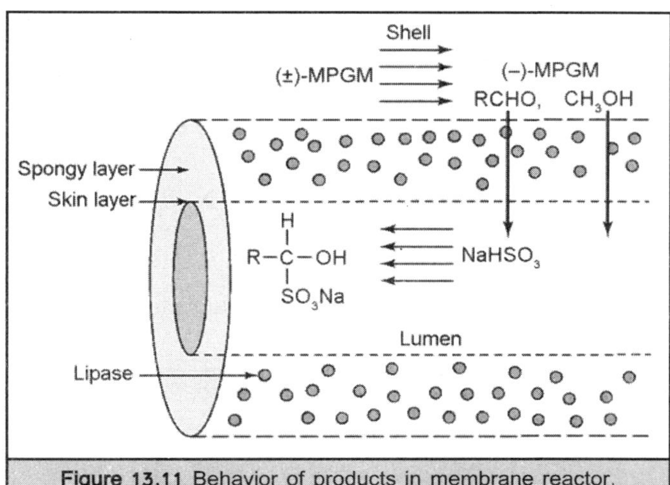

Figure 13.11 Behavior of products in membrane reactor.

Optimization

The residual concentration of (+)-MPGM decreases with reaction time and reaches less than 1% of initial concentration of (+)-MPGM during the first 15 h. The degradation of (–)-MPGM in merely trace amounts is observed. Thus this hydrolyzing rate of the lipase obeys the following first-order reaction kinetics:

$$r_S = -dC_S/dt = k_S C_S$$

$$r_R = -\ dC_R/dt = k_R C_R$$

where r_S and r_R are reaction rates for (+)- and (–)-MPGMs (M/h); C_S and C_R are (+)- and (–)-MPGM concentrations (M); k_S and k_R are reaction rate constants (h^{-1}); and t is reaction time (h). k_S and k_R are calculated to be 0.23 h^{-1} and 0.007 h^{-1}, respectively. The ratio of k_S to k_R is about 33, indicating that the lipase from *S. marcescens* has a high enantio-selectivity on hydrolysis of (+)-MPGM.

Optimization of the asymmetric hydrolysis reaction with the following parameters was investigated using the reaction rate constant k (h^{-1}) as the indication of reaction. This hydrolyzing reaction is influenced by pH, temperature, phase ratio, and immobilized enzyme amount. The optimized conditions are pH 8.5; temperature 20°C; phase ratio (=aqueous:toluene phase) 2:1; immobilized enzyme amount 0.8-1.6 x 10^5 unit/m^2 (based on esterase activity). In the industrial application of the membrane bioreactor, one of the most important factors is the stabilization of enzyme.

The reaction rate constants for the asymmetric hydrolysis of (±)-MPGM exponentially decrease with time, and the half-life of enzyme activity is 127 h. This half-life is much lower than for enzymes employed for the industrial production of various amino acids. However, one is about 30 times higher than that in the emulsion reactor. In this membrane bioreactor, one is able to remove inactivated enzymes by a simple back flash. When the enzyme activity in the membrane reactor lowers, one can immobilize new enzyme without regeneration of the membrane. As a result, it has become possible to achieve repeated runs of the enzyme reaction for periods up to a month. When the regeneration of the membrane after the every eight successive batch runs is repeated 10 times over a period of 4 months, the water flux rate of the old membrane is the same as that of new membrane. Thus the production method of (–)-MPGM using the membrane reactor can be industrialized.

Table 13.5 Comparison of reactor performances for laboratory- and pilot-scale membrane reactors

Reactor membrane area	*Laboratory scale*	*Pilot scale*	
	0.75 m^2	*7.8 m^2*	*57 m^2*
L/D (–)	3.9	5.5	4.4
Loaded enzyme (units/m^2)	0.8 x 10^5	1.1 x 10^5	0.9 x 10^5
k_S value (h^{-1})	0.23	0.23	0.22
Half-life of enzyme activity (h)	127	137	144

Scale-Up

Advantages of the production process of (–)-MPGM using the membrane bioreactor are as follows: (i) the reaction and products separation are achieved simultaneously, and crystalline (–)-MPGM with a high yield of 40–43% and optical purity of 100% e.e. is obtained by just concentrating the toluene solution; (ii) the working time taken to isolate crystalline (–)-MPGM is reduced in comparison with that of the emulsion bioreactor, (iii) stabilization and recovery of the enzyme are achieved in laboratory-scale experiments. On the basis of these results, the scale-up experiments were started using membrane modules having an effective membrane area of 7.8 m^2 and 57 m^2. Both membranes were magnified from that of the laboratory scale on the basis of geometrical analogy.

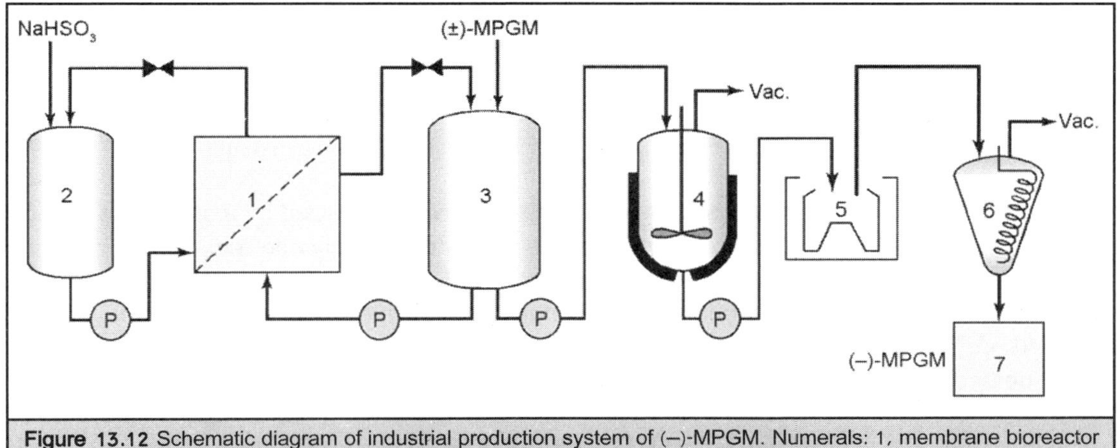

Figure 13.12 Schematic diagram of industrial production system of (–)-MPGM. Numerals: 1, membrane bioreactor skid; 2, aqueous solution tank; 3, organic solution tank; 4, crystallizer; 5, centrifuge; 6, dryer; 7, product container.

The basic strategy of scaleup of the membrane reactor is as follows: (i) the flow rate of supplied fluids to the membrane area is kept constant, (ii) when a lot of modules are used in series, a pressure gradient with respect to flow direction among the modules is not allowed to form. We have tried the reaction in pilot-scale experiments under a liquid line rate comparable to that in the laboratory-scale experiments. The functions of pilot-scale membrane reactors are identical with those of laboratory-scale membrane reactors. From the result, it is believed that scale-up should proceed very smoothly without any trouble. On the basis of the obtained data, we made a design for an apparatus producing (–)-MPGM. It is estimated that the membrane bioreactor using membrane modules having the effective membrane area of 57 m^2 produces about 50 kg (–)-MPGM m^{-2} $year^{-1}$. The pilot-plant experiments were carried out in collaboration with Sepracor, Inc., and this technique was commercialized in 1993.

14

ENZYMATIC REVERSE MICELLES

Reverse micelles are fine dispersions of water in organic solvent stabilized by a surfactant molecule. The main property of this system is the coexistence of two completely different microphases in an apparently macrohomogeneous solution. The nature of the surfactant molecules, known as amphiphiles, induces molecular organizates, among which reverse micelles are thermodynamically stable and are formed spontaneously. The solutions are optically isotropic because one solvent is finely dispersed into the other, forming microdroplets that have a diameter on the order of 100 .

PRODUCTION OF REVERSE MICELLES

The formation of the reverse micelles depends on the relative concentrations of water, surfactant, and organic solvent and can be explained by the energy change caused by dipole–dipole interactions between the polar head groups of the surfactant molecule. This mechanism can be understood by taking into consideration the values of the interfacial tension between water and oil (the term *oil* refers to any hydrophobic solvent). This interfacial tension (with values varying from 30 to 60 dyn/cm) causes a reduction of the surface between the two solvents, which results in their separation. The addition of surfactants induces the formation of micelles because of the reduction of the interfacial tension down to very low values (10^{-5} dyn/cm).

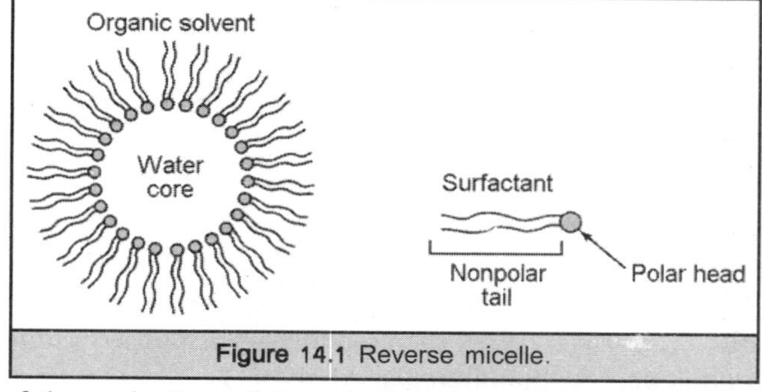

Figure 14.1 Reverse micelle.

The amphiphilic nature of the surfactant molecule causes its self-association in solution. In a reverse micellar solution, the polar head of the surfactant is directed toward the aqueous core of the micelle, and the hydrophobic tail extends into the continuous organic phase. Depending on the hydrophilic–lipophilic balance (HLB) of the surfactant (i.e, the rate of the weight percentage of hydrophilic groups to the weight percentage of hydrophobic groups in the surfactant molecule), the interface will curve spontaneously toward water or the oil phase, favoring the formation of water-in-oil or oil-in-water structures, respectively.

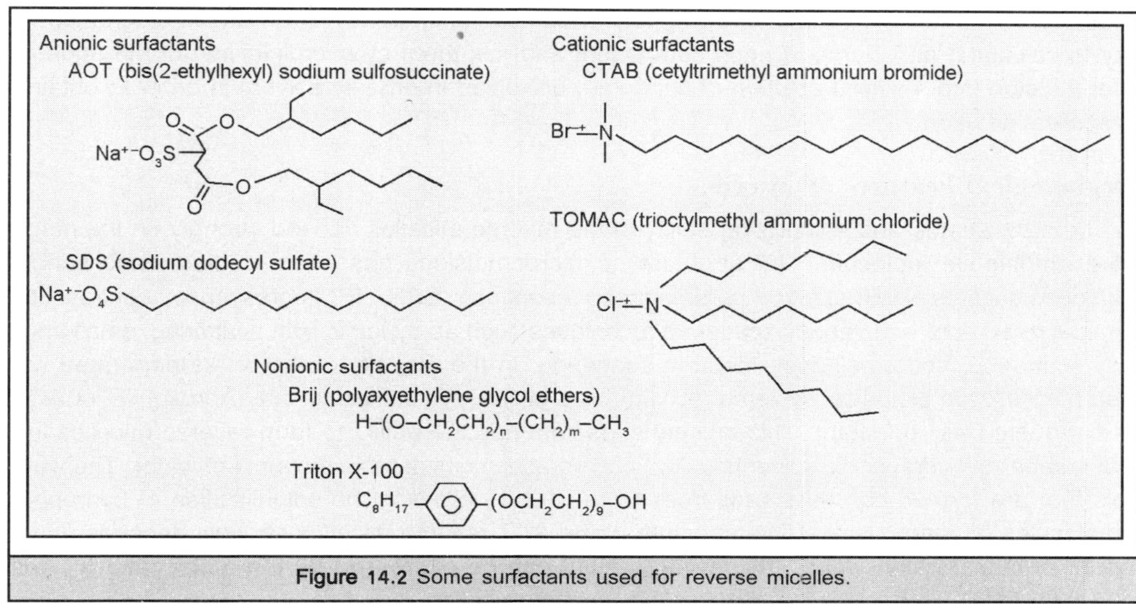

Figure 14.2 Some surfactants used for reverse micelles.

Various types of surfactants have been used as amphiphiles to formulate reverse micellar systems, including natural membrane lipids and artificial surfactants. The most commonly used surfactants are the anionic AOT [bis(2-ethylhexyl) sodium sulfosuccinate]; the cationic CTAB (cetyltrimethyl ammonium bromide); various nonionic surfactants of the Brij, Triton and Tween series; and the zwitterionic biological membrane-forming phospholipids. They can be grouped in three main categories: anionic, cationic, and nonionic, depending on the nature of their polar moiety. Cationic micelles are very small ($[H_2O]/[\text{surfactant}] \leq 3$). To increase the micelle size, and thus the amount of water solubilized, a cosurfactant is included in the system, usually an aliphatic alcohol, the cosurfactant being partitionated between the interface and the apolar phase. Anionic surfactants

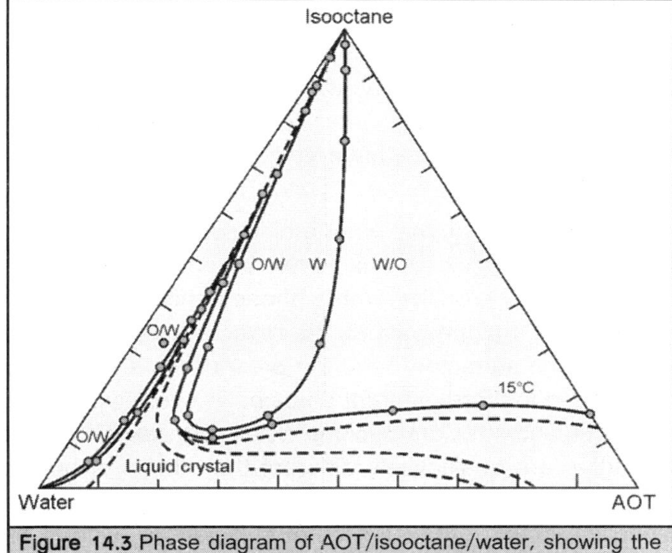

Figure 14.3 Phase diagram of AOT/isooctane/water, showing the reverse micelles region. O = oil; W = water.

form micellar aggregates of large size ($[H_2O]/[\text{surfactant}] \leq 60$). There is a surfactant concentration under which the reversed micelle formation does not occur the critical micellar concentration (CMC). The CMC depends on the chemical structure of the surfactant, solvent, temperature, and pressure.

The solubilization properties of surfactants are often expressed by a three- or four-component phase diagram, after determination of regions of optical transparency. From these diagrams, it is easy to calculate the volume of each component and mix them by microinjection of the aqueous buffer solution into a stirred solution of surfactant dissolved in organic solvent in order to obtain a transparent solution.

Morphological Features of Micelle

The size, shape, and solubilizing ability of the reverse micelles depend strongly on the nature of the amphiphile molecule. The structure of microemulsions has been studied using various techniques such as NMR, electron paramagnetic resonance (EPR), IR, fluorescence spectroscopy, chemical relaxation, and various scattering techniques such as dynamic light scattering, small-angle X-ray scattering, and small-angle neutron scattering. In the majority of the works performed with reverse micelles in biological systems, sodium(bis-2-ethylhexyl)-sulfosuccinate (Aerosol-OT or AOT) has been used as surfactant. This anionic surfactant has the ability to form reverse micelles in a great number of hydrophobic solvents (oils), incorporating considerable amounts of water. The water pools that are formed can have radii from 15 to >100 T, allowing the solubilization of hydrophilic biomolecules of various sizes. The structure of the AOT reverse micelles strongly depends on the amount of incorporated water. The water content can be expressed by the molar ratio w_0, with $w_0 = [H_2O]/[AOT]$, reflecting the sizes of the reverse micelles.

Many experimental studies using scattering techniques have shown the existence of empirical relations between w_0 and the radius of the reverse micelles (whole microdroplet), for example, $r(nm) = 1.5 + 0.175\ w_0$. Also, for a given w_0 value, the increase of the respective quantities of water and AOT lead to an increase of the number of micelles. It has been well established from various structural studies that the AOT reverse micelles are spheres formed by a monolayer of AOT molecules surrounding a water core. The water molecules that are located in the close vicinity of the AOT membrane are quite immobilized and possess different properties as compared to bulk water. This is because these water molecules serve to hydrate the polar heads of the surfactant.

On the contrary, the water molecules that are situated in the center of the sphere have almost the same properties as bulk water. By increasing w_0, that is, increasing the amount of water, the overall behavior of the water phase tends to that of normal water. In general, the physical characteristics of the water in the reverse micelles strongly depends on the value of w_0 and on the nature of the surfactant, and the presence of these differently organized water molecules can be categorized in three different domains in every type of reverse micellar system: *free water,* whose properties and structure become closer to those of bulk water as w_0 increases; *bound water,* whose properties are qualitatively different than those of bulk water because of hydrophilic interactions with the polar head groups of the surfactant; and *surfactant tails,* which penetrate the apolar solvent. When the amount of water is not sufficient to hydrate the surfactant molecules, there is not free water and reverse micelles have only two domains: bound water and surfactant.

Dynamics of Reverse Micelles

Reverse micelles are not rigid, impenetrable spheres; they are dynamic entities. In addition to the normal Brownian motions, a number of relaxation processes characterize the monomer, that is, the individual surfactant molecules participating in the micellar aggregate and the water molecules. In contrast to micelles formed in the aqueous solution, the reverse micelles are electrically neutral in their exterior shell, so that no Coulombic repulsions occur, and collisions are frequent. The collision of reverse micelles can be predicted from simple diffusion theory to be on the order of nanoseconds. Not all collisions result in mixing of the aqueous pool.

When micelles collide with each other, they may build a transient dimer with a communication channel, which permits the exchange of materials. Exchange of the contents of reverse micelles is believed to occur by means of a collision–fusion–fission process. The characteristic time for exchange of solubilizates between water pools in AOT reverse micelles has been found to be 10^{-4} to 10^{-6} seconds. The exchange is a function of droplet size and temperature, decreasing with increasing droplet size at a given temperature, and increasing with increasing temperature at a given w_0. The rate of exchange for reverse micelles stabilized by the surfactant alkylarylsulfonate is much slower, of the order of hours, whereas that of CTAB reverse micelles is on the order of microseconds.

pH and Ionic Strength in the Reverse Micelles

The pH of the solubilized water in reverse micelles can differ form that of the bulk water, as a consequence of the high concentration of its electric charges caused by the ionization of the polar head groups of the surfactant molecule. For example, in an anionic surfactant reverse micellar system, a decrease the apparent pH of the water within the reverse micelles is observed, and this difference is larger for low w_0. In studies concerning the role of salts, when salts are present in the aqueous pool of the micelles, it has been found that the ionic strength, the type of salt, and the type of buffer used can affect the size as well as the shape of the reversed micelles. These two factors, pH and ionic strength, are of particular importance for the encapsulation of enzymes inside reverse micelles in their application in biocatalysis and liquid–liquid extraction of proteins.

ROLE OF ENZYMES IN REVERSE MICELLES

Enzymes represent novel routes to a number of important products, ranging from β-blockers to synthetic cocoa butter. Although they have outstanding catalytic properties, including regioselectivity, stereoselectivity, and enantioselectivity, which can be applied for the preparation of compounds not available by chemical means, their use in organic synthesis is quite limited. The main reason is that most organic compounds of commercial interest are very sparingly soluble in aqueous media used in conventional enzymology. Water is a poor solvent for nearly all applications in industrial chemistry. Biocatalysts—enzymes, microbial cells, and others—are traditionally applied to aqueous reaction systems, including water-soluble reactants. Since the recent development of biotechnology, much wider applications of biocatalysts have been demanded, for example, bioconversions of biological and xenobiotic compounds having lipophilic or hydrophobic characters. In such bioconversion reaction systems, it is indispensable to introduce organic solvents into the system to improve the poor solubility of the reactants in water. Furthermore, in an attempt to apply biocatalysts for the conversion of petrochemicals, the substrates and products are themselves organic solvents in such biochemical reactions. The recent realization that enzymes can function also in organic solvents has given the synthetic potential of enzymes a powerful boost for many processes.

Other uncommon biocatalysis media are supercritical gases such as $CO_2(CO_2SF)$, liquid membranes, and reverse micelles. The activity and specificity of the enzymes in these uncommon environments are changed. The enantiomeric excesses are normally increased. Several substrates (e.g, lipophilic ones) are better soluble in these media and downstream processing is often easier. Various enzymes (such as alcohol dehydrogenase and esterases, to mention some of them), catalyzed asymmetric redox reactions, stereospecific trans-esterification, regioselective esterification of sugars, and others have been studied in nonaqueous systems.

Use of Enzymes

The use of reverse micelles to solubilize enzymes in organic solvents has attracted considerable interest in the past decade. A large variety of enzymes, including cofactor-dependent enzyme, have been successfully encapsulated in reverse micellar systems, retaining their catalytic activity. The enzyme molecules can be entrapped in the reverse micelles, avoiding direct contact with the organic medium, potentially limiting their denaturation. The interior of the reverse micelles acts like a microreactor that provides a favorable aqueous microenvironment for enzyme activity as well as an enormous interfacial area through which the conversion of hydrophobic substrates can be catalyzed. Increasing the interfacial area is of great technological interest since this results in the increase of the number of substrate molecules available to react. When biomolecules such as enzymes are solubilized in reverse micelles, they can be localized in the different microcompartments, depending on their chemical structure or relative solubility in these microphases. The degree of hydrophobicity of the biomolecule plays an important role to the preferential localization among the various micro-environments (water core, bound water, surfactant membrane, and organic solvent).

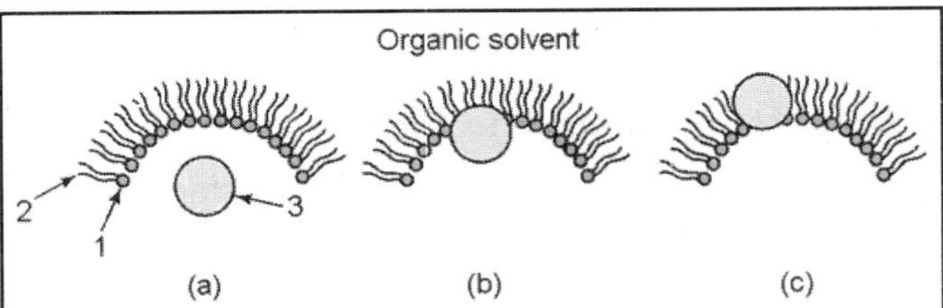

Figure 14.4 Enzyme localization in reverse micelles depending on the degree of hydrophobicity of the molecule. (a) Hydrophilic; (b) Amphiphilic, and (c) Hydrophobic enzyme. 1, hydrophilic head of the surfactant; 2, hydrophobic tail of the surfactant; and 3, enzyme molecule.

Hydrophilic enzymes are entrapped in the water core, whereas others with an amphiphilic character can be anchored to the surfactant layer, penetrating even in the continuous organic phase. Recent studies relate the enzyme localization in the different domains of the reverse micelles with their selectivity for various substrates with a different degree of polarity. This hypothesis at least stands for lipase-catalyzed esterification reactions. In addition, this interaction between the enzyme molecule and the interface can explain the different enzyme behavior found in several cases with respect to bulk water.

Encapsulation of enzymes

The most common technique used for the encapsulation of enzymes in the polar core of reverse micelles is the injection method. The protein already solubilized in a concentrated stock aqueous (or buffered) solution is added to the surfactant-containing organic solution upon stirring. With this method, the solubilization is almost instantaneous, and it is possible to control the amount of water present in the system. Another method is to contact the protein solution with the micellar solution upon stirring. Phase transfer will occur, depending on the charge of distribution of the enzyme, surfactant charge, and pH and ionic strength of the aqueous phase. With this method, the enzyme solubilization is slow, but a large amount can be solubilized with minimal values of w_0. A third way of introducing the enzyme is to add the dry enzyme into the empty reverse micelles while aggitating. This technique causes denaturation of the enzyme to a degree. The injection and dry addition methods are commonly used in biocatalytic applications, the latter being suited to hydrophobic proteins. The phase-transfer technique is the basis for extraction of proteins from aqueous solutions. In addition to the method of protein inclusion to the reverse-micelles, other variables that can influence protein solubilization are the size and isoelectric point of the protein, the size of the reverse micelles, the nature of the surfactant used, and the pH and ionic strength of the aqueous solution.

Factors affecting enzyme solubilization

For both the injection and phase contact methods, proteins with a net charge opposite to the surfactant charge and a low ionic strength will favor solubilization. As far as the size of protein molecule is concerned, their solubilization may occur even when the radius of the molecule is larger than the micellar polar core radius. The protein itself probably stabilizes the enlarged micelle, constructing a protective surfactant shell with a certain amount of water. Data support the possibility that there is not a definite maximal molecular size upon which encapsulation in surfactant aggregates does not occur. The hydrophilic character of large proteins may have a significant role on the formation of reversed micelles. It seems that the hydrophilic proteins will favor the formation of surrounding surfactant shell. This fact could account for the solubilization of large entities, even when the total charge of the macromolecules is identical to the charge of the surfactant heads, thus excluding the formation of micelles via direct electrostatic interactions between the protein and the surfactant layer. This type of interaction within a micellar system can be particularly important for the phase transfer of proteins, applied to the liquid--liquid extraction of proteins and other molecules via reverse micelles. The various parameters affecting this process are discussed in more detail in the section on liquid–liquid extraction.

Spectroscopic techniques

Enzyme-containing reverse micellar solutions are optically clear. Thus, spectroscopic techniques, such as ultraviolet (UV), fluorescence, circular dichroism (CD), EPR, and NMR, are able to be used for monitoring the conformation changes that may occur in enzyme molecules upon their solubilization in reverse micelles. Moreover, in this system it possible to precisely control the range of water microenvironments (interfacial/bulk), the size of the micelles by varying the water/surfactant molar ratios, and the role of the chemical structure of the surfactant. Generally, a more ordered structure has been identified by CD for myelin basic proteins and poly(α-aminoacids) in AOT reverse micelles and trypsin and α-chymotrypsin in lecithin reverse micelles. UV spectra of enzymes in reverse micelles show little change from aqueous solutions.

Fluorescence spectra in reverse micelles are red shifted for some enzymes and blue shifted for others. An increase in quantum yield has been observed with α-chymothrypsin, trypsin, pepsin, lysozyme, and a decrease in quantum yield has been observed for lactate dehydrogenase. ^{1}H-NMR studies of some peptides and small proteins show that met-encephalin became more folded, and pancreatic secretory trypsin inhibitor became highly flexible. Furthermore, EPR studies have been performed for investigation of structural changes to α-chymotrypsin, serum albumin, and cutinase upon solubilization in reverse micelles. The information derived by several studies shows that proteins can be categorized according to the perturbation of their structure induced by reverse micelles: conformations that are not changed or only slightly changed after incorporation into reverse micelles; conformations that are altered in reverse micelles but unaffected by the amount of water present in the system; and conformations that are altered and that also change as a function of the water/surfactant molar ratio.

Enzymes in Reverse Micelles

The first report on enzymes retaining their catalytic activity in reverse micellar systems was in 1978. Today the number of active enzymes described in reverse micelles has increased to more than 50, including hydrolases, dehydrogenases, and oxidases used in single or multienzyme catalyzed reactions. Enzymatic catalysis in microemulsions has been used for a variety of reactions, such as synthesis of esters, peptides and sugar acetals, trans-esterifications, various hydrolysis reactions, and steroid transformations.

Table 14.1 Examples of bioconversions in reversed micellar systems

Type of reaction	*Enzyme*	*Reverse micelles*
Reduction of decanal	Alcohol dehydrogenase	AOT/isooctane
Polymer synthesis	Horseradish peroxidase	AOT/isooctane
Reduction of ketosteroids	Alcohol dehydrogenase	AOT/isooctane
Oxidation of polyunsaturated fatty acids, linoleic acid	Lipoxidase	AOT/isooctane
Cholesterol oxidation	Cholesterol oxidase	CTAB/octanol/octane
Oxidation of aliphatic alcohols	Alcohol dehydrogenase	AOT/octane
Peptide synthesis	α-Chymotrypsin	AOT/isooctane
Hydrolysis of olive oil	Lipase	AOT/isooctane
Inter- and *trans*-esterification	Lipase	AOT/isooctane
Bioluminescent assay	Luciferase	Brij-96/octane
Fatty acid esterification with alcohols	Lipase	AOT/isooctane
Polyol–fatty acid esters	Lipase	AOT/isooctane

Enzymatic reactions in reverse micelles are found to obey normal Michaelis-Menten kinetics, but some attention should be given to the way of expressing the kinetic constant K_m. K_m is the substrate concentration where the reaction rate is half the maximum rate. The substrate concentration in a reversed micellar system can be based on the total volume of the system or the water-pool volume of the micelles, but the latter is very difficult to measure. Other factors that have to be taken under consideration is the compartmentalization of the enzyme and substrates between the various microdomains of the reverse micelles and the degree of hydrophobicity of substrate, product, and enzyme molecule, which is related to its localization in the micelles. Several theoretical models have been developed to explain enzyme kinetics in reverse micelles. They are based on different assumptions, such as all droplets in the medium are equal, and there are not any changes in the enzymes kinetic parameters (K_m and V_{max}) during their entrapment in the reverse micelles.

The nondiffusional models are based on the fact that the differences in the kinetic constants with respect to bulk water are caused by conformational changes produced in the reverse micellar microenvironment. Another important parameter obtained from Michaelis-Menten kinetics is the K_{cat}, the turnover number, which is related to maximum reaction velocity. Because K_{cat} is concentration independent, its determination is more straightforward. Turnover numbers for enzymes in reverse micelles are usually dependent on the amount of water present in the system, which is expressed by the molar water-to-surfactant ratio, *w*0. This dependence is often a bell-shaped curve, explained by the existence of an optimal inner cavity for catalytic activity, which corresponds to the size of the enzyme. Maximum K_{cat} is sometimes higher than that in aqueous solutions, giving rise to the concept of superactivity. This effect can be explained by a greater reactivity of the structured water in the micelle, increased rigidity of the enzyme, and enhanced substrate concentration at the enzyme surface. Thus, the first parameter that should be considered when studying the enzyme activity in reverse micelles is the amount of the water in the system (w_0).

The electricproperties of the substrate and surfactant can often have a marked effect on the K_m and K_{cat} constants. When the charges of substrate and surfactant are the same, the differences of K_{cat} and K_m values with those found in water are not significant, but when the charges are different, partition effects are usually observed because of the increase of the substrate concentration near the vicinity of the surfactant layer, hampering their access to the enzyme, which is surrounded by water. α-Chymotrypsin and trypsin encapsulated in AOT and CTAB show a decrease in the second-order rate constant (K_{cat}/K_m). The specificity of the enzymes can also be altered because of the distribution of the substrate between the system phases or interactions with the surfactant layer. An example is the oxidation of aliphatic primary alcohols, catalyzed by alcohol dehydrogenase encapsulated in an AOT/octane system.

In the reverse micellar system, octanol ($n = 8$) is the optimum substrate, whereas in water the preferred substrate is butanol ($n = 4$). It is profound that these parameters are more accurately determined using the active enzyme concentration, which can be obtained typically by active site titration, for example, based on fluorescence techniques for the active site titration of α-chymotrypsin and lactate dehydrogenase. Another parameter to take into account is the pH. When the enzyme is entrapped in reverse micelles, significant shifts of the pH profile usually occur, because of the heterogeneous distribution of protons inside the polar core of the micelle. Larger shifts of optimal pH occur for small values of w_0. The shifts reported in literature for systems with anionic surfactants are of one to two unities.

Fat Splitting Lipases

A particular case of enzymatic studies in microemulsions is that of lipases. Among the enzymes so far investigated, lipase is one of the most advantageous because it is stable, inexpensive, and widely used in the development of various applications in the detergents, oils and fats, dairy, and pharmaceutical industries. In addition to hydrolysis of triglycerides to glycerol and free fatty acids, lipases can be used in esterification and transesterification reactions in low-water content media. This catalytic process is heterogeneous and can be favored by the use of microemulsion. Lipases are active almost exclusively near interfaces in a classical heterogeneous procedure. There are various reports in the literature concerning studies on the hydrolytic and synthetic action of lipases in different types of microemulsions.

Table 14.2 Lipase-catalyzed reactions in microemulsions

Lipase source	*Microemulsion system*
	Hydrolytic reactions
Pancreas	AOT/octane
R. arrhizus	AOT/octane
R. delemar	AOT/isooctane
	CTAB/isoctane
	$C_{12}E_4$/decane
	Lecithin/isooctane
Rhizopus sp.	AOT/isooctane
	$C_{12}E_3$/isooctane
Candida rugosa	AOT/isooctane
Chrom obacterium viscosum	AOT/isooctane
Candida cylinracea	Lecithin/cyclooctane
	Synthetic reactions
Rhizopus delemar	Phosphatidylcholine/hexane
	$C_{12}E_3$/nonane
	$C_{12}E_4$/decane
	AOT/isooctane
	AOT/hexane
Rhizopus arrhizus	AOT/isooctane
Chrom obacterium viscosum	AOT/heptane
Candida cylindracea	Brij-35/olive oil
	AOT/isooctane
Penicillium simplicissimum	AOT/isooctane

Hydrolytic reactions

Lipase catalyzes the hydrolysis of long-chain aliphatic esters to glycerol and free fatty acids. Several hydrolytic reactions performed by lipases in microemulsion systems have been reported, such as the hydrolysis of triolein and tributyrin by pancreatic lipase in an AOT/octane microemulsion system. The same system has been used for the hydrolysis of palm kernel olein by *Rhizopus arrhizus* lipase. Hydrolysis of triolein and tributyrin has been reported with the use of *Rhizopus delemar* lipase in AOT/isooctane, CTAB/isooctane, and $C_{12}E_4$/decane systems. Olive oil hydrolysis by *Candida rugosa* lipase and glycerolysis of triolein by *Chromobacterium viscosum* lipase in AOT microemulsion systems have been performed. A bioreactor system has been designed that is suitable for continuous glycerolysis of fats and oils, using liposomes as a matrix for adsorbing enzyme and AOT reverse micellar systems, and as a supplier of interfacial area for glycerolysis. Palm oil and trimyristine hydrolysis by *Rhizopus* sp. lipases for the production of monoglycerides and diglycerides with a good yield has been also reported in AOT and $C_{12}E_3$ based systems. Several parameters play an important role in lipase-catalyzed hydrolytic reactions in microemulsion systems.

The reverse micellar structural characteristics, such as the size, shape, and solubilizing ability, which depend strongly on the nature of the polar moiety of the amphiphile used, strongly affect the enzyme activity. Anionic, cationic, and nonionic types of surfactants have been used in various hydrolytic reactions. Among the different systems studied, the anionic surfactants seem to be more suitable for the hydrolytic activity of lipase, as compared to the cationic ones, whereas nonionic systems do not seem adequate for this type of reaction. Other important parameters affecting lipase activity that have been studied in detail by several groups are pH, temperature, water content (w_0), fatty acid chain length of the triglycerides used as substrates, and enzyme stability. In studies concerning the hydrolysis of triolein by *Rhizopus delemar* lipase, the pH optima were reported to be 6.5 in the anionic systems and 5.8 in the cationic systems, considering as pH values the values of the buffered enzyme solutions used for the preparation of the various micro- emulsion systems.

The pH optimum for the same reaction performed in classic heterogeneous biphasic systems was found to be 5.6. This difference is obviously related to the fact that the pH values within the water-dispersed phase do not necessarily correspond to that in the bulk water. The difference observed between AOT and CTAB systems is possibly the result of the different amounts of bound water and the nature of the ions present in the water pools of each system. In the case of AOT systems, it has been reported that 8 to 10 water molecules are indispensable for hydrating each Na^+ and sulphonate group, whereas for CTAB, only 1 to 3 water molecules per Br^- and ammonium head are enough. In other studies on AOT-formulated microemulsion systems, a pH optimum of 7.1 was reported for lipase from *Candida rugosa* 7.0 for lipase from *Rhizopus* sp.; and a maximum in the range 6.0–7.5 for lipase from *Chromobacterium viscosum*.

Temperature also plays an important role in lipase activity in microemulsions; in the case of anionic systems, the optimum temperature is reported to be 30°C; in cationic systems, 22.5°C. However, at temperatures lower than 15°C and higher than 40°C, the structure of these systems may be considerably altered. Water content in microemulsions appears to be the most important factor that determines the reaction rates. The molar ratio w_0 (w_0 = $[H_2O]/[\text{surfactant}]$) values are used to express the water content of the system. Several reports concerning the activity of lipase from various sources as a function of $w0$ shows a maximum in the range of 9 to 11. The same pattern is also observed for other enzymes studied in the same microemulsion systems, such as chymotrypsin and pyrophosphatase.

However, when reverse micelles based on lecithin were used, the maximum reaction rate was observed at an extremely low water content (w_0 = 2.2). In studies concerning the effect of the fatty acid chain length as well as the type of microemulsion system on the substrate conversion rate, it was found that for microbial origin lipases, triolein was a slightly better substrate as compared to tributyrin, but pancreatic lipase cleaves triolein 15 times more efficiently than tributyrin. Also, the enzyme in anionic microemulsion systems shows a higher affinity as compared to the cationic systems, whereas nonionic systems do not seem adequate for hydrolytic reactions. One serious problem for the application of lipase to triglyceride hydrolysis is the stability of the enzyme in the microemulsion systems. The size of the microemulsion microdroplet seems to play an important role on the enzyme stability. It has been reported that lipase stability is improved as w_0 decreases. Also, the presence of various additives, such as $CaCl_2$ and casein, considerably increases the stability of lipase in AOT microemulsion systems.

Synthesis and transestrification

The hydrolytic reaction catalyzed by lipase is reversible because the enzyme can also catalyze the synthesis and transesterification of triglycerides, under certain conditions. This catalytic process is of great technological interest and may be favored by the use of microemulsions. The lipophilic substrates and products can be solubilized in the continuous organic phase, whereas lipase remains active in the water core of the microemulsion droplets. In this case, specific synthetic reactions can be catalyzed, under mild conditions. The synthesis of triacylglycerol by *Rhizopus delemar* lipase has been reported in a phosphatidylcholine reverse micellar system. The esterification of glycerol with oleic acid, catalyzed by *Chromobacterium viscosum* lipase in glycerol-in-oil–type microemulsions of AOT in heptane, and the synthesis of various fatty acid esters in water-in-oil microemulsions, stabilized by CTAB, have also been reported. Similar systems have been used for the transesterification of palm oil with stearic acid.

Synthesis of heptyloleate and the esterification of various triglycerides by *Candida cylindracea* lipase in a microemulsion-like system has been reported using Brij-35 as surfactant. The esterification of various alcohols with fatty acids by *Rhizopus delemar* lipase has been performed using nonionic microemulsion systems stabilized by $C_{12}E_4$, with significantly fast reaction rates also reported. In this work, it was shown that pH ranging from 3.45 to 8.55 does not affect the reaction rates in this type of microemulsion system. Also, studying the effect of the system's water content showed that when $w_0 < 5$ the conversion is considerably slow, but for $w_0 > 7$ a 90% conversion is attained in 4 h. Esterification reactions of alcohols with fatty acids and synthesis of polyols have been reported using *Rhizopus delemar* and *Candida cylindracea* lipase in AOT microemulsion systems.

Recent studies have been reported on the activity of lipases from *Rhizopus delemar, Rhizopus arrhizus,* and *Penicillium simplicissimum* in esterification reactions of various aliphatic alcohols with fatty acids in microemulsions formulated by AOT in isooctane. The results showed a remarkable selectivity of the lipases regarding the chain length and the structure of the substrates. *P.*

simplicissimum lipase presented higher reaction rates in the esterification of long-chain alcohols as well as secondary alcohols. Primary alcohols had a low reaction rate, and tertiary alcohols had a very slow rate of esterification.

Long-chain fatty acids were better catalyzed as compared to the shorter ones. *R. delemar* and *R. arrhizus* lipases showed a preference for the esterification of short-chain primary alcohols, but the secondary alcohols had a low rate of esterification and the tertiary could not be catalyzed. Medium-chain fatty acids were also better catalyzed than the long ones.

Fluorescence quenching measurements suggest that there are not any significant changes of the shape and size of the microemulsions caused by the substrates, which could have been related to the enzyme selectivity. In extensive studies using a large number of substrates with different degrees of polarity as well as spectroscopic techniques, this selectivity appears to be related to the localization of the enzyme molecule within the micellar microstructure, because of the hydrophobic/hydrophilic character of the protein and not of a specificity of the enzyme itself. This hypothesis can further be supported if one takes into consideration the structural differences of these lipases.

Michaelis-Menten kinetics

Lipase-catalyzed hydrolysis of triglycerides in reverse micelles follows Michaelis-Menten kinetics. The release of fatty acids as a function of reaction time at different enzyme concentrations and fixed substrate concentrations (20–40%, v/v, where $[S] \gg K_m$) is proportional to the enzyme concentration. The K_m and V_{max} parameters of the Michaelis-Menten kinetic equation can be determined from the Lineweaver-Burk double-reciprocal plot of the lipase activity only as apparent ones. To find a true K_m, the substrate concentration at the enzyme surface must be used. This requires knowledge of the substrate partitioning between various microphases, as the water pool, the interface, and the bulk organic solvent. The partition coefficient is difficult to be determined experimentally. Another constant obtained from Michaelis-Menten kinetics is the turnover number, K_{cat}, which is related to the maximum reaction velocity.

Determination of K_{cat} depends on accurate measurement of active-enzyme concentration in the microemulsion system. Usually, the total enzyme concentration and not the active-enzyme concentration is used for the calculation of K_{cat}. The latter can be achieved by active site titration studies. Kinetic studies of the esterification reaction on the reaction of ethylene glycol and fatty acids and of lauric acid and (–)menthol, catalyzed by lipase in AOT/isooctane/water microemulsion systems, have been described. In the second case, the authors have shown that the kinetics of this synthesis follow a ping-pong Bi-Bi mechanism. According to this mechanism, lipase reacts with lauric acid to form the lipase–lauric acid complex.

The complex is then transformed to a carboxylic–lipase intermediate and water is released. This is followed by an attack of (–)menthol to this intermediate, forming the (–)menthyl laureate. The values of all apparent kinetic parameters were determined, and a model concerning this type of reaction involving two hydrophobic substrates was proposed. According to this, the partitioning of the two hydrophobic substrates between the continuous organic phase and the reverse micelles must be taken into account to determine the real kinetic constants.

Significance of Reverse Micelles

The potential applications of lipases in biotechnological processes have been the subject of increased scientific, economic, and industrial interest and efforts in recent years. Regioselectivity, stereospecificity, substrate specificity, and low energy consumption are some of the features that make lipase-catalyzed processes an alternative to conventional nonenzymatic reactions. Special attention is given to the transformation of fats and oils as well as the preparation of synthetically useful enantiomers. Among the systems under investigation, the encapsulation of lipases in reverse micelles seems to present several advantages such as greater interfacial area, lower mass transfer limitations, and enhancement of catalytic activity. The development of a bioreactor that enables reactions in a semibatch or continuous mode allowing the separation of substrates, products, and enzymes is a critical demand in reversed micellar technology today.

Recently, two types of reactors have been developed for the glycerolysis of olive oil by lipases. In the first case, the bioreactor, abbreviated as LRM system, uses liposomes for adsorbing the enzyme and reversed micelles as a good supplier of interfacial area for the glycerolysis reaction. According to this system, the lipase was adsorbed on the liposomes prepared in a simple manner, and reverse micelles containing the immobilized enzyme were formed by injecting appropriate amounts of glycerol and predetermined amounts of water in AOT/isooctane system. Olive oil in isooctane is fed at controlled flow rates into the bioreactor by pressurizing the reservoir with nitrogen gas. The reaction takes place at 37°C. The products are obtained from the outlet in a single phase, but the microemulsions are not broken. A microporous membrane is used to separate products formed (oleic acid, monoolein, and diolein) and unreacted substrate (triolein) from the microdroplets containing immobilized lipase. The membrane, made of polysulfone, is about 12.5 μm thick, and the micropores have a maximum diameter of 25 nm. This system presents a long operational stability. The second bioreactor system is a membrane bioreactor for simultaneous lipolysis and bioproducts separation in a reversed micellar system.

The main purpose of this type of bioreactor is to retain lipase behind the membrane where reaction occurs and to recover the products on the permeate side. The membrane is a ceramic one, resistant to organic solvents, made of a zirconium oxide layer over a porous carbon support with a 10,000 molecular weight cutoff. The inside diameter is 6 mm; the wall thickness, 2 mm; and the effective length, 20 cm. The internal permeation area is 38 cm^2. The reaction mixture is recirculated from the thermostated vessel into the reactor by means of a gear pump flowing tangentially to the membrane. Part of the solution is permeated through the wall of the semipermeable membrane as a result of the hydrostatic pressure difference, and it is recycled to the reactor. All connecting piping is solvent resistant. Enzymatic catalysis in reverse micelles has also been performed mostly in batch-type reactors in various other processes. Some examples are the synthesis of peptides by α-chymotrypsin in reverse micelles in a hollow fiber reactor, or the production of tryptophan by tryptophanase in a continuous reversed micellar reactor.

Enzymatic Liquid–Liquid Extraction

The application of reverse micelles in the extraction and isolation of enzymes is a promising technique at the downstream biotechnological processes, having interesting advantages over existing processes for large-scale recovery of extracellular enzymes. The traditional process of transfer proteins into organic solvents often causes irreversible denaturation of proteins and loss of biological function. A reverse micellar extraction combines the potential for concentration and purification of the enzymes through transfer between aqueous and organic micellar phases. Enzymes can be extracted without loss of activity. The recovery of enzymes from the aqueous phase involves two steps: forward and back extraction. In the first step, the desired enzyme is transferred selectively from the aqueous phase to the reversed micellar phase, and in the second step, the enzyme is subsequently recovered from the reversed micellar phase by extraction with a second aqueous phase, the micellar phase being recovered for further extractions.

Table 14.3 Examples of enzyme extraction using reversed micelles

Enzyme	*Reversed micellar system*
Trypsin	AOT/isooctane, octane, heptane
Extracellular alkaline protease	AOT/isooctane
α-Amylase	TOMAC/isooctane
Separation of a mixture of cytochrome *c*, ribonuclease, lysozyme	AOT/isooctane
Lysozyme, α-chymotrypsin, pepsin	AOT/isooctane
Isocitrate dehydrogenase, β-hydroxybutyrate dehydrogenase, glucose-6-phosphate dehydrogenase	CTAB/octane/exanol
α-Chymotrypsin	NR_4^+/cycloexane AOT/octane/water/glycerol
Ribonuclease A	AOT/isooctane

The rates of forward transfer from the aqueous to the micellar phase are extremely fast. Upon steering, for example, only a few seconds are necessary for the transfer of cytochrome *c* into a system of AOT/isooctane. The process of back extraction into the new aqueous phase is much slower and requires constant agitation. In ideal conditions, the back extraction of cytochrome *c* can be achieved in several minutes. The distribution of an enzyme between the reversed micellar and the aqueous phase is influenced by several factors such as pH, ionic strength, and surfactant type (anionic, cationic, or nonionic). The pH determines the ionization state of the surface-charged groups of the protein molecule. The solubilization is favored when attractive electrostatic interactions occur between the protein molecule and the surfactant head groups; in this case, the overall charge of the surfactant is opposite to the overall charge of the protein. This implies that, for cationic surfactants, solubilization of the protein in reverse micelles is favored at pH values above the isoelectric point (pI) of the protein, whereas the opposite is true for anionic surfactants. Together with the pH, the ionic strength plays a very important role in protein recovery. The presence of salts in the aqueous phase can alter the equilibrium size of the surfactant aggregates.

A higher ionic strength will decrease the size of the micelle, and this effect can be stronger for larger ions, which causes shorter electrostatic interactions. Thus, it is possible to achieve a selective transfer based on size exclusion effects. The attraction between the surfactant and the protein is weaker, and this contributes to exclude the protein from the micellar core. When the ionic strength is very low, phase transfer does not occur. The transfer of proteins from an aqueous to a micellar phase requires a minimum of ionic strength in the aqueous phase. In addition to the surfactant charge, other surfactant-dependent parameters have to be taken under consideration, such as the size of the reversed micelles that are formed, the energy required to enlarge the reverse micelles, and the charge density on the inner surface of the reverse micelles. The liquid–liquid extraction technique in general is well known, and apparatus and scale-up rules have been established for numerous applications, including the use in recovery processes for antibiotics and organic acids from fermentation broths.

The forward and back extractions can be performed in a continuous mode, with the reverse micellar phase circulating between the two extraction units. No specialized equipment is required. The process developed for conventional liquid–liquid extraction is, in principle, also suitable for this application. The most important types of equipment are mixer/settler, agitated column, centrifugal, and membrane extractors. At the present time, process development has centered on the use of mixer/settlers and membrane extractors.

CONCLUDING REMARKS

So far, the results derived from studies with reversed micelles show that the use of these systems can reach a considerable biotechnological relevance and be extended to a wide range of applications. In the field of bioconversion, the enzyme solubilization in reversed micelles clearly offers advantages over other techniques, including the very fast reaction rates or the small quantity of enzyme to be used in comparison to other approaches of enzymology in extreme conditions. However, there are a few problems that must be solved in order to take full advantage of this system, such as the need to remove the excess surfactant from the products and to develop appropriate reactors, which will allow the continuous production and in situ recovery of the bioproducts applying new separation techniques, for example, membrane polymers. Another promising field for reverse micellar systems application is in the downstream processing for the separation of other bioproducts besides proteins and peptides, such as amino acids, nucleotides, nucleic acids, solvents, organic acids, antibiotics and steroids.

INDUSTRIAL PROTEASE

Industrial enzymology is an important branch of bioprocess technology. Enzymatic processes permit natural raw materials to be upgraded and finished. Using enzymes as a functional ingredient in detergents, laundry and dishes, for example, are cleaned in an efficient, environmentally mild, and energy-saving way. The most used industrial enzymes are *proteases* for hydrolysis reactions within the detergent, dairy, protein, pet food, feed, and leather tanning industries.

WORK DONE

In detergents proteases are the most widely used enzymes. In laundry detergents, protein stains such as grass, blood, egg, and human sweat are removed through proteolysis. In automatic dish-washing detergents, proteases secure the removal of proteinaceous food films, which are a particular problem with glassware and cutlery. Enzymes offer alternative ways of making products previously made using conventional chemical processes. Proteases, in particular, offer attractive ways of performing hydrolysis reactions needed for most tasks within the mentioned industries. Historically, proteases were used in ancient Greece for the production of cheese, as referenced in Greek epic poems dating from about 800 B.C. The development of the submerged fermentation technique has resulted in tremendous progress in the field of industrial enzymology. Due to their efficiency, specific action, ability to work under mild conditions, high purification, and standardization, proteases are ideal catalysts for the food industry. Simple equipment can be used, and because moderate temperature and pH values are required for the reactions, only a few by-products influencing taste and color are formed. Furthermore, reactions using pro- teases are easily controlled and can be stopped when the desired degree of conversion is reached.

Consumption of Proteases

Worldwide consumption of proteases amounted to approximately $600 million in 1996. The total growth (including nonfood applications) in volume of the enzyme business from 1974 to 1986 was

10–15% per year, although no major new applications emerged in this period. From 1980 to 1990 the growth was estimated as 5–10% per year, and from 1992 to 1996 as 8–10%. The principal producers of industrial proteases are Novo Nordisk A/S and Genencor International, Inc. These companies provide more than 90% of the total volume of enzyme products to the detergent enzyme market.

Characteristics of Proteases

In most countries, regulatory status classification, and labeling of enzymes are determined in accordance with existing schemes for chemicals. Many enzyme types are listed on chemical inventories, for example, European Inventory of Existing Commercial Chemical Substances (EINECS) in the European Union and Toxic Substances Control Act (TSCA) in the United States. In some cases, enzymes are considered natural substances exempt from listing. In other cases enzymes are regulated by specific biotechnology products legislation. The Association of Manufacturers of Fermentation Enzyme Products (AMFEP) has defined *good manufacturing practice* (GMP) for microbial food enzymes. This practice is generally followed for detergent enzymes as well. The most important element is to ensure a pure culture of the production organism. When an enzyme is used for a nonfood and non-feed industrial technical application, its regulatory status is determined by its properties as a naturally occurring substance. These properties determine the classification and consequent labeling in accordance with existing schemes for chemicals.

Applications of Food Enzymes

The application of enzymes in food processing is governed by food laws. Within the European Union large parts of food laws have been harmonized by directives and regulations. For general purposes, the Joint FAO/WHO Expert Committee on Food Additives (JECFA) and the Food Chemicals Codex (FCC) have made guidelines available for the application of enzymes as food additives. AMFEP and the Enzyme Technical Association (ETA) work nationally, as well as internationally, for harmonization of regulations with regard to enzymes. AMFEP members ensure that the enzymes used in food processing are obtained from nonpathogenic and nontoxicogenic microorganisms, that is, microorganisms that have clean safety records, with no reported cases of pathogenicity or toxicosis attributed to the species in question. When the production strain contains recombinant DNA, the characteristics and safety record of each donor organism contributing genetic information to the production strain are assessed.

The majority of food enzymes are used as processing aids and have no function in the final food. In this case they need not be declared on the label, because they will not be present in the final food in any significant amount. Some enzymes are used as both processing aids and food additives. When used as additives, they must be declared on the food label. GMP is used in the microbial food enzyme industry. The key issues in GMP are the microbial control of the microorganism selected for enzyme production, the control and monitoring systems ensuring pure culture and optimum enzyme productivity conditions during fermentation, and control of the hygienic conditions throughout recovery and finishing of the enzyme preparation. Commercial enzyme products are usually formulated in aqueous solutions or processed to nondusty dry products (granulates or microgranulates). Both types of preparations must be formulated to be suitable for the final application. Requirements mainly regarding the storage stability (e.g., enzyme activity stability, microbial stability and physical stability) and formulations as significant quality parameters of industrial enzyme products must be taken into consideration by both the producer and the user.

Threshold Limit Values

Like many other proteins foreign to the human body, enzymes are potential inhalation allergens. Inhalation of even small concentrations of a foreign protein in the form of dust or aerosol can stimulate the body's immune system to produce antibodies. In some individuals, increased concentrations of antibody–enzyme protein complexes can trigger increased concentrations of histamine. The latter compounds can cause hay fever–like symptoms, such as watery eyes, runny nose, and sore throat. When exposure ceases, these effects cease also. Enzymes must be inhaled to present a risk of causing sensitization, which may lead to an allergic reaction. Working environments in which enzymes are used are therefore subjected to extensive monitoring to confirm that *threshold limit values* (TLV) for atmospheric enzymes are not exceeded. In many countries, the TLV for enzymes is based on the proteolytic enzyme subtilisin, and is stated as 0.00006 mg/ m^3 of pure crystalline subtilisin in air.

Successful experience in controlling enzyme exposure in detergent manufacturing and in protecting workers' health has been reported by Procter & Gamble. As early as 1971, a *National Research Council* (NRC) report concluded that consumers of enzyme-containing laundry products did not develop respiratory allergies. Further studies over the years have confirmed that enzyme-containing laundry and dishwashing detergents are safe for consumer use. Use of proteases as aids in food processing virtually never represent safety problems for consumers. Because liquid enzyme preparations are normally used in processing plants in very small concentrations, dust- or aerosol-containing enzymes are usually not released from process equipment based on closed systems.

DEVELOPMENT OF NEW PROTEASES

It is difficult to tell which applications spurred the development of *new proteases* in the early days. Besides for production of foods such as cheese, beer, and shoyu (soy sauce), proteases have been developed for the tanning industry, and in the past the detergent industry has been the driving force in development. Scientific interest in proteolytic action on different food proteins has resulted in the development of new proteases for both extraction processes and production of functional ingredients for the food industry. By far the largest tonnage production of protease is based on microbial sources. Recombinant DNA methods and protein engineering are today's means to develop important proteases for both food and nonfood uses. Proteases are classified according to their source of origin (animal, plant, microbial), their catalytic action (*endopeptidase* or *exopeptidase*), and the nature of the catalytic site (*active site*).

They are characterized by common names and trade names, typical pH range, and preferential specificity. Based on a comparison of active sites, catalytic residues, and three-dimensional structures, four major protease families are recognized: serine, thiol, aspartic, and metalloprotease.

The serine protease family contains two subgroups: the chymotrypsin-like and the subtilisin-like proteases—the most important industrial products. Many commercial proteases are mixtures of the different types. This is especially the case for pancreatin, papain (crude), and some proteases from *Bacillus amyloliquefaciens, Aspergillus oryzae, Streptomyces,* and *Penicillium duponti,* mainly used within the food area. The enzymes *Alcalase, Esperase, Savinase,* and *Trypsin* are serine proteases, while the enzyme *Neutrase* is a metalloprotease having Zn^{++} in its active site. Furthermore, this enzyme is stabilized by Ca^{++}.

Durazym and *Everlase* are protein-engineered variants of *Savinase*. The milk-clotting enzyme product *Rennilase* or *Fromase* is an aspartic protease. Examples of thiol (or cysteine) proteases are plant proteases from papaya latex (papain), from pineapple stem (bromelin), and from fig latex (ficin). Practical selections of proteases are made on the basis of performance and stability tests carried out under certain standard conditions chosen in conjunction with detergent formulators.

Table 15.1 Merchant market detergent proteases

Producer	*Alkaline*	*High alkaline*	*High bleach stability*	*Cold water*
Genencor International	Maxatase	Purafect	Maxapem	Properase
	Optimase	Maxacal	Purafect OxP	
		Opticlean	Opticlean plus	
Novo Nordisk A/S	Alcalase	Esperase	Durazym	New developments
		Savinase	Everlase	

Utility of Protein Hydrolysis

Proteases catalyze the hydrolytic degradation of the peptide chain, as shown for reactants and products in Figure 15.1. In aqueous solutions and suspensions of protein the equilibrium lies so far to the right that degradation rather than synthesis of larger molecules is thermodynamically favored.

$$-\underset{R_1}{\overset{H}{\underset{|}{\overset{|}{C}}}}-\overset{O}{\overset{\|}{C}}-N-\underset{H}{\overset{R_2}{\underset{|}{\overset{|}{C}}}}- + H_2O \;\underset{}{\overset{\text{Protease}}{\rightleftharpoons}}\; -\underset{R_1}{\overset{H}{\underset{|}{\overset{|}{C}}}}-COO^- + {}^+H_3N-\underset{H}{\overset{R_2}{\underset{|}{\overset{|}{C}}}}-$$

Figure 15.1 The hydrolysis reaction: Enzymatic cleavage of a peptide bond.

When a protease acts on a protein substrate a catalytic reaction actually consists of three consecutive reactions. This reaction mechanism is simplified by scheme shown in Figure 15.2.

$$E + S \underset{k_{-1}}{\overset{k_{+1}}{\rightleftharpoons}} ES \xrightarrow{k_{+2}} EP + H-P' \xrightarrow[+H_2O]{k_{+3}} E + P-OH + H-P'$$

Figure 15.2 A catalytic mechanism of proteolysis: E, enzyme, S, substrate, P and P', resulting peptide, k, reaction velocity constant.

1. Formation of the Michaelis complex between the original peptide chain (the substrate) and the enzyme
2. Cleavage of the peptide bond to titrate one of the two resulting peptides
3. A nucleophilic attack on the remains of the complex to split off the other peptide and to reconstitute the free enzyme

The rate-determining step is the acylation step characterized by the reaction velocity constant k_{+2}. The ratio k_{-1}/k_{+1} is equal to the Michaelis constant k_w for serine proteases. In other words classical Michaelis–Menten kinetics are valid for protein hydrolysis. In detergent applications the substrate and enzyme concentrations are very small. Strangely enough, the *E/S*% ratio is of the same order of magnitude.

Degree of Hydrolysis

The technical discipline of enzymatic hydrolysis of food proteins has been developed over the past 20 years to become a significant area of modern food processing. The number of literature references has grown enormously. Descriptions of the principles, methods of controlling the hydrolysis reaction, ways of carrying out industrial processing, and characterization of functional properties of proteins treated with proteases have been intensively researched and developed at NovoNordisk A/S. A great deal of this work was introduced by Adler-Nissen. The most important parameters of hydrolysis reactions are *S* (percent protein in the reaction mixture), *E/S* (enzyme–substrate ratio in activity units per kilogram protein), pH, and temperature. Together with the specificity and properties of the enzyme itself, these parameters are responsible for the course of reaction on a given protein raw material.

The quantitative criterion of a proteolytic reaction is the degree of hydrolysis, which is calculated from determination of the number of peptide bonds cleaved and the total number of peptide bonds in the intact protein. As a quantitative measure for the hydrolytic reaction, hydrolysis equivalents are applied, termed *h*. The unity of *h* is equivalent peptide bonds cleaved per kilogram protein. The term *h* is used for kinetic investigations of protein hydrolysis, but in other usages the derivative quantity *DH* (degree of hydrolysis) is preferred. *DH* is calculated from the following equation:

$$\mathrm{DH} = \frac{\mathrm{h}}{\mathrm{h_{tot}}} \times 100\%$$

h_{tot} is estimated on the basis of the amino acid composition of the protein. For most food proteins, the average molecular weight of amino acids is about 125 g/mol, making h_{tot} about 8 g equiv/kg protein (calculated as 6.25 x *N*). Table 15.2 gives more exact figures for Kjeldahl conversion factors f_N and for h_{tot} for some common proteins. The degree of hydrolysis of enzyme-treated proteins determines the properties of relevance to food applications. Therefore, it is of the utmost importance that the degree of hydrolysis can be measured while the reaction is going on. Only in this way is it possible to stop the reactions at a well-defined stage, when the desired property of the product has been obtained. Within detergent applications (laundry or dishwashing), no estimates of this parameter have been reported, presumably because the substrate concentration is considerably low and therefore difficult to measure. When sufficient substrate is available, relatively simple analytical tools that can be applied directly during the reactions are often used. Some of them are the pH-stat technique, pH drops, osmometry, viscosimetry, and chemical determination of free amino groups, using, for example, a measurement of liberated α-amino groups by reaction with trinitrobenzenesulfonic acid (TNBS), whereby a colored derivative of the peptide is formed.

Table 15.2 Kjeidahl conversion factors and content of peptide bonds for various food

Protein	*Kjeldahl conversion factor, f_N*	*h_{tot} equiv/kg ($N \times f_N$)*
Casein	6.38	8.2
Whey protein isolate	6.38	8.8
Meat	6.25	7.6
Fish muscle	6.25	8.6
Egg white	6.25	approx. 8
Soy meal, concentrate and isolate	6.25	7.8
Cotton seed	6.25	7.8
Red blood cells	6.25	7.8
Wheat protein	5.7	8.3
Gelatin	5.55	11.1

The TNBS method has been adapted for use for hydrolysates of food proteins, which usually are partly soluble (i.e., the sample should be dispersed in sodium dodecyl sulphate (SDS)). The *O*-phthaldialdehyde and the ninhydrin reaction with liberated α-amino groups may also be used, but with some limitations. The pH-stat technique and the osmometry technique are quicker and less labor-demanding than the "chemical" methods. When simultaneous solubilization of protein occurs as an effect of the protein hydrolysis measurements of soluble nitrogen, for example, by Kjeldahl analysis or a simple detection of soluble dry matter by measuring of the increase of refraction index (°BRIX) can be used to describe the hydrolysis reaction.

pH-Stat Technique

Originally, the *pH-stat method* was developed at the Carlsberg Laboratory. It is based on the principle that pH is kept constant during hydrolysis by means of automatic titration with a base, when the hydrolyses are carried out under neutral to alkaline conditions. When the hydrolysis is carried out under acidic conditions, (e.g.,byuse of pepsin at pH 3) the titration must be made with acid. The high buffer capacity of proteins means that at extremely pH values (i.e., pH > 11 or pH < 3) the pH stat is inoperable.

The free carboxyl and free amino groups found after hydrolyses will be more or less ionized depending on the pH of the hydrolysis reaction. Working in the pH range of 6–9.5, the amino groups will be partially protonated. This means that the hydrolysis of proteins in this pH region is accompanied by a release of H^+, as already noted by Sorensen. Consequently, at pH values above

7.5–7.8 (the p*K* values at 25°C), the amino group will be less than half protonated, but the carboxyl groups will be fully dissociated. This leads to a net release of 0.5–1 mol H^+ for each mole of peptide bonds cleaved. In the pH range of 4–6.5, amino groups and carboxyl groups will be approximately equally dissociated. Therefore, in this pH range the pH state is not accurate. *DH* is calculated on the basis of the titration equations as follows:

$$DH = \frac{h}{h_{tot}} \times 100\% \qquad ...(1)$$

$$DH = B \times N_b \times \frac{1}{\alpha} \times \frac{1}{MP} \times \frac{1}{h_{tot}} \times 100\% \qquad ...(2)$$

where *B* is base consumption in mL (or L), N_b is normality of base, *a* is average degree of dissociation of the α-NH_2 groups, *MP* is mass of protein ($N \times f_N$) in g (or kg), *h* is hydrolysis equivalents in mequiv/g protein (or equiv/kg protein), and h_{tot} is total number of peptide bonds in the protein substrate (mequiv/g protein or equiv/kg). The degree of dissociation is

$$\alpha = \frac{10^{pH-pK}}{1+10^{pH-pK}} \qquad ...(3)$$

p*K* is the average p*K* value of the α-amino groups liberated during the hydrolysis, and it can be determined from a direct assay of these amino groups by use of the TNBS methods. p*K* also varies significantly with temperature because the ionization enthalpy of the amino group is considerable. Inserting the ionization enthalpy (+45 kJ/mol) in the Gibbs–Helmholtz equation, it was found that p*K* changed about 0.23 pH units for a change of 10°C in the hydrolysis temperature. Table 15.5 gives results of calculations of the degrees of dissociation α from equation 3 for various pH values as a function of the temperature using the previously described method.

Osmometry Technique

The *osmometry technique* can be used to calculate the degree of hydrolysis of proteins when no soluble components are added during the reaction (i.e., in ranges of the pH scale where the pH-stat principle is not functioning). A freezing-point osmometer can record the osmolality of a sample in a few minutes on a sample taken directly from the hydrolysis mixture. By drawing samples as a function of time during a hydrolysis reaction, a good reaction curve can be constructed. With the osmometer ΔT is measured and converted to osmole though the simple proportionality $\Delta T = K_f \times$ osmolality.

From the increase in osmolality ΔC, *DH* is calculated by use of the following equation:

$$DH = \frac{\Delta C}{S\% \times f_{osm}} \times \frac{1}{\omega} \times \frac{1}{h_{tot}} \times 100\% \qquad ...(4)$$

where ΔC is the increase in osmolality measured in milliosmol/kg H_2O, *S*% is the protein substrate concentration ($N \times 6.25$), w/w%, *w* is the osmotic coefficient ($\omega = 0.96$ for most actual concentrations of protein), and f_{osm} is the factor to convert percent to grams per kilogram H_2O, which can be calculated using the following equation:

$$f_{osm} = \frac{1000}{100 - D\,\%} \qquad ...(5)$$

where $D\%$ is percent dry matter present in the reaction mixture, and h_{tot} is total number of peptide bonds in the protein substrate (mequiv/g protein or equiv/kg).

Table 15.3 Degree of dissociation (α values)

	T(°C)					
	40	*50*	*60*	*70*	*75*	*80*
pK	*7.3*	*7.1*	*6.9*	*6.7*	*6.6*	*6.5*
pH						
6.5	—	0.20	0.29	0.39	0.44	0.50
7.0	0.33	0.44	0.55	0.67	0.71	0.76
7.5	0.61	0.71	0.80	0.86	0.89	0.91
8.0	0.83	0.89	0.93	0.95	0.96	0.97
8.5	0.94	0.96	0.97	0.98	0.99	0.99

Kinetic Aspects of Hydrolysis

The pH-stat technique has been utilized at Novo Nordisk to study the proteolytic degradation of hemoglobin, for example, and to understand the reaction mechanism. In discussions of initial proteolysis, Linderstr□m-Lang referred to the reaction of a protease on native hemoglobin molecules as one-by-one, indicating that a particular protease molecule degraded one substrate molecule at a time. No appreciable amounts of intermediary products will be present. The reaction mixture will consist of native proteins and end products only. In another reaction mechanism discussed, the native protein molecules were rapidly converted into intermediary forms, which degraded more slowly to end products—the zipper reaction. As seen on the hydrolysis curves made by using Alcalase at 50 and at 55°C in the pH stat, the slight bending at low *DH* values of the blood-cell fraction hydrolysis curves indicates a thorough initial degradation to small peptides (a zero-order reaction). This degradation is close to the ideal one-by-one reaction.

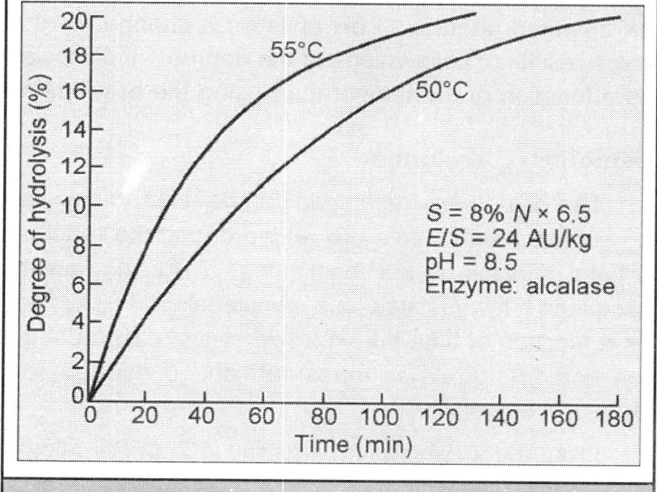

Figure 15.3 Hydrolysis curves for a hemolyzed red blood cell function.

The hydrolysis proceeds rapidly with little decrease in velocity until the substrate concentration is so low that first-order kinetics take over. A gel chromatogram of the soluble part of the hemoglobin hydrolysate at *DH* 10% and *DH* 20% fully confirms the degradation pattern of the hemoglobin substrate. At *DH* 10%, the hydrolysate consists of high molecular weight material and small peptides in accordance with the one-by-one mechanism. At *DH* 20% the high molecular weight material has completely disappeared, as predicted by Linderstrom-Lang as early as 1952. This is an important factor behind the application of ultrafiltration in the industrial process for separation of the hydrolysate from precipitated hemin, called the sludge. If the hydrolysis was not of the one-by-one type, large peptides would be present also, and consequently the yields and fluxes would have been much lower. One should be careful not to draw too simple conclusions about kinetic models on enzymatic hydrolysis on even a fairly simple and well-defined substrate as hemoglobin. At least 50 different peptide compounds are present under the single peak in the gel chromatogram of *DH* 20%.

Proteolysis

When the protein structure is modified by *proteolysis*, the various effects are considered either positive or negative, depending on whether the effect is wanted or unwanted. The developments within enzymatic hydrolysis have often been within both these criteria. The uses and importance of wanted effects including increased solubility, reduced surface tension, increased emulsification, increased foaming properties, and detergency of textile fibers and hard surfaces have been studied. Unwanted effects, such as increased bitterness and off-flavor developments, have been key problems to solve within the food area. During an enzymatic hydrolysis of proteins, the size character of the soluble peptides produced is changed. Key indices such as %*DH*, protein solubility index (PSI), and protein solubility in 0.8 M trichloroacetic acid (TCA index) increase, whereas peptide chain length (PCL) is reduced.

Inhibitory Mechanism

Some proteins are natural inhibitors of proteases. Trypsin and chymotrypsin inhibitors of legume and cereal grains have thus been reviewed and studied. Protease inhibitors for bacterial proteases, such as serine proteases (e.g., *subtilisin*) have been purified and characterized from legume seeds and potatoes. The probably best-known and most-studied protease inhibitors are the trypsin inhibitors of soybeans. The nutritional significance of the antitryptic activity of soybeans, soybean foods, and feed materials has been known for a long time. The way to inactivate these inhibitors has been an inherent part of most manufacturing processes for processes based on soybean material. Addition

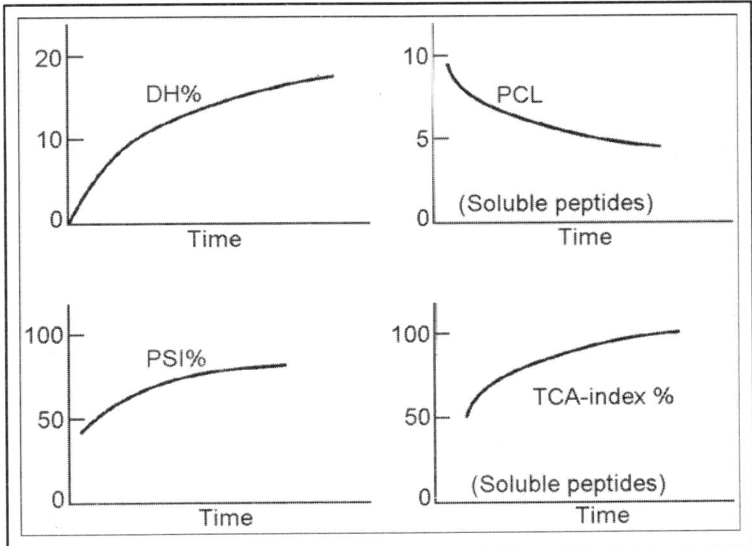

Figure 15.4 Quantitative illustration showing the time dependence of key indices during enzymatic hydrolysis of protein.

of protease inhibitors is known to protect proteases in detergents during storage. In powder detergents, where proteases are granulated and encapsulated, all enzymes are stable in the absence of water. In liquid detergents, the aqueous environment is challenging enzyme stability, and inhibitors are necessary to prevent autodigestion of the proteases themselves or to prevent other enzymes from being degraded.

Polyols such as glycerin and boric acid have been used in combination to inhibit protease activity and reduce autodigestion. Borate works as a reversible protease inhibitor by interacting with groups in the active site of the enzymes. Once in the wash water, these inhibitors are diluted and enzyme activity increases. Protein hydrolysate products and amino acids are also known to be able to stabilize proteases by binding to the active site of the enzyme. In liquid formulations used in the food industry, the way to stabilize proteases against loss of catalytic activity by denaturation and against microbial growth is by combining low water activity and a reduced pH.

IMPORTANCE IN DETERGENTS

Over the past 30 years, the importance of enzymes in detergents has changed from being minor additives to being key ingredients. The main areas where the use of proteases has grown are household laundry, automatic dishwashing, and industrial and institutional cleaning. I&I covers laundry, dishwashing, hard-surface cleaning, cleaning-in-place, and membrane cleaning. All commercial proteases from the early 1990s are *serine endoproteases* produced by *Bacillus* strains. Some have a maximum activity in the high pH range (pH 8–12) and others have a maximum activity in the pH range 7–10. Protease performance is influenced by factors such as detergent pH, ionic strength, wash temperature, wash time, detergent composition, and mechanical handling. Both performance and stability are influenced by some detergent surfactants and some bleach systems.

Heavy-duty powdered detergent formulations may vary across geographic regions, with a tendency for higher content of builders in North America than in Europe and Japan. The ratio of anionic to nonionic surfactant usage also varies from region to region. Furthermore, the content of surfactants is higher in Japan than elsewhere. The remarkable difference between detergent formulations for laundry and dishwashing is a much lower content of surfactants in dishwashing powder. The variable compositions in different regions should be seen in terms of the laundry practices common in North America, Europe, and Japan. The ratio of wash liquor to fabric load, temperatures, the usual dosage of detergents, and water hardness around the world are significantly different, as are detergent compositions. Most of these parameters may influence the detergency of proteases in a washing machine or automatic dishwasher. Particularly in laundry, the water hardness (i.e., the calcium concentration) is of concern. Calcium-ions are known to stabilize some enzymes and destabilize others. Usually, sufficient amounts of enzymes are present in the powders to secure the desired detergency power.

Laundry Applications

Proteases are usually evaluated by wash trials of test pieces soiled with milk, blood, or grass. Commercial pre-soiled test pieces are available from a number of research and test institutes. In laboratories, washing performance studies are carried out at relevant temperatures using Terg-Otometers, which simulate the U.S. top-loading type of washing machine, and Launder-O-meters, which simulate European drum-type machines. Launder-O-meters to some degree simulate washing mechanics and temperature profiles, including initial temperatures, heating rate, and maximum temperatures. Washing results depend on a number of factors, including detergents, water, type of soils, textiles, washing machine (temperature, time, mechanics, and number of rinses). Dose-response trials may be made in ordinary washing machines with ballast laundry and artificially soiled fabrics.

A sufficient amount of standard swatches may be fixed onto the ballast cloth. Evaluation of the effects on the test pieces may be made visually or by measuring the reflectance of light remitted at 460 nm when illuminating the test pieces. The intensity of reflected light % R (% remission) can thus be measured. The value $\Delta R = R_{washed} - R_{unwashed}$ is a measure of the total detergency. $\Delta R_{Enz} = R_{washed} - R_{washed\ without\ enzyme}$ reflects the enzyme's effect. Using a particular standard test textile swatch, EMPA 117 (cotton and polyester), containing blood, milk, and carbon black, the optimal effect of the enzyme was almost reached at a dosage of about 0.2% of the Savinase 6.0 T formulation. In practice, a dosage of about 1% is used to compensate for the detergency factors mentioned earlier. At $\Delta R < 22$ the fabric was clean. This means that the maximum measured R value was almost identical to that of an unsoiled test piece.

The data shows that the performance is better at 60°C than at 40°C until a dosage of about 3% Savinase 6.0 T (granulate) in the tested detergent was used. Stored detergents in open packages may become humid. At storage conditions of about 55% relative humidity, the bleaches (percarbonate or perborate) become activated. After some weeks of storage the residual enzyme activity may become reduced. However, stability of proteases in activated bleach systems has been improved significantly in the past few years by the use of genetic and protein engineering. Using these techniques, oxidation-sensitive regions of amino acids in protease molecules have been replaced by more stable amino acids.

Role in Dishwashing

A standard laboratory procedure for evaluating the cleaning efficacy of enzymes, mainly proteases and amylases, has been published. To test the efficiency of proteases, stainless steel plates are soiled with a baked egg–milk film. After washing light reflectance values, R, are measured directly on the plates. Calculation of percentage of removed, protein film, $RPF\%$, is shown by the formula

$$\text{RPF}\,\% = \frac{(R_{\text{after wash}} - R_{\text{before wash}})}{(R_{\text{clean plate}} - R_{\text{before wash}})} \times 100\,\% \qquad \ldots(6)$$

The soil-removal values are affected by the temperature at which the protein film has been baked. The effect of the baking temperature is using a citrate-based model detergent.

Synergism and Cleaning

Proteinaceous soils often contain complexes formed between proteins, fats, and carbohydrates. Thus it may be advantageous to combine enzymes to obtain a synergism during the hydrolysis of the protein. This has been demonstrated by the addition of *lipase* to protease in a model hydrolysis experiment. A homogeneous freeze-dried model substrate called *burned whole milk* was prepared from whole milk stirred in a pressure cooker at 120°C for 30 min. This substrate was tested for protein degradation using combinations of *Esperase* and *Lipolase*.

Surprisingly, it was found that when Esperase was combined with Lipolase, a synergistic effect on the release of free amino groups was found (TNBS method). Esperase used without Lipolase resulted in a lower solubilization of the burned whole milk. Using fresh milk, high levels of Lipolase did not greatly improve the degradation of the protein. A practical utilization of this hydrolysis synergism was demonstrated for removing burned-on soil on a heat exchanger used for high pasteurization of whole milk. A 2- to 3-mm layer of burned milk was removed using a protease and lipase at pH 8 (phosphate buffer), 50°C for 2 h. Scale was removed using 0.5% nitric acid after the enzyme treatment. During the enzyme cleaning, surfactants (soaps) were produced in situ from the fat, and emulsifiers and foaming agents were formed from the proteins. This was demonstrated by measuring the surface tension of a reaction mixture as a function of the consumed amount of base in a pH-stat trial using lipase and protease.

A complete cleaning-in-place (CIP) program using enzymatic cleaning of heat exchangers is estimated to be economical and useful when there are problems with conventional chemical cleaning processes. Membrane cleaning was tested in the laboratory and pilot plant using a pure enzyme–containing medium in dairy applications. Spiral-wound modules are very sensitive to harsh chemicals and in general not easy to clean by use of traditional cleaning products. Protease and lipase used at 40–60°C, pH 7–10, for 10–120 min under recirculation at high speed seem to solve the problems. In dairy plants enzymatic CIP are usually carried out according to the following scheme: rinsing → enzymes → rinsing → acid → rinsing.

Mechanism of Enzymatic Cleaning Mechanisms

To understand the mechanism of performance of proteases during washing of cloth it is important to know the molecular binding properties of the surfaces. Issues such as adsorption and desorption of stains to textile, composition of dirt substrate, specificity of the enzyme, net charge, and isoelectric properties of the enzyme in relation to the pH value of the detergent solution must be considered. Research within this field may involve activity tests on soluble and insoluble substrates and wash trials using test textile swatches with common stains, such as milk, blood, grass, and egg. Detergent proteases hydrolyze the insoluble proteins in stains into fragments that can be easily repeptide-protein fragments from hydrolysis with specific proteases could be removed more efficiently from the surface than small hydrophobic peptides. The simple analytical tools mentioned earlier may be used for evaluation of the protein hydrolysis reaction during a washing process. The one-by-one reaction, described earlier, may, together with the absorption theory, be part of the explanation of why the surfaces may be cleaned in a short time (15–45 min).

Protein Hydrolysates

Enzymatically produced *protein hydrolysates* made by-processes similar to those described in the following can, in combination with anionic surfactants, modify viscosity, foaming ability, foam stability, and cleaning efficiency for fat on textile fabric. Such protein hydrolysates have been acylated (condensated) with fatty acids of various lengths and have been found to be powerful amphiphiles in the sense that they can lower surface and interfacial tension, enhance wetting and foaming ability, enhance skin compatibility, and reduce viscosity. Such products are produced from wheat gluten and are used in shampoos and cosmetics. Enhanced emulsifying properties were demonstrated in laundry and cosmetic products, such as bath soap, conditioning shampoo, and cream hand cleanser, when enzymatic soy protein hydrolysates were used to partly replace detergent components. *N*-acyl protein hydrolysates made from enzymatic hydrolysates of keratin by acylation with fatty acid chlorides may be used as detergents during scouring of raw wool or other wool-cleaning operations or as fiber-protective agents during wool dyeing and bleaching.

APPLICATIONS OF PROTEASES IN FOOD PROCESSING

Hydrolysis of proteins with enzymes is an attractive means of obtaining better functional and nutritional properties of food proteins from vegetable origin or from by-products (e.g., from slaughter-houses). Industrial baking of biscuits, conversion of milk to cheese, and fermentation of soy sauce are examples of applications of proteases to produce food.

Table 15.4 Functional properties of proteins in foods and their applications

Property	*Application*
Emulsification	Meats, coffee whiteners, salad dressing
Hydration	Dough, meats
Viscosity	Beverages, doughs
Gelation	Sausages, gel desserts
Foaming	Toppings, meringues, angel food cakes
Cohesion binding	Textured products, dough
Textural properties	Textured foods
Solubility	Beverages

Role in Baking

The proteins of wheat flour form a network with disulfide bridges. The structure of gluten becomes stronger as more sulfur bridges are present. Oxidizing agents, such as bromate and dehydroascorbic acid, strengthen the dough in this manner. By hydrolysis of the gluten fraction of wheat, neutral bacterial endoproteases may weaken a flour that is too strong or may provide the plastic properties required in a dough to be used for biscuits. In biscuit manufacturing, either sodium meta-bisulfite or a neutral bacterial protease, such as Neutrase is used to weaken the gluten, so that the biscuits, cookies, or crackers keep their shape and size.

Role in Dairy Products

Milk is clotted to cheese by the application of proteases. Proteases of various kinds are used for acceleration of cheese ripening, for modification of the functional properties, and for modification of milk proteins to reduce allergic properties of cow milk products.

Obtaining cheese

For industrial cheese production, the milk must be standardized with respect to protein and fat content. This is normally achieved by blending milk batches, skimming (by centrifugation) to remove fat, or adding fat. The standardized milk is normally pasteurized at 72°C for 15 s. Subsequently, the milk is cooled to 30°C before being transferred to the cheese tank. Starter culture is added to the pasteurized milk to initiate fermentation. Rennet (the milk-clotting enzyme) and calcium chloride are added to promote the milk protein clotting reaction.

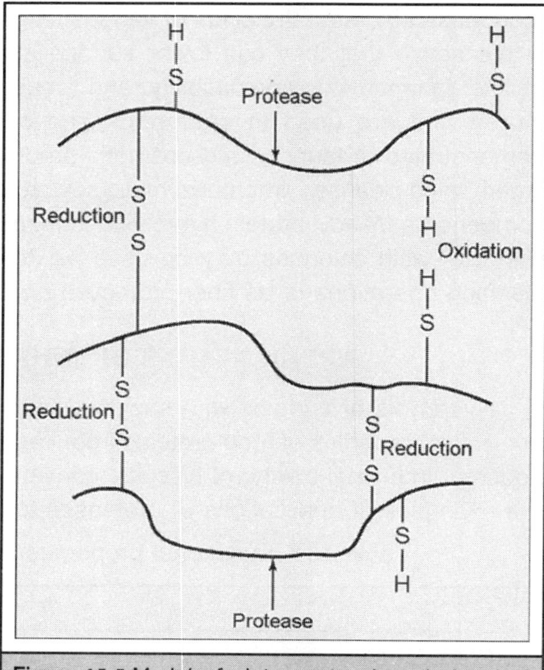

Figure 15.5 Model of gluten network and chemical and enzymatic modifications.

The milk-clotting effect of rennet is due to a limited hydrolysis of the *j*-casein surrounding the protein micelles. Thereby, the micelles become electrostatically discharged and able to aggregate with the help of calcium and phosphate ions and can form a network entrapping the fat micelles. A gel structure is thus formed. After about 25 min at 30°C, the gelled milk (the coagulum) is cut or stirred to promote syneresis, or exudation of the whey. Syneresis is further promoted by heating the curd–whey mixture. Specialized equipment for drainage of the whey is used. After drainage, the cheese is pressed, and lactic acid fermentation continues during pressing and subsequent storage. The cheeses are then steeped in brine. The fermentation of lactose continues until all lactose has been fermented, so that after removal of the brine the cheese is free of lactose. The cheese is then ripened during an appropriate period of storage.

Role in rennet

Rennet, also called *rennin,* is a mixture of *chymosin* and *pepsin*. Rennet is extracted from the gastric mucosa of the abomasum of young mammals (e.g., calves and lambs). The pepsin-to-chymosin ratios differ in different rennet preparations because chymosin is present only in the stomachs of unweaned mammals and is later replaced by pepsin. Thus, the content of pure chymosin depends on the age and species of the animal. Chymosin rapidly splits κ-casein into a glycomacropeptide and *para-κκ*-casein by selectively cleaving the 105–106 bond between phenylalanine and methionine. Pepsin and other proteolytic enzymes are less specific and can give rise to a number of degradation products, which tend to taste bitter.

Substitutes of rennet

Microbial rennets from a number of producers are marketed under various trade names (e.g., *Fromase, Hannilase, Marzyme,* and *Rennilase*). The enzyme costs for treating the same volume of milk are considerably lower than using standard rennet. Microbial rennets are produced by submerged fermentation of selected strains of fungi, often *Rhizomucor miehei*. The properties of microbial rennets have proven similar to those of chymosin, and only slight modifications of the cheese-making technique are made in practice. The gene for calf chymosin has been cloned into selected bacteria, yeasts, and molds. *Chymosins* made with recombinant DNA techniques on *Kluyveromyces lactis, Escherichia coli,* and *Aspergillus nidulans* are now commercially available.

Flexible Properties

Protein structure can be modified to improve solubility, emulsification, and foaming properties. Chemical modification is not desirable for food applications because of harsh reaction conditions, nonspecific chemical reagents, and difficulties of removing residual reagents from the final product. On the other hand, enzymes provide several advantages, including fast reaction rates, mild conditions, and, most importantly, high specificity. There is also a need for gentle methods that modify the food itself, in order to limit the use of additives. Over the years many different protein raw materials have been used, with different objectives.

Examples of extraction processes with enhanced yields include production of soya milk, recovery of scrap meat, cleaning of bone from slaughterhouses, recovery of gelatin, and production of meat (flavor) and yeast extracts. Furthermore, processes such as evaporation offish or meat stick water, rendering of fat, deskinning offish roe, and tenderizing of meat are assisted by proteases. Functional food ingredients from protein sources are made by use of proteases, including soluble protein hydrolysates used for nutrition, and foaming and emulsifying ingredients. Examples of such products are isoelectric soluble soy protein (ISSPH), egg white substitute of soy protein, emulsifier of soy protein, soluble wheat gluten, foaming wheat gluten, blood cell hydrolysate, whey protein hydrolysates, casein hydrolysates, soluble meat proteins, and gelatin hydrolysates.

Values of Hydrolysates

A limited hydrolysis of soy protein isolates was made at a very early stage with *Aspergillus oryzae* protease at varying ratios of enzyme to substrate. The *DH* was not directly measured, but from the data given by Puski in 1975 the maximum *DH* value obtained was about 5%. It was shown that whipping and emulsifying capacities were improved. The first application of the pH-stat technique for a controlled production of enzymatically hydrolyzed soy proteins was by use of Alcalase at pH 8 and Neutrase at pH 7. The isoelectric solubility, the foaming capacity, and the emulsifying capacity were all increased. If the reactions were continued to *DH* values higher than 5%, the foaming and emulsifying capacities again decreased. Also, a significant difference of the functional capacities was observed, depending on which of the two different pro- teases was applied. Alcalase had a much better performance, in general.

The reason for this effect could be due to the fact that foams and oil emulsions are stabilized better by peptides having a higher content of nonterminal hydrophobic amino acids than by peptides having terminal hydrophobic amino acids. In further research on functional ingredients based on enzymatic modification, improved products were developed with other raw soy materials or when physical treatments, such as ultrafiltration, were used during purification of the active components responsible for the specific effect. When a soy protein concentrate was hydrolyzed to *DH* 3–5%, with Alcalase, a functional protein hydrolysate with high foam expansion and remarkably good foam stability was obtained. On the other hand, the emulsifying capacity was considerably lower than if soy isolate was modified the same way. Using a native soy protein isolate produced by ultrafiltration of an aqueous extract of defatted soybean meal, the shape of the hydrolysis curve was remarkably different compared to that obtained for acid-precipitated protein.

The hydrolysis curves are shown for acid-precipitated soy protein isolate (denatured and nearly insoluble in water) and ultrafiltered soy protein isolate (native and soluble in water). It appears that the ultrafiltered protein isolate is hydrolyzed more slowly than the acid-precipitated protein. This is due to the compact molecular structure of the ultrafiltered protein, which is still in its native stage. The degree of denaturation of a protein substrate has a profound influence on the kinetics of proteolysis, as has been known for a long time. Using this soy protein isolate, functional protein hydrolysate with better foam expansion and better foam stability was obtained compared to those from the acid-precipitated protein. If the small peptides in this hydrolysate were removed by ultrafiltration, the foam stability was improved even further. The use of ultrafiltration before or after the enzymatic hydrolysis was investigated in pilot-plant scale in a number of process combinations. The high molecular weight hydrolysates had excellent whipping properties; furthermore, they were also able to heat-coagulate, so that they could substitute for egg white 100% on a protein basis in a meringue batter.

Soya proteins

Soya protein concentrate is produced by leach of defatted soybean grits using a water and alcohol mixture of 20–40% H_2O in ethanol. By this process the protein denatures so that the protein solubility is reduced and does not function adequately. A way to increase the solubility might be to enzymatically hydrolyze the separated protein before the alcohol is stripped off. To test if a protease such as Alcalase could operate in relatively high ethanolic environments, a series of hydrolysis curves were made under various conditions. The alcohol reduced the enzyme's effect considerably, but even when the percent DH was low, an improvement could be obtained on the protein solubility index. The practical utilization of these data could be to add water to an alcohol-precipitated protein sediment from a soy concentrate factory and to perform a protein hydrolysis before drying. A product having improved functional properties may thereby be produced.

Stepwise Applications

Enzymatic processing methods have become an important area for new processes used by the food industry to produce a large and diversified range of products based on protein raw materials. The principal advantages offered by enzymatic processes are the enzymes' specific mode of attack and the mild temperature and pH conditions required. Gentle treatment of the raw material is achieved, and only a few by-products that may affect the taste and smell of the final products are produced. Considering a typical enzyme application in details, several effects influence the enzyme reaction and the final product. The scheme shows a selection of parameters that can be varied when optimizing and developing new enzyme applications. Inactivation of the enzymes is of particular concern. In most enzymatically treated food products, active enzyme is undesirable because of changes that could appear in the final product. Therefore an efficient method for irreversible inactivation of the enzyme must be developed. Usually an efficient heat treatment is used, but neutral and alkaline proteases are also effectively inactivated when pH is adjusted to 4 or below.

Engineering aspects with regard to the up-scaling of inactivation must be taken into account for enzyme treatments carried out in production scale. Examples of inactivation procedures are discussed in the following. Purification of the products made enzymatically in which the enzymatic method does not produce the product in situ may in most cases be wanted. The selection of unit operations and the understanding of the performance of downstream processes are therefore also important parts of the application technology.

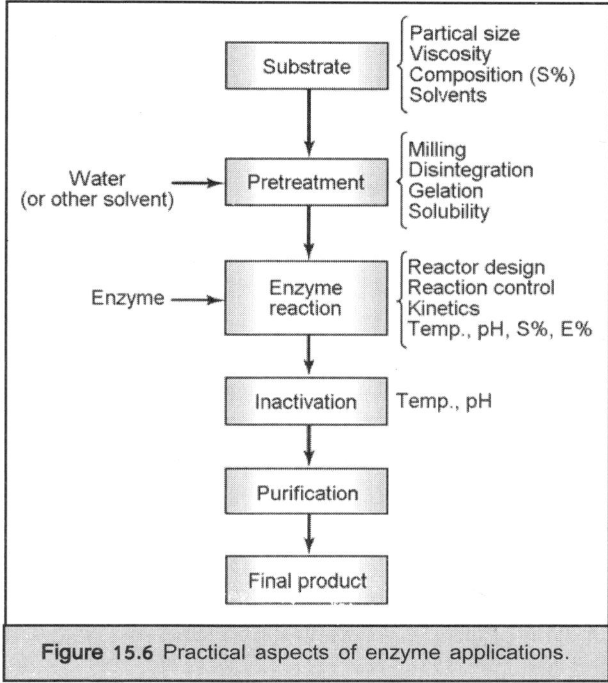

Figure 15.6 Practical aspects of enzyme applications.

PRACTICAL APPLICATIONS OF PROTEASES

Hydrolysis and Inactivation

If the reaction is carried out at a temperature above the enzyme's denaturation temperature, the protease is inactivated simultaneously during the hydrolysis. Examples of processes using this principle are shown for modification of wheat gluten and, as an alternative to rendering, for upgrading fresh bones from slaughterhouses to valuable products. The meat protein hydrolysates from the latter process are used as functional ingredients in meat products or as a clear, soluble meat extract for soups and seasonings. The cleaned bones make an excellent raw material for production of gelatin.

Role of wheat gluten

After a certain time, which seems to be dependent on the enzyme dosage used, the curves end at a plateau, showing that no further hydrolysis is occurring. Two product types from wheat gluten are considered. The first is soluble wheat gluten. High protein solubility may be obtained at higher degrees of hydrolysis. A *DH* value of about 10% results in solubilization of more than 90% of the gluten. A 100% soluble, bland-tasting wheat gluten hydrolysate with high yield may be recovered by centrifugation and concentration. A completely soluble wheat gluten hydrolysate is desirable for application as protein enrichment of food (e.g., soft drinks). Also, such protein hydrolysates may be excellent raw material for production of enzymatically produced hydrolyzed vegetable protein (HVP) for flavoring using, for example, exopeptidases. Nonfood applications were mentioned earlier. The second type of wheat gluten is whipping wheat gluten. The optimal whipping properties have been found at a degree of hydrolysis of 2–3%. The active whipping protein is recovered by centrifugation and drying. A whipping protein product is required for baked goods and for different types of candy.

Two-step process

A two-step enzyme process in which functional meat extracts are produced by the first enzymatic extraction and cleaned bones by the second enzymatic extraction can be performed. The fresh bone material (beef or pork) is crushed and mixed with hot water. The enzymes *Neutrase, Alcalase, Esperase,* and *Flavourzyme,* used alone or in combination, are added, and the slurry is agitated and extracted batch- wise or continuously at 60–75°C at neutral pH (6.5–7). Poultry material may also be used, but the bone material is not in general used for gelatin production. Bone cleaning trials have been carried out in Novo Nordisk's pilot plant for process development. Using a bone crusher, 250 kg of beef bones were milled to a particle size of approximately 20 mm. This material was poured into 250 kg of 80°C water and the temperature was adjusted to 65–70°C. Hereafter the enzymatic hydrolysis reaction was carried out under powerful stirring. Samples, drawn as a function of time, were centrifuged for 3 min before determination of °BRIX (dry matter), osmolality, and pH.

The total reaction mixture was sieved and the liquid phase was pasteurized, defatted by centrifugation, concentrated, and dried. The bone material was washed and dried. Within the 20 min of reaction the °BRIX value and the osmolality were still increasing. This indicates that the enzyme was not completely inactivated at this temperature (–70°C). The value mOSM/°BRIX is a measurement of the molecular size of the soluble phase. In the beginning, when the content of soluble protein is low, mOSM/°BRIXis high because of a high content of low molecular weight compounds. Later, this value stabilizes because no further degradation occurs when the protein is solubilized from the surface of the bone material. The one-by-one reaction system seems to be confirmed. Inactivation of the enzyme Esperase may simultaneously be achieved at a higher temperature. Centrifuging at 90°C during defatting of the hydrolysate inactivates the enzyme and pasteurizes the product.

Treatment of meat protein with esperase

Based on the osmolality difference, ΔC in equation 4, *DH* can be calculated and compared with those values from hydrolyses of lean beef. Hydrolyses using *Esperase* were carried out at 55, 65, and 75°C on minced beef. Both pH-stat trials and hydrolyses followed by osmo-meter were done at temperatures in the high region. The pH was adjusted to 7.5 so there was no appreciable hydrolysis of triglycerides. The percent *DH* was higher for raw meat, and the highest *DH* value for raw meat was obtained at 55°C. For both the raw and cooked meat the highest solubilization was found at 65°C. This was also seen in pH-stat trials. The enzyme dosage based on protein was doubled for the bone trial as compared to the meat trial, yet a high protein-solubilizing effect of

Esperase was seen in both cases. But, surprisingly, a lower percent *DH* secured a high solubility for the raw meat than for the meat on bones.

Table 15.5 Bone or meat treated with esperase at 65–70°C

	Bone			Raw meat		
Time (min)	*% DH*	*pH*	*% Soluble protein*	*% DH*	*pH*	*% Soluble protein*
0	0	7.14	1.2	0	7.5	1.8
1	5.7	6.82	2	0.1	7.33	2.5
5	7.7	6.81	3.2	1	7	3.4
10	9.3	6.67	3.7	2.1	6.85	4.1
15	10.3	6.67	4.2	2.5	6.77	4.5
20	11.1	6.68	4.3	2.8	6.71	4.8

Membrane Processes

Membrane processes are used within many areas of industrial enzymatic hydrolysis of proteins.

Application of ultrafiltration processes

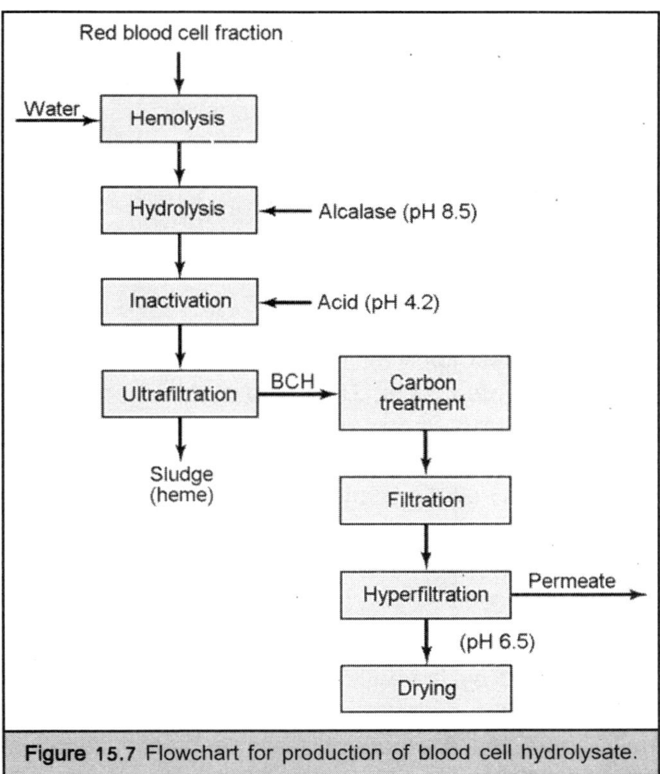

Figure 15.7 Flowchart for production of blood cell hydrolysate.

Examples of the application of ultrafiltration processes in various process combinations for production of highly functional protein hydrolysates of soy protein were mentioned earlier. Principally, the technique of removing small peptides from total hydrolysates may be used on many substrates other than soy. Thereby, excellent foaming and coagulating protein products may be isolated (purified). The first process step after inactivation is normally a centrifugation carried out to remove insoluble material, which may reduce the whipping expansion and the foam stability. To purify the supernatant as effectively as possible, ultrafiltration can be carried out in three steps. In the first step, the concentrate reaches a protein content of about 10%. Then, a diafiltration process is performed for washing out small, low-function peptides. A final ultrafiltration can bring the protein content up to approximately 20%. Finally, the concentrate can be spray-dried. The efficiency of these processes and the overall yields are dependent on the actual ultrafiltration equipment used, the membrane cutoff value, and the character of the hydrolysate (enzyme used, *DH*%, pH, protein type).

Blood cell hydrolysis

Low molecular weight hydrolysates are produced from raw materials that are hydrolyzed to a high degree of hydrolysis. Examples of such products are the mild-tasting decolorized BCH developed from the hygienically collected red cell fraction of slaughterhouse blood and the isoelectric soluble soy protein hydrolysate (ISSPH). These two examples demonstrate how the membrane processes can be used for different purposes. In the BCH process, the pH is 4.2 and insoluble hem product may be removed by centrifugation or ultrafiltration. Comparison of the centrifugation process and the ultrafiltration process shows that the ultrafiltration process was most economical when new plants were installed. Ultrafiltration equipment from various suppliers has been tried with technical success.

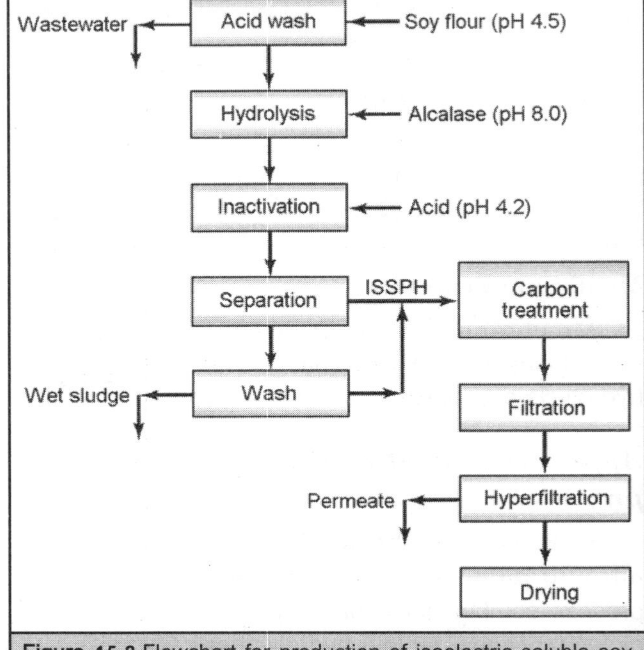

Figure 15.8 Flowchart for production of isoelectric soluble soy protein hydrolysate (ISSPH).

Application Trials

The earliest application trials were for pumping of meat cuts. Using up to 3% BCH on a final meat product basis, nutritional protein equivalence was obtained, yields were increased, costs were reduced, toughness of meat was reduced, and the fat to protein ratio was reduced. In a study using BCH in low concentrations in meat, it was demonstrated that BCH has a polyphosphate-like character in water-added meat systems. BCH has also been compared to sodium triphosphate in luncheon meat heat-treated at 75°C and 108°C, respectively.

During production of the soy protein hydrolysate, the clear filtrate after carbon treatment is concentrated by hyperfiltration as an alternative to evaporation. A falling film evaporator can concentrate the hydrolysate to above 50% dry matter. Hyperfiltration can only be used efficiently up to about 25% dry matter but allows the simultaneous removal of salt if nanofiltration membranes are used for the concentration. Thus, about 75% of the salt in ISSPH was removed by direct concentration and 93% if a diafiltration was included in a pilot-plant trial on a selected cellulose acetate membrane.

Ultrafiltration system

The *membrane reactor* is an ultrafiltration system in which a high concentration of hydrolytic enzyme is confined. High molecular weight substrate is fed continuously to the reactor, and the low molecular weight products are removed simultaneously as they permeate through the membrane. Ideally, a steady state is reached in which the degradation of substrate is carried out indefinitely with high efficiency and negligible loss of enzyme. A considerable number of publications on the use of membrane reactors for enzymatic hydrolysis of proteins appeared 10–15 years ago. However practical use of the system has not been widespread. One of the reasons may be that the economical advantage of the membrane reactor depends critically on the need for purging of the reactor. A small part of the substrate is always nondegradable. Therefore inert material will rapidly build up in the reactor, causing mechanical problems. A considerable purge is thus necessary if the concentration of this material is to be kept at a reasonably low level. This has a drastic, negative influence on the yield of product. Also, the enzyme loss during purging is considerable. Kinetic investigations of protein hydrolysis in the membrane reactor have been presented and a mathematical model established. Based on these calculations, it was found that the concomitant loss of enzyme in certain cases may lead to an overall enzyme consumption close to that of the batch hydrolysis process.

Unpleasant Activities

Unpleasant bitterness has been observed for some protein hydrolysates. This problem can be managed by proper selection of the hydrolysis parameters and the enzymes used. Bitterness is a complex problem that can be influenced by the following variables:

1. Hydrophobicity of the substrate, hydrolysates become bitter if amino acid side chains containing hydrophobic groups become exposed due to hydrolysis of the protein
2. High *DH* and, therefore, the concentration of soluble hydrophobic amino acids and their peptide chain lengths
3. The specificity of the protease, to some extent, determines the bitterness
4. Separation steps applied in downstream processing, for example, centrifuging at isoelectric pH and carbon treatment
5. Masking effects from other components in the protein hydrolysate; organic acids such as malic acid and citric acid exert a significant masking effect in soy protein hydrolysates that can be utilized in the final formulation of the food, where the hydrolysate is incorporated

In addition, recent recognition of the following effects points out other solutions to solve bitterness problems in production of protein hydrolysates with high *DH* values. A debittering of protein hydrolysates can be performed by a selective extraction or removal of hydrophobic peptides to a 2-butanol phase. This knowledge can be used to explain why it was surprisingly possible to produce a non-bitter *DH* 20% soy protein hydrolysate from full-fat soybean meal using Alcalase. This *DH* 20% hydrolysate was fully acceptable from an organoleptic point of view. The oil served as the hydrophobic extraction medium and eliminated the bitter peptides. The oil was simultaneously recovered by centrifugation. Reaction parameters such as pH and temperature, by which the specificity of the proteases are changed, also play an important role in managing bitterness. Furthermore, combinations of enzymes can change the overall mode of degradation of a protein so that bitterness is reduced. The application of exopeptidases is a generally recognized way to remove bitterness of high-*DH* hydrolysates.

Bitterness Problems

Animal raw materials considered useful for protein hydrolysis usually do not contain many in situ bitter compounds of significance. Bitterness can appear from rancid traces of fat or from extended hydrolysis of proteins. Oilseeds contain a variety of phenolic compounds, in particular, bitter-tasting phenolic acids. Some volatile compounds are hydrophobic and bind to the protein phase, especially to denatured protein. Off-flavors may be released by enzymatic hydrolysis of such proteins, some of which, but usually not all of which can be removed. A solution to such problems has been described for industrial production of ISSPH.

The white flakes were washed at pH 4.5 with water in a four-step extraction process to produce a soy concentrate with a low level of beany off-flavor. This soy concentrate was immediately hydrolyzed with Alcalase to a specific degree of hydrolysis. The hydrolysis reaction was terminated by lowering the pH, and the supernatant was recovered by centrifugation at the isoelectric point. The hydrolysate was carbon treated, concentrated, and used for protein-fortified fruit juices and for extension of cured, whole meat products. This was possible because the ISSPH had a bland taste and no bitterness.

Enzymatic Tenderization of Meat

The use of the vegetable proteinases papain, ficin, and bromelain for enzymatic tenderization of meat has been known for many years. These enzymes have a relatively strong potency on muscle tissue components such as collagen and elastin. Microbial proteinases have also shown the ability to tenderize meat, but not to the same extent as the vegetable proteinases. The most successful meat, tenderizing enzymes are those having the ability to hydrolyze the connective tissue proteins as well as the proteins of the muscle fibers. The enzymes are applied to the meat by sprinkling the enzyme powder on a thin slice of meat, dipping the meat in an enzyme solution, or injecting an enzyme solution into the meat. A method has been developed in which the enzyme is introduced directly into the circulatory system of the animal shortly before slaughter or after stunning the animal to cause brain death. Whatever method is used, the difficulty in using external enzymes to obtain an even distribution of the enzyme throughout the tissues seems to be in the control of the hydrolysis reaction.

Flavoring Techniques

Proteins in their natural state do not contribute chemically to formation of flavor in foods. The products of the hydrolysis of proteins, peptides, and amino acids are much more reactive, as are other breakdown products in general. Hydrolyzed protein, mainly produced by the use of hydrochloric acid, is one of the largest contributors to flavoring produced from protein. Using various sources, many savory products are on the market. The hydrolysis with hydrochloric acid usually takes place in the presence of triglyceride fat material. It is well known that chloro- and dichloropropanols are formed during acid hydrolysis under harsh conditions, although they are found in small quantities in the final food products. Concern for the safety of the products of hydrochloric acid hydrolysis of proteins has led to investigations into the safety of these chlorocompounds. Glutamic acid, as monosodium glutamate (MSG), is by far the most consumed single flavor enhancer of proteinogenic origin. About 270,000 tons are produced per year.

Glutamates are known as the fifth basic taste (in addition to sweet, sour, salty, and bitter). This taste is called *umami* by Japanese and savoriness by Americans. MSG is used at concentrations of 0.2–0.8% in a variety of foods, such as soups, broths, sauces, gravies, flavoring and spice blends, canned and frozen meats, poultry, vegetables, and finished dishes. To be able to reduce the intake of sodium and to produce in situ flavor enhancers in protein hydrolysates, peptidoglutaminases are of interest for industrial purposes.

Through their reactions with taste sites, peptides may give flavors that are bitter, sweet, salty, or *umami*. Also, sour taste and astringency have been attributed to peptides isolated from protein hydrolysates or by synthesis. Many studies have been made of the effects of flavor in dairy products, meat and fish products, and yeast extracts. The direct production of flavor products can be made with proteases used alone, or in combination with fermentation processes. Chemical hydrolysis of protein and carbohydrates combined with heat treatments form Maillard reactions that allow many flavors to be produced. Soy and wheat hydrolysates are used for flavoring and flavor enhancement of soups and seasonings; meat hydrolysates for addition of flavor to meat products, soups, bouillon, and sauces; and milk and cheese hydrolysates for production of cheese flavors.

Vegetable Proteins in Flavoring

Enzymatic hydrolysis usually leads to products that are light in color and have a much less pronounced meaty or savory flavor than, for example, yeast extracts. Meat flavor and similar condiments are often produced from many sources, and their production often involves a natural microbial fermentation. Among these can be included the Asian products based on soy protein. Wheat and maize gluten are also hydrolyzed to flavor intense peptides. The content of free glutamic acid is important for the *umami* flavor. A way to exhibit flavor enhancement characteristics by using mild acid hydrolysis of the protein, followed by enzymatic hydrolysis, has been described. The preferred protein for producing savory flavors is wheat gluten.

The mildly acidic hydrolysis is done to obtain a deamidation in the first step. This step was carried out for 1 h at about 95°C at a hydrogen ion concentration of about 1.0 M. The enzymatic hydrolysis was preferably carried out by use of an *Aspergillus oryzae* protease (Prozyme 6 from Amano) having both endoproteases and exoproteases. The overall degree of hydrolysis was 50–70%. The conditions during the acid treatment were mild enough to avoid the formation of substantial amounts of monochloropropanols.

Flavourzyme used in combination with Alcalase may obtain a degree of hydrolysis of 60–70% for hydrolysis of soy protein isolate. Except for the deamidation step, the same degree of hydrolysis could thus be obtained without the acid treatment. An enzymatically hydrolyzed soy protein, so-called eHVP, has been developed by Miwon C., Ltd. This product is suitable for special uses, such as infant diets, sports nutrition, and medical diets. eHVP gave no limitations of dosage in formulation as a flavor enhancer.

Animal Proteins in Flavoring

Products that have a high flavor of meat extract are used in soups, sauces, and prepared meals. Proteinaceous material is recovered with proteases from coarse and fine scrap-bone residues from mechanical fleshing of beef, pig, turkey, or chicken bones. The flavor intensity depends on the content of free amino acids and peptides and their reaction products. Reactions that develop flavor are, for example, Maillard reactions between reducing sugars and amino acids, thermal degradation due to the Maillard second stage, and deamination, decarboxylation, or degradation of cysteine and glutathione.

The latter reaction can give rise to a large number of volatile compounds of importance to aroma and taste. For the production of protein hydrolysates from meat, the first step is an efficient solubilization of the product. For this purpose the bacterial serine proteases and metalloproteases are very efficient. Unless certain treatments are undertaken, meat hydrolysates are usually found to be bitter when the degree of hydrolysis is above 10%, which is needed for sufficient solubilization.

A method comprising a series of steps using raw meat hydrolyzed with a specific combination of neutral and alkaline proteases has been developed. This meat hydrolysate exhibits excellent organoleptic properties and can be used as a meat- flavored additive in soup concentrates. The degree of hydrolysis above 20% did not show bitterness when this specific combination of enzymes was used. This effect may be due to the preferential proteolytic specificity being favorable when metalloprotease and serine protease are used simultaneously.

A method for further enhancement of the flavor of a natural beef-juice concentrate is hydrolysis using endoproteases (metalloprotease, serine protease, or combinations) followed by treatment with an exoprotease such as Flavourzyme or by incubation with a culture of a food-grade microorganism that is capable of producing exoproteases (e.g., aminopeptidase).

ROLE IN LEATHER INDUSTRIES

The processing of skin and hides for leather has been based on enzymes since 1908, when Otto Rohm patented the first standardized bate based on pancreatic enzymes. Leather chemistry research helped to improve understanding of the bating process and at the same time spurred on developments to improve leather processing. The stages in the processing of hides to leather include curing, soaking, liming or dehairing, deliming, bating, pickling, and tanning. The benefits of using enzymes in the different stages of leather manufacturing are mainly reduction of process time, increased opening of the fibrous structure, improved clarity of the surface, increased softness, increased uptake of water, improvement of area yield, and reduction of the need for chemicals. The animal hide is composed of three main layers.

The grain is the outer layer in which the hairs are embedded. This layer also contains the small glands that produce oil and sweat. The corium represents the principal part of the hide. To convert raw hide into leather a great deal of interfibrillar material must be removed from this layer.

Collagen fibers, elastin, and proteoglycans, such as dermatan sulfate, can be regarded as a kind of cement, that binds collagen fibrils. The connective tissue of the underside of the hide contains the blood vessels, fat tissue, and some meat proteins. Raw or cured hides must be properly soaked to obtain a satisfactory hydration and removal of unwanted materials, including meat and fat. Bacterial or pancreatic alkaline proteases are normally preferred for hydrolysis of interfibrillar proteins.

Role in Dehairing, Liming, and Bating

During liming, the highly charged dermatan sulfate may be removed from nearly each collagen fibril by use of alkaline-stable proteolytic enzyme preparations such as NUE 0.6 MPX from Novo Nordisk A/S. An accelerated dehairing process and an up to 40% reduction of the requirement for sulfide have been measured.

Furthermore, increased strength of grain leather and increased area yield are useful results. Before the bating process, the hides are delimed with ammonium sulfate or ammonium chloride. When proteases are applied a thickness similar to that of hairless pelt is reestablished. The early preparations proposed by Rohm were pancreatic trypsins. Some of the microbial proteases are used today as bating enzymes to make the hides soft and supple before tanning.

MORE APPLICATIONS OF PROTEASES

A number of applications of proteases have been mentioned. The emphasis of this article has been on my research over the years and on the most significant industrial applications, as seen from an enzyme producer's point of view. Some applications of proteases have not been mentioned in detail. This is the case, for example, with regard to the use of enzymes in fish processing, where proteases are used for production offish sauce, for salted matjes herring, for deskinning of fish, for production of protein hydrolysate as a cryoprotective agent, and for roe caviar purification.

Using a glutamic acid–specific protease from *Bacillus licheniformis*, the viscosity of a solution of whey protein concentrate could increase dramatically during hydrolysis due to production of plastein aggregates. An increased water-binding capacity and increased viscosity during hydrolysis of mechanically deboned meat was also reported.

Extending hydrolysis to smaller peptides reduced viscosity again. A practical application of these findings remains to be further developed. Use of proteases as processing aids for reduction of viscosity of stickwater before and during evaporation in a fish-meal plant or a meat-rendering plant could reduce the fouling on heat surfaces. The capacity of the evaporators was then increased due to the reduction of the overall heat transfer coefficients and because a greater amount of dry solids could be obtained.

Evaporation of raw fruit water from potato production represented an insurmountable problem, because gelling due to the protein components occurs at about 15% dry matter. It was obviously necessary to add proteolytic enzymes, as mentioned earlier, for concentration of stickwater. However, it was surprisingly found that no effects of the enzymes could be observed until heat treatment, preferably jet cooking to at least 125°C using a holding time of about 3 min, had inactivated a powerful protease inhibitor in the raw fruit water.

CONCLUDING REMARKS

Proteases are by far the most researched enzymes for industrial bioprocessing. The major uses of these enzymes are divided among many applications, all based on hydrolysis as the catalytic concept. Similarities in the catalytic mechanism within the widespread uses of proteases for detergents, for modification of food proteins, for better utilization of waste fractions of the food and feed industry, and for production of basic food have been shown. No doubt more development within proteolysis, as broadly defined, will be seen in the future. To meet challenges within the detergent industry and the food–feed industry, discovery of new protease molecules must be made through protein engineering and recombinant DNA techniques.

16

RECOMBINANT PROTEINS HETEROGENEITY

Recombinant proteins produced by animal cell expression systems can be subject to numerous posttranslational modifications that can result in a structurally heterogeneous product. *Glycosylation* is the most extensive of all posttranslational modifications, and for proteins destined for therapeutic use, this source of molecular diversity may be of particular significance. For example, distinct protein glycoforms may possess different properties such as biological activity, receptor binding, stability, immunogenicity, and clearance rate in vivo. The carbohydrate components of glycoproteins can also play crucial roles in protein folding, oligomer assembly, and secretion processes. Recombinant protein heterogeneity therefore represents an important issue for the biopharmaceutical industry and regulatory authorities, with regard to product characterization and definition, consistency, potency, and purity.

Authorities such as the U.S. Food and Drug Administration are now demanding increasingly sophisticated carbohydrate analyses as part of the "well-characterized" product or process validation. Therefore, within the biotechnology/biopharmaceutical industries there is now considerable interest in technologies enabling recombinant glycoprotein analysis and control of glycoprotein heterogeneity. Although there has been much progress in the field of glycobiology in recent years, with a plethora of published information on biological properties of carbohydrates, methods of glycan analysis, cellular biosynthetic pathways, and the molecular cell biology of the key enzymes involved, this article highlights key bioprocess factors that may influence the glycosylation of a recombinant protein produced by an animal expression system.

Carbohydrate Compounds in Recombinant Proteins

Although there is considerable diversity in the types of protein–carbohydrate conjugates that occur naturally, recombinant glycoproteins produced by eukaryotic expression systems are mainly associated with N-linked glycans (N-glycosidic bond to the amide group of an asparagine residue within the consensus sequence Asn-X-Ser/Thr) or O-linked glycans (O-glycosidic bond to the hydroxyl side group of either a serine or threonine residue). Carbohydrates are also components of the glycophosphatidyl inositol (GPI) anchor used to secure some proteins to cell membranes. The molecular diversity of recombinant glycoproteins arises from two major sources: variable glycosylation site occupancy, also called macroheterogeneity (i.e., the presence or absence of glycans), and microheterogeneity (variation in carbohydrate structures at individual glycosylation sites).

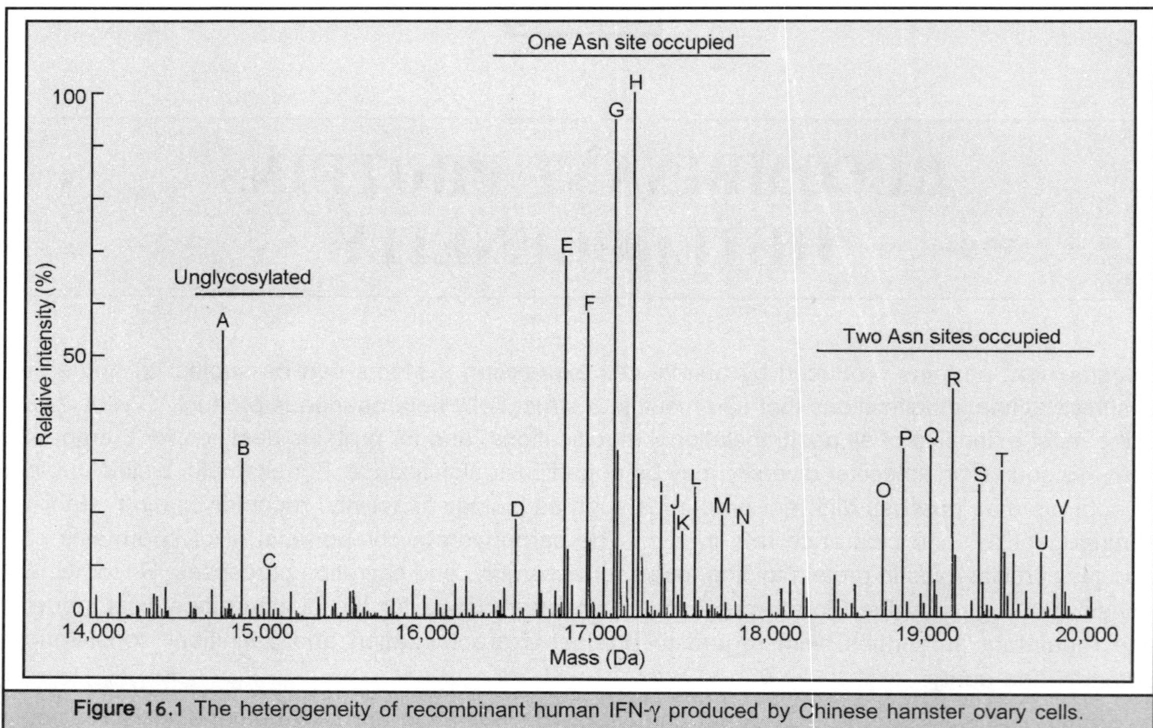

Figure 16.1 The heterogeneity of recombinant human IFN-γ produced by Chinese hamster ovary cells.

In the case of the former, the presence of sequence motifs or amino acids necessary for glycosylation within the polypeptide sequence does not guarantee their glycosylation. For example, at the cellular level N-glycosylation site occupancy at Asn-X-Ser/Thr sequons may be related to several factors including oligosaccharyldolichol substrate availability, oligosaccharyl transferase activity, rate of protein synthesis, competition with protein folding, and other cotranslational events. Protein structural or polypeptide sequence factors may also be significant; for example, the target glycosylation sequon Asn-X-Thr has been found to be 40 times more susceptible to glycosylation by oligosaccharyl transferase than the Asn-X-Ser sequence. Recombinant glycoproteins should most appropriately be considered as populations of individual "glycoforms," as illustrated for recombinant IFN-γ produced by Chinese hamster ovary (CHO) cells. In general terms, the relative abundance and distribution of protein glycoforms within a glycoprotein population are a function of the following:

1. The host cell and expression system employed
2. The interaction of the host cell and recombinant glycoprotein product with the bioprocess environment

N-Linked Glycans

All mature N-linked glycans contain a common trimannosyl core sugar structure, a pentasaccharide consisting of three mannose (Man) residues and two *N*-acetylglucosamine (GlcNAc) residues (Manα1-6(Manα1-3)Manβ1-4GlcNAcβ1-4GlcNAc-(Asn)). This can form part of simple oligomannose structures or be extensively modified by the addition of other GlcNAc, galactose (Gal), fucose (Fuc), and sialic acid (*N*-acetyl- or *N*-glycolyl-neuraminic acids; NeuAc or NeuGc) residues. N-linked chains can be classed into three main subgroups: high-mannose type, complex type, and hybrid type. High-mannose-type glycans contain α-linked mannose chains bound to both arms of the trimannosyl core. Complex-type chains contain no extra mannose residues and have α-linked GlcNAc residues bound to the trim annosyl core, with Gal and NeuAc residues added subsequently. Commonly, there may be α1-6- or α1-3-linked Fuc residues added to the core (Asn-linked) GlcNAc. Hybrid-type chains contain both α-linked Man residues bound to the α1-6-linked Man arm of the trimannosyl core, while complex-type chains are bound to the α1-3-linked Man arm. However, numerous possible variations are commonly found. Following are the basic sources of *N*-glycan heterogeneity:

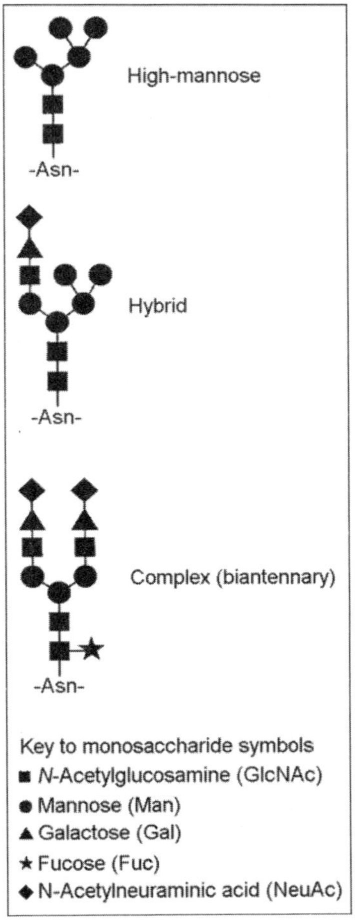

High-mannose

-Asn-

Hybrid

-Asn-

Complex (biantennary)

-Asn-

Key to monosaccharide symbols

■ *N*-Acetylglucosamine (GlcNAc)

● Mannose (Man)

▲ Galactose (Gal)

★ Fucose (Fuc)

◆ N-Acetylneuraminic acid (NeuAc)

Figure 16.2 Common N-glycan structures found on glycoproteins.

1. The degree of branching of complex type chains (number of antennae). For example, the complex chain contains two branching chains from the trimannosyl core and is thus termed a biantennary-complex-type chain. Typically, complex N-linked glycans may be binantennary, triantennary, or tetraantennary.
2. The presence or absence of terminal sialic acid residues linked to antennae.
3. Variation in the type of sialic acid; N-acetyl and N-glycolyl variants are the most common.
4. The presence or absence of a 'bisecting' GlcNAc attached to the trimannosyl core β-linked mannose (which prevents branching)
5. The presence or absence of terminal poly-*N*-acetyllactosamine repeats (Galβ1-4GlcNAcβ1-3).
6. Unusually, sulfation at GlcNAc residues or phosphorylation at Man residues.

O-Linked Glycans

O-linked glycosylation is extensive in structural proteins such as proteoglycans. Although the majority of O-linked glycans are bound to either a serine or threonine residue core, small glycan structures can also be O-linked to the side chain of hydroxylysine or hydroxyproline. *O*-Glycans may occur as a single *N*-acetylgalactosamine (GalNAc) residue, disaccharides NeuAcα2-6GalNAc/ Galβ1-3GalNAc, or larger glycans of up to 10 residues attached to the core GalNAc containing Gal, Fuc, and GlcNAc-residues.

Biological Synthesis of Glycan

The synthesis and elongation of glycan structures is an ordered, rather than random, process that begins in the rough endoplasmic reticulum and is completed in the *trans*-Golgi network. In the case of *N*-glycans, efficient transport of recombinant glycoprotein through the constitutive secretory pathway of eukaryotic cell expression systems is a prerequisite for high-level production of recombinant protein and modification of N-linked oligosaccharides. In the case of the latter, the ordered cellular localization and activity of glycosidase and glycosyl transferase enzymes define the sequence and structure of *N*-glycans associated with the secreted protein. This dynamic process is mediated by the trafficking of secretory vesicles between distinct subcellular compartments and by molecular interactions within those compartments. N-glycosylation may itself be necessary for efficient secretion of some proteins.

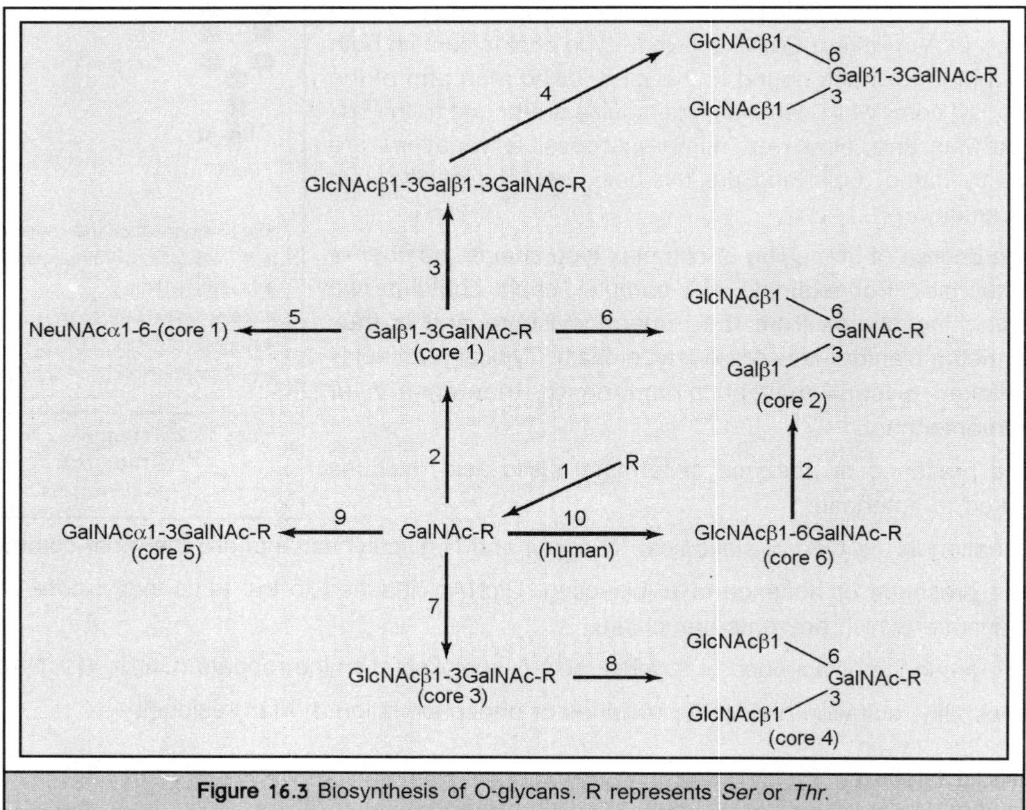

Figure 16.3 Biosynthesis of O-glycans. R represents *Ser* or *Thr*.

Recent studies have shown that secreted or membrane-targeted proteins become concentrated in an intermediate compartment (IC) between the endoplasmic reticulum (ER) and the *cis*-Golgi. Although the nature of this compartment remains obscure at present and may not be solely a vesicular transport system but a microtubule-directed system, distinct molecular markers of the IC, such as the membrane-bound mannose-binding lectin ERGIC53 or homologs such as p58, which cycles between the ER, IC, and *cis*-Golgi, may function to specifically regulate the sorting of glycoproteins. Biosynthesis of N-linked glycans begins with the cotranslational en bloc transfer of a large, high-mannose oligosaccharide terminating in glucose (Glc) residues ($Glc_3Man_9GlcNAc_2$) from a dolichol pyrophosphate–linked precursor to an appropriate Asn residue in the nascent polypeptide chain. This process, catalyzed by the oligosaccharyl transferase (OST) complex, occurs in the lumen of the ER.

This is followed by the subsequent removal of the three terminal glucose residues via an α-glucosidase, a quality-control process that has been shown to play an important part in the correct folding and assembly of some secreted proteins through interactions with the molecular chaperones calnexin and calreticulin. Subsequent trimming of mannose residues by ER and Golgi-specific α-mannosidases is followed by addition of terminal GlcNAc, Gal, and sialic acid and core fuc residues by glycosyltransferases utilizing nucleotide-sugar substrates imported into the lumen of Golgi compartments by specific transporters (antiporters). Control of glycosylation reactions is complex and relies upon a multiplicity of molecular, cellular, and physiological interactions such as relative distribution of enzyme activities, substrate availability, and rate of protein transport. In some circumstances, local protein structure may serve to modify glycan processing. For example, it is well documented that terminal galactosylation and sialylation of IgG *N*-glycans in the Fc region at Asn297 is restricted by steric hindrance. However, in general, key control points that contribute to *N*-glycan heterogeneity are those catalyzed by the following:

1. ER and Golgi mannosidase I, GlcNAc transferase I (conversion of high-mannose to complex-type glycans)
2. GlcNAc transferases II, IV, V (degree of branching of *N*-glycans)
3. GlcNAc transferase III (bisecting GlcNAc)
4. α-2,3 and α-2,6 sialyltransferases (degree of sialylation and linkage)
5. CMP-NeuAc-4-hydroxylase (conversion of NeuAc to NeuGc)

Mathematical modeling of this complex process has recently been attempted to permit comparison of strategies to obtain specific protein glycoform distributions. The biosynthesis of O-linked glycans is similar to that of *N*-glycans and undergoes similar competition and regulation of its glycosyl transferases. It has been well reviewed by Schachter and Brockhausen, and four of the basic core structures and known biosynthetic pathways are outlined in Figure 19.3. Although many of the glycosyl transferases involved in *N*-glycan processing have been identified and cloned, most of the *O*-glycan transferases, with the exception of the core 2 GlcNAc-transferase, have not yet been cloned. This is in part due to the low percentage of glycans present on recombinant proteins that are O-linked.

REGULATION OF PROTEIN GLYCOSYLATION

Underpinning our ability to control protein glycosylation is the necessity to monitor and in some cases completely characterize the recombinant glycoprotein product. More routine profiling of

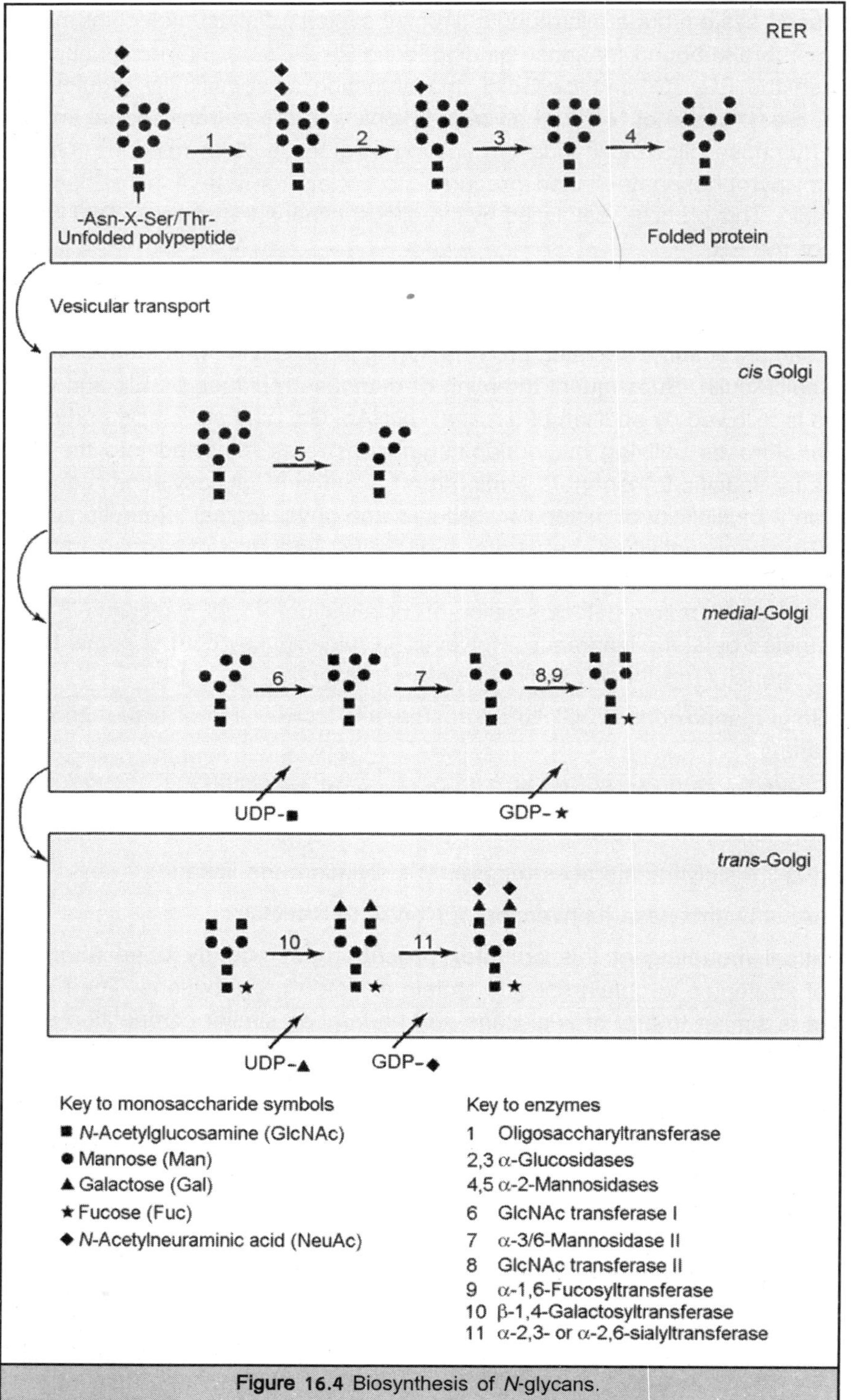

Figure 16.4 Biosynthesis of *N*-glycans.

glycoform populations may also be useful to ensure reproducible product quality, batch-to-batch consistency, and product stability. Recent advances in analytical technology now permit relatively rapid and detailed assessment of glycoprotein and glycan structure by a variety of different methodologies, yielding information of varying complexity. The choice of analytical strategy employed for characterization of a recombinant glycoprotein will depend on factors such as level of structural information required (e.g., site-specific glycosylation, oligosaccharide structure or monosaccharide composition, etc.), sample composition, and critically, the amount of protein available for analysis. Although some techniques, such as mass spectrometry, are now proving indispensable for many applications, it is a combination of complimentary techniques that will often provide the total analytical solution. The principal techniques currently employed are highlighted in the following sections.

Mass Spectrometry

Increasingly prevalent among current analytical strategies are mass spectrometric methods. Since the early 1980s, the emergence of "soft" ionization methods for biopolymer characterization by mass spectrometry (MS), such as fast atom bombardment (FAB), electrospray ionization (ESI), and matrix-assisted laser desorption/ionization (MALDI), has provided a formidable weapon in the armory of the analytical biochemist.

Fast atom bombardment mass spectroscopy

FAB-MS and its variant, liquid secondary ion MS (LSI-MS) have an upper mass limit of ~15 and ~25 kDa, respectively, and thus can be used only to directly characterize smaller glycoconjugates such as glycopeptides. Because molecules tend to fragment in the ion source, this technique has been employed successfully to provide detailed sequence, linkage, and branching pattern information for N- linked glycans; however, derivatization of the carbohydrates by permethylation, peracetylation, or periodate oxidation is frequently necessary to obtain strong signals and provide linkage information. This technique is not generally amenable to rapid analysis of mixtures of glycopeptides or oligosaccharides, and successful operation of instrumentation and interpretation of spectra requires substantial specialist training.

Matrix-assisted laser desorption/ionization mass spectroscopy

MALDI-MS is a relatively simple technique. Biopolymers (theoretically up to 300 kDa) are cocrystallized on a metal surface in the presence of a molar excess (typically >5000:1) of a low molecular weight, strongly UV absorbing matrix. Biopolymer and matrix ions are desorbed by a short laser pulses (typically at 337 nm, 5 ns), and the gas phase ions are electrostatically directed into a time-of-flight (TOF) mass analyzer. Analyte resolution (width of a peak at half-height divided by the mass/charge (m/z) ratio) using MALDI TOF-MS (typically <500) is ±0.1% for continuous ion extraction MALDI or >0.01% for MALDI with delayed ion extraction. This technique can be used to identify and characterize heterogeneous samples of glycopeptides, and released oligosaccharides (enzymatically or chemically), and for the simultaneous sequencing of oligosaccharides in conjunction with exoglycosidase digestion. Recent reports highlight the use of MALDI-MS to rapidly monitor the site-specific glycosylation of a recombinant glycoprotein by interfacing it with HPLC-based peptide mapping. There are few examples of MALDI-MS glycoform profiling using whole glycoproteins, a result of the extensive heterogeneity of natural and recombinant glycoproteins. However, glycoproteins of less than 20 kDa, with perhaps only a single occupied glycosylation site, maybe amenable to more detailed analysis. For example, variably sialylated forms of CHO-cell-derived recombinant interleukin 4 (~17 kDa) have been observed by Tsarbopoulos et al.

Electrospray ionization mass spectroscopy

ESI, typically in combination with a single- or triple-quadrupole mass analyzer, is capable of determining the molecular weight of whole recombinant glycoproteins or glycopeptides up to 100 kDa, with a mass accuracy of ±0.01% at a resolution routinely in excess of 1,000. Ionization of biopolymers in volatile solvents results in the production of multiply charged molecular ions (averaging 1 proton/kDa), and conditions are sufficiently mild that noncovalent interactions can often be observed because covalent bonds are not disrupted. The resulting mass spectrum of multiply charged molecular ions is then "deconvoluted" to yield a single zero-charge peak. The advantage of ESI-MS is that it may be coupled on-line with liquid chromatography (LC) or capillary electrophoresis (CE) to analyze complex samples such as proteolytic digests at low picomole to femtomole sensitivity. The increased resolution and softer ionization afforded by ESI-MS makes this technique the most attractive method for direct analysis of recombinant glycoprotein populations because individual glycoforms can often be resolved. For example, heterogeneous glycoform populations of recombinant human interferon γ have been analyzed by James et al.

ESI-MS interfaced with LC has proved to be the method of choice for direct, sensitive analysis of protein digests. Selective, on-line identification of glycopeptides within these complex mixtures can be achieved by the use of operating conditions that enhance the formation of oxonium fragment ions (e.g., $HexNAc^{+}$ = 204 Da), such as collision-induced dissociation. Isolated glycopeptides may be analyzed in detail by ESI-MS directly or in combination with exoglycosidase digestions to provide a more extensive site-specific structural analysis of N- and O-linked oligosaccharides. Although ESI-MS may be interfaced with CE for analysis of glycoproteins and glycopeptides, some CE separation protocols rely on acidic buffers, which are compatible with ESI-MS but unsuitable for separation of sialylated glycoconjugates.

Liquid Chromatography

Liquid chromatographic techniques have been extensively employed to separate underivatized and derivatized oligosaccharides released enzymatically or chemically (e.g., by hydrazinolysis) from glycoproteins. These approaches can most conveniently be described as being based on oligosaccharide size or charge. Low-resolution separations based on hydrodynamic volume, such as those employing gel permmeation chromatography, were utilized in combination with limited exoglycosidase digestion of oligosaccharides to provide structural information. However, this approach has largely been superceded by the development of high-resolution normal-phase separations using amide-based media, in combination with simple procedures to fluorescently label oligosaccharides at the reducing terminus. This technology may also be used in combination with exoglycosidase sequencing and for analysis of glycoproteins in polyacrylamide gels. Charge-based separations, distinguishing between differently sialylated glycans, have typically involved resolution of borohydride-reduced oligosaccharides by high-pH anion exchange chromatography with pulsed amperometric detection (HPAE-PAD) (76), although effective separation of fluorophore-derivatized glycans has also been demonstrated with amine-based columns now available commercially.

Capillary Electrophoresis

CE is proving itself to be a versatile analytical tool for (glyco)protein analysis both by itsel and in combination with other analytical techniques such as mass spectrometry. There are now several examples of glycoprotein analysis by CE, from analysis of intact glycoproteins, glycopeptides and released oligosaccharides, to individual monosaccharides. These examples employ a variety of separation mechanisms (e.g., capillary zone electrophoresis, micellar electrokinetic capillary chromatography, capillary isoelectric-focusing) designed to enhance the resolution of the specific analyte(s) in question.

Gel Electrophoresis

Separation of released oligosaccharides in polyacrylamide gels after derivatization with fluorescent labels at the reducing terminus provides a simple, inexpensive method for profiling oligosaccharide populations. Limited structural information is obtained in combination with exoglycosidase digestions.

REGULATORY MECHANISM

Given the ability to detect and analyze glycoproteins in detail, how can recombinant protein glycosylation be controlled? There are five main options currently available:

1. Choice of host cell
2. Genetic engineering of the host cell
3. Genetic engineering of the recombinant glycoprotein
4. Control of the host cell environment
5. Remodeling of protein glycans in vitro

Selection of Host Cell

Different host cell types are known to vary in their ability to glycosylate recombinant proteins. As there is now a profusion of expression systems available (mammalian, insect, or yeast cells in culture; transgenic animals and plants), initial choice of expression system is crucial and will, to a large extent, govern the type of glycosylation pattern obtained. Common bacterial expression systems such as *Escherichia coli* have no capacity to glycosylate proteins in either N- or O-linked conformations.

Yeast

Most yeast strains tend to hypermannosylate (the addition of a large number of mannose residues to the core trimannosyl oligosaccharide) recombinant proteins. We may infer that this modification would speed clearance of the glycoprotein from the bloodstream by hepatic and macrophage mannose receptors, and it has been shown to reduce the efficacy of recombinant hepatitis B vaccine. It is possible to use mutant yeast strains (e.g. mnn-9 or the temperature-sensitive ngd-29) that do not hypermannosylate recombinant proteins. In these strains, N-glycosylation is confined to core oligosaccharide residues containing up to eight mannose residues.

Insect cell lines

The first steps in *N*-glycan biosynthesis appear to be highly conserved throughout all eukaryotic cells, including insect cells, such as those commonly derived from the lepidopteran insect *Spodoptera frugiperda* (*Sf*9). For example, Kawar et al. recently demonstrated that *Sf*9 cells encode and express an α1,2-mannosidase, similar to other eukaryotic enzymes, that releases mannose from Man_9, $GlcNAc_2$. In addition, an α1,2-mannosidase has been isolated from virus-infected *Sf*21 cells; it acts preferentially on Man-6 oligosaccharides. The authors suggest that this is a pivotal step in glycan processing that determines whether complex or hybrid structures are formed. However, numerous studies have shown that baculovirus-infected insect cells differ markedly from mammalian cells in subsequent glycan processing capability. These perceived limitations in glycosylation have hindered the wide acceptance of insect cell systems for large-scale production of human therapeutics. Whereas mammalian cells are generally capable of complex-type *N*-glycan processing, insect cells produce proteins with truncated glycans, notably trimannosyl core structures ($Man_3GlcNac_2$) plus or minus fucose.

The molecular basis for the accumulation of trimannosyl *N*-glycans on proteins secreted by some insect cells (notably *Sf*-derived cell lines) has recently been elucidated. After initial transfer of *N*-acetylglucosamine onto Man-5 by a GlcNAc transferase I identified in insect cells, subsequent removal of mannose and *N*-acetylglucosamine by a Golgi-specific α-mannosidase II and β-*N*-acetylglucosaminidase occurs, yielding the trimannosyl core structure. In some insect cell lines, specifically those derived from *Estigmena acrea*, β-*N*-acetylglucosaminidase activity is absent, and consequently, *N*-glycans bearing terminal *N*-acetylglucosamine residues are present. Further *N*-glycan processing is cell line specific. For example, several studies report the inability of *Spodoptera*- (*Sf*9), *Estigmena*-, and *Trichoplusia*-derived cell lines to sialylate *N*-glycans. However, *Mamestra brassicae* and *Manduca sexta* cells have been reported as being capable of sialylating the *N*-glycans of recombinant human plasminogen. This group also reported that *Sf*21 cells were capable of synthesizing complex *N*-glycans, but that this was dependent upon time postinfection—complex *N*-glycans appearing at later stages of the infection cycle.

Mammalian cell lines

Although phylogenetically more similar to humans and generally capable of complex-type oligosaccharide synthesis, commonly used cell lines derived from rodent tissues maintain some key differences in glycan processing. Firstly, murine cell lines (e.g., hybridomas, mouse–human heterohybridomas, C127 cells, and NS0 cells) express the enzyme α-1,3-galactosyltransferase (α-1,3-GalT), which leads to the synthesis of glycans terminating in Galα1,3-Galβ1,4-GlcNAc motifs. As more than 1% of human serum IgG is directed against this epitope, its association with a therapeutic protein would be undesirable, resulting in rapid clearance. In humans, apes, old world monkeys, and CHO cell lines the α-1,3-GalT gene has become inactivated through frameshift mutations. Secondly, in contrast to human cells, murine cell lines have a tendency to confer *N*-glycolylneuraminic acid (NeuGc), an oncofetal antigen and immunogenic derivative of *N*-acetylneuraminic acid (NeuAc).

The hydroxylase enzyme that converts CMP-NeuAc to CMP-NeuGc is cytosolic and has recently been cloned. NeuGc has been found to be associated with the oligosaccharides of an IgM produced by a human–mouse hybridoma, where it was twice as prevalent as NeuAc. Similarly, comparison of glycans associated with a chimeric IgG produced by CHO and NS0 cells revealed modification of NS0-derived glycans with NeuGc, and not NeuAc, which was linked to enhanced rate of plasma clearance.

Proteins produced by CHO cells have a low NeuGc content, typically 1–2% of total sialic acids, and are generally not immunogenic, although higher levels of substitution with NeuGc may elicit an immune response. CHO and baby hamster kidney (BHK) cell lines lack the functional α-2,6-sialyltransferase (ST) enzyme found in human and murine cells and synthesize exclusively α-2,3-linked terminal sialic acids. This alteration has been addressed by transfection of CHO and BHK cells with an α-2,6-ST gene, as discussed later. Lastly, although not of immunological significance, another variable *N*-glycan modification that may be associated with recombinant proteins derived from CHO cells is the addition of poly-*N*-acetyllactosamine repeats (Galβ1-4GlcNAcβ1-3-) to terminal Gal residues on the antennae of complex oligosaccharides.

Transgenic animals

There are very few studies describing the glycosylation of recombinant proteins produced in the mammary gland of transgenic animals. The few reports that exist indicate that glycan processing is altered. James et al. compared the site-specific glycosylation of recombinant human interferon-γ (IFN-γ) produced by CHO cells, *Sf*9 insect cells, and the mammary gland of transgenic mice. In the latter case, *N*-glycans at one site (Asn_{97}) were predominantly truncated or of the high-mannose type, whereas *N*-glycans at the other site (Asn_{25}) were of the complex, sialylated variety similar to those associated with IFN-γ from CHO cells.

Similar data has recently been reported for recombinant antithrombin produced in transgenic goat milk, where one of three N-linked sites (Asn_{155}) in the transgenic-derived protein was unaccountably associated with high-mannose-type glycans, not complex glycans as in the natural human protein. Furthermore, the recombinant antithrombin contained a GalNAc for Gal substitution on some *N*-glycans, was highly fucosylated and less sialylated than natural human antithrombin, and contained both NeuAc and NeuGc.

The total extent of glycosylation of a recombinant protein produced by transgenic animals may also differ from that of its natural counterpart. A recombinant erythropoietin fusion protein produced in transgenic mouse and rabbits was less glycosylated overall than the molecule produced in CHO cells. In addition, recombinant human bile salt–stimulated lipase (BSSL) derived from transgenic mouse milk exhibited a significantly lower degree of O-glycosylation than was found for the native enzyme or for recombinant CHO or C127 cell–derived BSSL. In both cases, however, biological activity was not compromised.

Transgenic plants

Transgenic plants are now being evaluated as a viable option for therapeutic protein production. However, not surprisingly, *N*-glycan processing in plant cells is substantially different from that obtained in mammalian cell systems. *N*-Glycans universally lack terminal sialic acid residues, and potentially allergenic residues such as core α1-3-linked fucose, xylose or β1-2-mannose residues have been detected. Recombinant erythropoietin (EPO) produced in transgenic tobacco cells was found to have no biological activity in vivo, possibly because of its lack of sialylation and consequently high clearance rate.

Manipulating Genes in Host Cell

As already outlined, the glycosylation of recombinant glycoproteins is generally host cell specific, being largely dependent on the complement of specific glycosyltransferases available for glycan processing. Thus, the majority of efforts to augment glycan processing in various host cells by genetic means has involved introduction of a cloned heterologous glycosyltransferase with a desirable processing activity that is absent in the parental host cell. It should be stated, however, that successful introduction of a new biosynthetic activity into a host cell by this route depends on the preexisting capability of the cell to synthesize the requisite nucleotide sugar substrate and the ability to transport it into the Golgi lumen using a specific nucleotide sugar transporter. For this reason, our ability to engineer the protein glycosylation capabilities of yeast and insect cells is restricted at present.

In the case of CHO and BHK cells, which both lack a functional α-2,6-sialyltransferase (α-2,6-ST), terminal sialylation of *N*-glycans has received the most attention. This is primarily for two reasons, (i) the potential for modulation of the pharmacokinetic behavior of a therapeutic protein product: the relationship between terminal sialylation of *N*-glycans and rate of protein clearance from the bloodstream (mediated by hepatic asialoglycoprotein receptors) is now well documented, and (ii) the production of recombinant proteins with "authentic" humanlike *N*-glycans consisting of a mixture of α2-3- and α2-6-linked sialic acids. However, experiments involving cotransfection of cells with cloned α-2,6-ST cDNA have met with variable success: although α2-6-linked sialic acids have been shown to occur in all cases, the absolute level (mole per mole) of recombinant protein sialylation was either not determined, unaffected, or increased in some transfectants but not others.

A soluble form of the α-2,6-ST expressed in BHK cells did not modify a model secreted recombinant glycoprotein, indicating that functional activity is dependent upon correct localization of the heterologous transferase. Recent reports highlight the potential use of CHO cells expressing heterologous fucosyl transferases for the production of bioactive oligosaccharides, specifically Golgi membrane and secreted forms of fucosyl transferase III for Lewis (a) synthesis, and fucosyl transferase VI for sialyl Lewis X synthesis.

In the latter case, expression of the introduced glycosyltransferase was controlled using antisense technology. There have also been recent attempts to engineer baculovirus-infected insect cell expression systems for improved recombinant protein glycosylation. This has been carried out directly, for example, via the stable expression of β1-4-galactosyltransferase in *Sf*9 insect cells, which resulted in the presence of corresponding terminal galactose residues on glycans associated with human tissue plasminogen activator secreted after recombinant baculovirus infection.

Alternatively, the glycosylation of proteins produced in insect cells may be improved by the use of baculovirus expression vectors employing the immediate-early promoter element, permitting the production of recombinant glycoproteins with more extensively processed N-linked oligosaccharides. There is only one example where the transgenic animal mammary gland has been engineered for altered glycan processing. Prieto et al. expressed human α-1,2-fucosyltransferase (α-1,2-Fuc T), which generates the H blood group antigen, in transgenic mice. Milk from these animals contained soluble α-1,2-Fuc T, as well as 2′- fucosyllactose (Fucα1-2Galβ1-4Glc) and modified glycoproteins containing the H antigen.

Manipulating Recombinant Proteins

There is now a well-documented link between local protein structure and glycosylation at a particular site, in other words, glycosylation of a protein produced in a single cell type can be different at individual glycosylation sites. This may arise due to steric hindrance of glycosyltransferases, as in the case of the Fc *N*-glycans of IgG, or other peptide–carbohydrate interactions. Thus, by utilizing protein engineering technology such as site-directed mutagenesis, the possibility arises that glycosylation sites associated with undesirable glycans (e.g., high-mannose) may be deleted, new sites may be created, or local protein structure may be redesigned to alter glycan processing at a glycosylation site.

Two current examples are highlighted: engineered forms of tissue plasminogen activator (tPA) and IgG. Recombinant human tPA produced by CHO cells has three potential N-glycosylation sites at Asn_{117}, Asn_{184} (variably occupied), and Asn_{448}, and one O-linked site at Thr_{61}, which is associated with a single fucose residue. *N*-Glycans at Asn_{117} are invariably of the high-mannose type, in contrast to the complex-type *N*-glycans at the other Asn sites, and are responsible for promoting the clearance of tPA from circulation via interactions with macrophage and hepatic mannose receptors.

Thus, second-generation tPA molecules, TNK-tPA, have been mutated to remove Asn_{117} (N117Q), with the introduction of a new glycosylation site, Asn_{103} (T103N). The latter mutation restores fibrin binding activity to the wild-type tPA molecule. Significantly, the TNK-tPA variant exhibits complex-type *N*-glycans at Asn_{103} and has a 10-fold increase in its serum half-life. Human IgG is glycosylated at Asn_{297} in the Fc region, predominantly with a mixture of variably galactosylated, minimally sialylated biantennary oligosaccharides.

Numerous studies have shown that variations in the structure of these glycans can serve to modulate antibody effector functions via noncovalent interactions between carbohydrate and local amino acid residues and Fc receptors.

Lund et al. mutated a number of amino acid residues proximal to the Asn_{297} of a recombinant IgG3 produced in CHO cells; specifically, residues 241, 243, 264, 265, and 301 were replaced with alanine in each case. These mutations resulted in increased galactosylation and sialylation relative to the wild-type oligosaccharide chains, although there was a reduction in antibody effector functions in some cases.

Contribution of Cellular Environment

For most recombinant proteins produced by mammalian cells in culture, the cell environment itself can contribute to product molecular heterogeneity. The following factors may all influence the structural diversity of glycans associated with recombinant therapeutic proteins or antibodies:

1. Serum
2. Ammonium
3. Extracellular glycosidases
4. Method of cell culture

Serum

In light of recent bovine spongiform encephalitis epidemics and the desire for control of a fully defined cell culture environment, there is a trend in industry to remove serum from culture media. However, the presence or absence of serum has been shown to affect protein glycosylation. For an IgG1 monoclonal antibody produced by hybridomas, Patel et al. demonstrated that ascites-derived protein differed from serum-free-culture-derived material only with respect to the content of sialic acid. IgG1 derived from culture in serum-containing media had an intermediate sialic acid content and a lower incidence of outer-arm galactosylation than the other two preparations.

However, for antibodies produced by CHO cells, *N*-glycan galactosylation may be improved in serum-containing medium. For other types of recombinant protein, serum content of the medium may also be significant. Adaptation of BHK-21 cells producing a recombinant mutant IL-2 from a serum-containing to a serum-free environment resulted in changes to *N*-glycan antennarity, sialylation, sialylation, and fucosylation, as well as an increase in the overall degree of protein glycosylation.

Ammonium

Increases in the concentration of the toxic catabolite ammonia during cell culture have been linked to various changes in protein glycosylation. Borys et al. demonstrated that exogenous additions of ammonium (as ammonium chloride) up to 9 mM decreased the overall glycosylation of recombinant mouse placental lactogen 1 produced by CHO cells in a pH-dependent manner, this effect being most pronounced at elevated pH (8.0). Subsequently, it was shown that ammonium ion concentrations above 2 mM specifically inhibited the sialylation of recombinant granulocyte-macrophage colony-stimulating factor produced by CHO cells. This effect was thought to be mediated by an ammonium-induced elevation of the acidic *trans*-Golgi compartment pH, resulting in reduced sialyltransferase activity in vivo, a concept originally explored by Thorens and Vassalli. In parallel with this work, it was demonstrated that an artificial increase in ammonium ion concentration induced an accumulation of UDP-*N*-acetylhexosamines in mammalian cells, with a concomitant decline in growth rate (169); in fact the "U ratio" (UTP:UDP-*N*-acetylhexosamine) was shown to be a sensitive indicator of cell growth rate.

Subsequently, increases in the antennarity of *N*-glycans associated with a human mutant IL-2 produced by BHK-21 cells were also correlated with an increased cellular concentration of UDP-*N*-

acetylhexosamines, as artificially induced by exogenous glutamine (7.5 mM) or ammonium (15 mM). In another report, similar increases in the cellular UDP-*N*-acetylhexosamine pool were induced by incubation of BHK cells with glutamine and uridine, which increased *N*-glycan antennarity but not sialylation. This suggests that the effect of ammonium on sialylation alone is due to another mechanism—probably changes in *trans*-Golgi pH, as discussed above. In BHK cells, it is now likely that ammonium ions elevate cellular UDP -*N*-acetylhexosamine levels by direct incorporation of ammonium into fructose-6-phosphate to yield a synthetic precursor of UDP-GlcNAc, glucosamine-6-phosphate; the reaction being catalyzed by glucosamine-6-phosphate isomerase.

Presumably UDP-GlcNAc transport into the Golgi is stimulated, increasing the availability of substrate for resident GlcNAc transferases III and IV involved in *N*-glycan branching. In practice, therefore, excessive ammonium concentrations in cell culture should be avoided. Alternatively, use of the glutamine synthetase (GS) expression system avoids this issue. Ammonia is utilized enzymatically by the transfected selectable marker, GS, and does not therefore accumulate during cell culture.

Extracellular glycosidases

Sialidase, β-galactosidase, β-hexosaminidase, and fucosidase have been identified in cell culture supernatant from CHO and other mammalian cell lines. At neutral pH the most active is sialidase, which has been purified from CHO cell culture supernatant. Gu et al. have demonstrated that sialylation of recombinant human IFN-γ produced by CHO cells decreased in parallel with loss in cell viability in batch culture. Therefore elimination of the sialidase from CHO cells constitutes a viable target for cell engineering. However, the fucosidase purified from CHO cell culture supernatant could not release fucose from the recombinant CHO cell–produced glycoproteins gp120 or soluble CD4 with the Fucα1-6GlcNAc linkage, or from human serum α(1)-acid glycoprotein with the Fucα1-3GlcNAc linkage.

Cultural methodology

The type of culture used to produce a recombinant protein affects its glycosylation. For example, during batch and fed-batch cultures of antibodies produced by NS0 cells, and during recombinant human IFN-γ production by CHO cells, there was a progressive increase in the proportion of high-mannose (predominantly $Man_5GlcNAc_2$) and truncated *N*-glycans associated with the recombinant protein. The increased appearance of high-mannose glycans suggests a change in the glycan processing capability of cells as a function of time in culture (in this case presumably affecting GlcNAc transferase I activity, by as yet unknown mechanisms). This may also be true for truncated *N*-glycans (lacking terminal monosaccharides), although at present it is more likely that these derive from the action of extracellular glycosidases, as previously mentioned. These considerations emphasize the necessity to harvest product from batch cultures prior to significant cell lysis.

Furthermore, As a means of ensuring reproducible product quality, batch-to-batch consistency, and stability, deleterious changes to recombinant glycoprotein product heterogeneity may be averted by implementation of appropriate monitoring technologies capable of rapid quantitative analyses of intact glycoprotein populations, enabling control of bioreactor operations (e.g., to avoid late harvest of undersialylated product). Few reports describe the effect of long-term continuous culture on the processing and modification of complex recombinant proteins.

Goldman et al. have shown quantitatively that both long-term fluidized-bed perfusion and batch cultures of CHO cells produce recombinant human IFN-γ with a consistent and highly comparable degree of glycosylation, monitored as glycosylation site occupancy and level of *N*-glycan sialylation. However, as previously mentioned, the sialylation of recombinant IFN-γ produced in stirred-tank batch culture declined significantly during the stationary and death phase. Differences in the sialylation of recombinant human tissue kallikrein produced by CHO cells from microcarrier culture and from a serum-free suspension cell recycle process have been reported, in which a reduced degree of *N*-glycan sialylation was observed with the microcarrier process.

Other factors

Other factors may affect glycosylation of recombinant glycoproteins produced by animal cells in culture to a limited extent, although such effects are likely to be host cell and recombinant protein specific. For example, mild hypoxia and elevated pCO_2 have minimal effects on the glycosylation of CHO cell–derived tPA. In perfusion culture of BHK-21 cells producing a recombinant mutant IL-2, nutrient limitations (glucose, amino acids, dO_2) led to short-term changes in macroheterogeneity, but microheterogenity was largely unchanged.

Synthesis of In Vitro Oligosaccharide

Advances in the chemical synthesis of oligosaccharides, notably using cloned glycosyltransferases, now permit the synthesis or modification of *N*-glycan-type structures. Thus, postharvest remodeling of *N*-glycans on purified proteins is now possible (e.g., to fully sialylate *N*-glycans for improved protein half-life in vivo, although cost constraints may limit the commercial application of this approach.

CONCLUDING REMARKS

The glycosylation of a potential recombinant therapeutic protein will contribute to a set of product characteristics relating to product efficacy, including bioactivity and/or pharmacokinetic behavior, although these will be protein specific. Initial choice of expression system is of crucial significance and will define the structure of oligosaccharides associated with individual protein glycoforms, although other factors implicit in large-scale bioprocess development may also be significant. At present, the most realistic goal may be a consistent, reproducible, humanlike (albeit heterogeneous) glycosylation pattern. However, an ideal production process would perhaps result in a single, defined protein glycoform, rather than a mixture of glycoform variants, each with varying characteristics.

This may be considered as analogous to chiral separation of small-molecule drugs, where there is a necessity for optically pure molecules for medical use. Such control of recombinant protein glycosylation may possibly be achieved at a number of levels: (i) genetic control of glycan processing in the host cell, (ii) engineering recombinant proteins for a defined glycosylation status, (iii) control of cell physiology in a bioreactor environment for enhanced production of specific glycoforms, and (iv) downstream processing to selectively recover specific glycoforms. Incorporation of any of these developments into a bioprocess must involve a consideration of clinical efficacy and potency, technical feasibility, cost of goods, and regulatory requirements. Central to these issues is our ability to quantitively monitor and completely characterize heterogeneous biological products. Glycoproteins can now be analyzed by a diverse array of techniques, yielding structural information of varying complexity. Recent advances in analytical methodology permit relatively rapid, detailed assessment of glycan structure, glycosylation site-occupancy, and particularly sialylation (extent, type of sialic acid, linkage). More routine fingerprinting of glycoform populations may also be useful to ensure reproducible product quality, batch-to-batch consistency, and stability.

17

ENZYME KINETICS

The kinetic behavior of enzymes has been studied in detail for a century, beginning with the classic work of Henri and Michaelis and Menten. The objectives have been threefold: to gain an understanding of the mechanisms of enzyme action; to illuminate the physiological roles of enzyme-catalyzed reactions; and, mainly in recent years, to manipulate enzyme properties for biotechnological ends. Experimental practice has been overwhelmingly dominated by the first of these aims; most experiments have been designed as if shedding light on the mechanism was the principal, or even the only, objective. However, although much valuable information has been obtained in this way, there are some important aspects of enzyme function that are obscured when working mainly with isolated enzymes in conditions far removed from those that exist in the cell. For this reason the latter part of this chapter will be devoted to a discussion of enzymes as they behave in complex mixtures.

MICHAELIS-MENTEN KINETICS

The starting point for any discussion of enzyme kinetics is the Michaelis–Menten equation, which expresses the initial rate v of a reaction at a concentration a of the substrate transformed in a reaction catalyzed by an enzyme at total concentration e_0:

$$v = \frac{k_0 e_0 a}{K_m + a} = \frac{Va}{K_m + a} \qquad \text{...(1)}$$

The parameters are k_0, the catalytic constant, and K_m, the Michaelis constant. The form shown in the middle is more fundamental than that on the right, but the second form, in which $k_0 e_0$ is written as the limiting rate V, is often used because the enzyme concentration in meaningful units is often not known. V is the limit that v approaches as the enzyme becomes saturated (that is, when a becomes very large) and K_m is the value of a at which $v = 0.5V$ (that is, at which the rate is half-maximal). The ratio k_0/K_m is called the specificity constant and given the symbol k_A: It is a more fundamental constant than K_m in the analysis of enzyme mechanisms (i.e., it has a simpler mechanistic meaning), but the equation is usually written in terms of V (or k_0) and K_m nonetheless.

The principal assumption implied by equation 1 is that the rate of the reverse reaction is negligible: This may be because the reaction is irreversible for practical purposes, but even with reversible reactions the reverse reaction can be made negligible by measuring the rate in the absence of products and by extrapolating the rate back to zero time, that is, by estimating the initial rate at a time when no products have accumulated. Even if there is no significant reverse reaction, products can still affect the rate of the forward reaction because product inhibition is a common phenomenon. Another assumption is that the reaction is allowed sufficient time to reach a steady state because all reactions pass through an initial acceleration phase known as the transient state. This phase is normally very brief (a few milliseconds), and in practice enzymes are usually studied under steady-state conditions.

Note, however, that this is made experimentally possible by working with extremely small enzyme concentrations compared with those that may exist in the cell. The use of very low enzyme concentrations has two important consequences: First, it normally means that the enzyme concentration can be neglected in comparison with the substrate concentration, and second it makes the steady-state rate sufficiently slow to be easily measured and the steady-state phase long enough to be meaningful. If high enzyme concentrations were used, the transient state would be as brief as before, but the steady state would also be very brief, so that there would be no period in which one could adequately treat the rate as constant.

Equation 1 may be derived from the following model

$$A + E \underset{k_{-1}}{\overset{k_1}{\rightleftarrows}} EA \underset{k_{-2}}{\overset{k_2}{\rightleftarrows}} EP \underset{k_{-3}}{\overset{k_3}{\rightleftarrows}} E + P \qquad \ldots(2)$$

which assumes that the reaction passes through an enzyme–substrate complex EA, which undergoes catalytic transformation to an enzyme–product complex EP, which then breaks down to form products. Although real enzyme mechanisms may be more complicated than this, every reaction passes through steps of substrate binding, chemical transformation, and product release. In simple introductory treatments the second and third steps are often treated as a single step, but conceptually they are clearly distinct. In reactions of more than one substrate, the steps do not necessarily occur in the order one might guess; that is, some products may be released before all substrates have bound, but the general principle that any reaction involves the same three kinds of step remains valid.

The Michaelis–Menten parameters can be defined as follows in terms of the rate constants shown in equation 2:

$$k_0 = \frac{k_2 k_3}{k_{-2} + k_2 + k_3};\ k_A = \frac{k_1 k_2 K_3}{k_{-1}k_{-2} + k_{-1}k_3 + k_2 k_3};\ k_m = \frac{k_{-1}k_{-2} + k_{-1}k_3 + k_2 k_3}{k_1(k_{-2} + k_2 + k_3)} \qquad \ldots(3)$$

Note that none of the three parameters has a simple transparent meaning. The interpretations commonly attributed to them depend on additional simplifying assumptions that are not always correct. For example, K_m is often said to be equal to the equilibrium constant k_{-1}/k_1 for dissociation of A from EA, but the expression in equation 3 does not take this form unless k_2 is very small. As there is no good reason for k_2 to be small, and indeed ideas of evolutionary optimization of enzyme function lead one to expect the opposite when the enzyme is acting on its natural physiological substrate, it follows that K_m should not, in general, be regarded as a measure of the equilibrium dissociation constant.

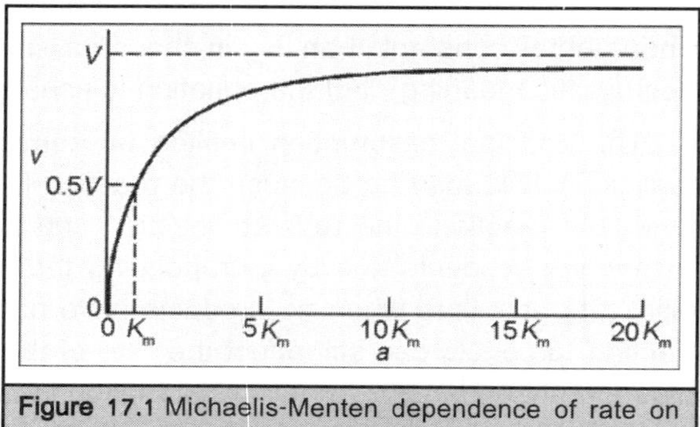

Figure 17.1 Michaelis-Menten dependence of rate on substrate concentration.

Despite these difficulties in providing a detailed mechanistic meaning to K_m, it does provide a measure of the tightness of substrate binding in the steady state, as it is quite correct to take a/K_m as equal to the ratio of the sum of concentrations of all enzyme complexes (i.e., both EA and EP) over the concentration of free enzyme. Similarly k_0 provides a valid measure of the capacity of the enzyme– substrate complex to react to give products, even if it cannot be interpreted as the rate constant for a unique step in the mechanism. The reason for the term *specificity constant* for k_A, that is, its relationship to enzyme specificity, will become clear once inhibition has been considered.

Hyperbolic Dependence

Equation 1 defines a hyperbolic dependence of rate on substrate concentration. The initial steep rise in *v* as *a* increases from zero is rapidly transformed into the phenomenon of saturation, whereby further increases in *a* produce smaller and smaller increases in *v*, which approaches, but does not reach or exceed, the limiting rate *V*. The rectangular hyperbola makes this type of plot inconvenient for estimating the values of the kinetic parameters (because the line does not approach the saturation limit closely enough at reasonable values of *a* to allow direct measurement of *V*). For this purpose, therefore, it is usual to transform equation 1 into one of the following three forms, which underlie the three straight-line:

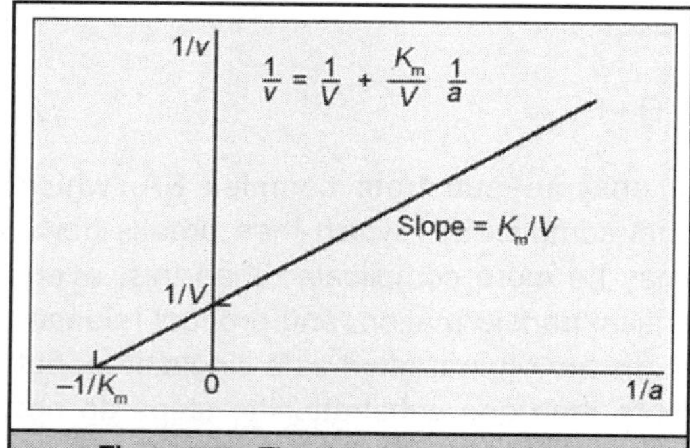

Figure 17.2 The double-reciprocal plot.

$$\frac{1}{v} = \frac{1}{V} + \frac{K_m}{V}\frac{1}{a} \quad \text{...(4)}$$

$$\frac{a}{v} = \frac{K_m}{V} + \frac{1}{V}a \quad \text{...(5)}$$

$$v = V - K_m\frac{v}{a} \quad \text{...(6)}$$

The double-reciprocal plot, based on equation 4, is the mostly widely used, but it is also the least satisfactory because it distorts the effect of experimental error to such an extent that it is difficult to form any visual judgement of where the best line should be drawn. The other two plots are better, and the plot of v against v/a has the particular advantage that the entire observable range of v values, from 0 to V, is mapped onto a finite range of graph; this makes it easy to judge by eye if an experiment has been well designed. On the other hand, it has the disadvantage that v, normally the less reliable measurement, contributes to both coordinates, and errors in v cause deviations along lines through the origin, rather than parallel with one or the other axis. In modern practice it is usually best to regard these plots as for illustration purposes only, and to use suitable computer programs for the actual parameter estimation. For this purpose it is not sufficient just to apply unweighted linear regression to the straight-line plots, as this suffers from the same statistical distortions as the plots themselves. The following two equations for calculating best-fit values of K_m and V give satisfactory results if the v values have uniform coefficient of variation (uniform standard deviation expressed as a percentage), as is usually at least approximately correct:

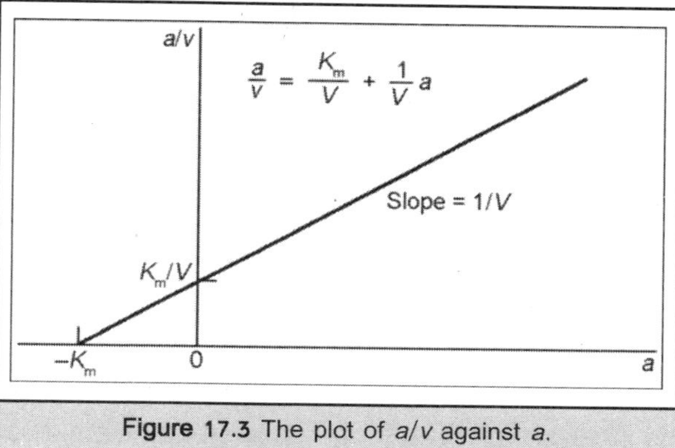

Figure 17.3 The plot of a/v against a.

$$K_m = \frac{\Sigma v^2 \Sigma(v/a) - \Sigma(v^2/a)\Sigma v}{\Sigma(v^2/a^2)\Sigma v - \Sigma(v^2/a)\Sigma(v/a)} \quad \text{...(7)}$$

$$V = \frac{\Sigma(v^2/a^2)\Sigma v^2 - [\Sigma(v^2/a)]^2}{\Sigma(v^2/a^2)\Sigma v - \Sigma(v^2/a)\Sigma(v/a)} \quad \text{...(8)}$$

Reactions with Two Substrates

Enzymes that catalyze reactions of a single substrate are only a small minority of all the enzymes known, but the Michaelis–Menten equation remains useful for examining the kinetics of the more common case of a reaction with two substrates and (often but not necessarily) two products, because such a reaction normally obeys Michaelis– Menten kinetics when only one substrate concentration is varied at a time. This is illustrated by the following typical equation for such a reaction:

$$v = \frac{k_0 e_0 ab}{K_{iA}K_{mB} + K_{mB}a + K_{mA}b + ab} \quad \text{...(9)}$$

Although at first sight appears quite different from equation 1, it can be arranged in the same form if one of the two substrate concentrations, for example b, is treated as a constant:

$$v = \frac{\left(\frac{k_0 b}{K_{mB} + b}\right) e_0 a}{\frac{K_{iA}K_{mB} + K_{mA}b}{K_{mB} + b} + a} \quad \text{...(10)}$$

The two fractions in this equation can be regarded as "apparent" values of the Michaelis–Menten parameters for A, that is, the equation can be written as

$$\nu = \frac{k_0^{app} e_0 a}{K_{mA}^{app} + a} \qquad \text{...(11)}$$

with

$$k_0^{app} = \frac{k_0 b}{k_{mB} + b}; \; k_A^{app} = \frac{(k_0/K_{mA})b}{(K_{iA}K_{mB}/K_{mA}) + b}; \; K_{mA}^{app} = \frac{K_{iA}K_{mB} + K_{mA}b}{K_{mB} + b} \qquad \text{...(12)}$$

Notice that the expressions for the apparent values of k_0 and k_A are both individually of Michaelis–Menten form with respect to *b*, whereas that for the apparent value of K_m is more complicated: This behavior is quite typical and is one of the reasons why k_0 and k_A should regarded as more fundamental parameters than K_m. More generally, the concept of apparent parameters pervades the analysis of simple cases in steady-state enzyme kinetics, being important for the study of reactions with multiple substrates and inhibition.

INHIBITION AND ACTIVATION

Inhibition

For most enzyme-catalyzed reactions, molecules exist that resemble the substrate closely enough to bind to the enzyme, but not closely enough to undergo a chemical reaction. Such a molecule is known as a competitive inhibitor and causes competitive inhibition, characterized by a rate equation of the following form:

$$\nu = \frac{k_0 e_0 a}{K_m(1 + i/K_{ic}) + a} \qquad \text{...(13)}$$

in which *i* is the concentration of the inhibitor and K_{ic} is the competitive inhibition constant. (The qualification "competitive" and the second subscript c are usually omitted if only this simplest kind of inhibition is being considered.)

Inhibitors can interfere with catalysis as well as with substrate binding. In the simplest case, an inhibitory term affects the variable term in the denominator of the Michaelis–Menten equation, instead of the constant term:

$$\nu = \frac{k_0 e_0 a}{K_m + a(1 + i/K_{iu})} \qquad \text{...(14)}$$

This is called uncompetitive inhibition, and the inhibition constant K_{iu} is the uncompetitive inhibition constant. This is important as a limiting case of inhibition, but in its pure form it is not at all common. Much more often one has mixed inhibition, when both competitive and uncompetitive effects occur simultaneously:

$$\nu = \frac{k_0 e_0 a}{K_m(1 + i/K_{ic}) + a(1 + i/K_{iu})} \qquad \text{...(15)}$$

There is no particular reason for the two inhibition constants K_{ic} and K_{iu} to be equal, and most of the mechanisms one might propose to account for mixed inhibition lead one to expect them to be different, yet the case where $K_{ic} = K_{iu}$ is often given an undeserved prominence in discussions of inhibition, largely because experiments done many years ago suggested that it was a more common phenomenon than it is. This is called noncompetitive inhibition and its rate equation is the same as equation 15, but with both K_{ic} and K_{iu} written simply as K_i.

All of these kinds of inhibition are conveniently discussed in terms of apparent Michaelis–Menten parameters. In the general case (equation 15), these are as follows:

$$k_0^{app} = \frac{k_0}{1+i/K_{iu}};\ k_A^{app} = \frac{k_A}{1+i/K_{ic}};\ K_m^{app} = \frac{K_m(1+i/K_{ic})}{1+i/K_{iu}} \quad \ldots(16)$$

Note that the first two expressions have the same form, and both simplify to independence of *i* in the event that one or other inhibition term is negligible. The expression for the apparent value of K_m is more complicated, especially when one considers how it varies with the different types of inhibition: It increases with the concentration of a competitive inhibitor, it decreases as the concentration of an uncompetitive inhibitor increases, it may change in either direction as the concentration of a mixed inhibitor increases, or it is independent of inhibitor concentration if the inhibition is noncompetitive. In general it is simplest to regard k_A as the parameter affected by competitive inhibition, negligibly so when the competitive component is negligible, k_0 as the parameter affected by uncompetitive inhibition, negligibly so when the uncompetitive component is negligible, and K_m just as the ratio of the two, so $K_m = k_0/k_A$.

The effects of the different kinds of inhibition on the common plots follows naturally from equation 16. Any competitive effect affects the apparent value of k_A, hence, it increases the slope of the plot of $1/v$ against $1/a$, it increases the ordinate intercept of the plot of a/v against a, and it decreases the abscissa intercept of the plot of v against v/a. Conversely, any uncompetitive effect increases the ordinate intercept of the plot of $1/v$ against $1/a$, increases the slope of the plot of a/v against a, and decreases the ordinate intercept of the plot of v against v/a. When both components of the inhibition are present, both kinds of effects occur. As an illustration we may consider just one example, the effect of competitive inhibition on the plot of $1/v$ against $1/a$: Plots made at various different inhibitor concentrations produce a family of straight lines intersecting on the ordinate axis, the lack of effect on the ordinate intercept being a direct consequence of the lack of effect on the apparent value of V.

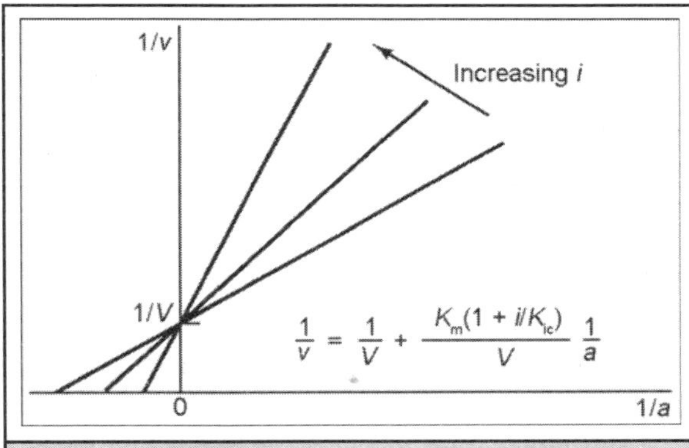

Figure 17.4 Effect of competitive inhibition on the double-reciprocal plot.

Specificity

Specificity is the most fundamentally important property of enzymes. Although one is often impressed by the catalytic effectiveness of enzymes, accelerating a reaction is not, in reality, difficult: Heating the reaction mixture in a sealed tube is an efficient way of accelerating virtually any reaction, essentially without limit. What is difficult is to accelerate one selected reaction without at the same time accelerating a great mass of unwanted reactions. What is important about an enzyme, therefore, is not that it is an excellent catalyst for a small set of reactions, but that it is an extremely bad catalyst—virtually without any activity—for all other reactions. In other words, the essential properties of enzymes are that they act under very mild conditions and are highly specific.

In the past, specificity was often assessed by comparing the kinetic parameters for different reactions measured in isolation from one another, which led to sterile arguments as to whether specificity was best measured in terms of k_0, K_m, or some combination of the two. This type of argument was resolved once it was realized that the only meaningful way of defining specificity is as a property of an enzyme that allows it to discriminate between substrates that are mixed together. The simplest way to consider this is with a model similar to that for competitive inhibition, except that one assumes that both molecules are capable of reacting. The equation for reaction of one substrate A in the presence of a competing substrate A' follows an equation similar to that for competitive inhibition (equation 13):

$$v = \frac{k_0 e_0 a}{K_m(1 + a'/K'_m) + a} \qquad \ldots(17)$$

with the inhibitor concentration replaced by *a'*, the concentration of A', and the inhibition constant by K'_m, the Michaelis constant for the reaction of A' considered in isolation. The rate of reaction of A' is given by the same equation with an obvious transposition of symbols:

$$v' = \frac{k'_0 e_0 a'}{K'_m(1 + a/K_m) + a'} \qquad \ldots(18)$$

It can then be seen that the ratio of rates is the ratio of substrate concentrations multiplied by the ratio of specificity constants:

$$\frac{v}{v'} = \frac{k_0/K_m}{k'_0/K'_m}\frac{a}{a'} = \frac{k_A a}{k'_A a'} \qquad \ldots(19)$$

This result, which is still less well known than its importance merits, is the reason for the term *specificity constant*. Note that although inspection of equation 1 suggests that *k*A is no more than the parameter that defines the rate at very low substrate concentrations, no assumption about the magnitudes of the concentrations was made in arriving at equation 19. It is thus valid at all concentrations, and the specificity constant measures specificity at all concentrations, not just low ones.

Activation

Activation is the opposite from inhibition, in which a reaction proceeds more rapidly in the presence of a particular molecule than in its absence. It is less common than inhibition, and discussion is complicated by the fact that a variety of quite different phenomena have been termed "activation." The most important of these is a confusion between true activation and the case where the activator is really a component of the substrate. Numerous ATP-handling enzymes are said to be activated by magnesium ions, when in reality the complex MgATP is the true substrate, that is, the species that reacts with the enzyme. Other metal ions, such as the zinc in a number of enzymes, may be true activators as they bind to the enzyme itself and confer catalytic properties on it. Another misuse of the term activation relates to coenzymes such as NAD in many dehydrogenases: Alcohol dehydrogenase, for example, may be said to be activated by NAD, but although consideration of the metabolic pathway in which the reaction occurs may make it convenient to regard ethanol as the substrate and NAD as the coenzyme, this is meaningless when the reaction is considered in isolation. So far as alcohol dehydrogenase is concerned, it catalyzes a reaction that requires two substrates, ethanol and oxidized NAD; the reaction will not proceed unless both are present, and neither has any more reason to be called the substrate than the other. When such improper uses of the term are excluded, there remain a number of enzymes for which the true inverse of inhibition occurs. In the simplest cases the equations are just the inverse of inhibition equations, with terms of the form i/K_i replaced by ones of the form K_x/x (for an activator X with activation constant K_x). However, the simplest cases constitute a smaller proportion of the whole than they do for inhibition. This is because whereas most inhibitors inhibit completely, in the sense that enzyme species with inhibitor bound retain no activity as long as the inhibitor remains bound, many enzymes subject to activation retain some activity in the absence of the activator. As a result, full analysis of activation is often more complicated than it is for inhibition, but this will not be discussed further here.

Irreversible Inhibition

The types of inhibition considered so far are examples of reversible inhibition; the inhibitor binds reversibly and catalytic activity returns when the inhibitor is released. Irreversible inhibition also occurs, in which the inhibitor either binds so tightly that for practical purposes it cannot be removed, or reacts with the enzyme and converts it irreversibly to a form that has no catalytic activity. These two cases are conceptually different, and the former is more correctly called tight-binding inhibition, rather than irreversible inhibition. However, they are not easy to distinguish in practice, and have similar practical effects and, hence, similar practical uses. Although irreversible inhibition has played a smaller part than reversible inhibition in the academic study of enzyme mechanisms, it has far greater industrial and pharmacological importance. This is because competitive inhibitors, the most common kind of reversible inhibitors, are almost completely ineffective in complete physiological systems, for reasons to be considered shortly.

By contrast, whenever irreversible (or tight-binding) inhibition occurs in a physiological system, it can be expected to have profound effects. Many toxic and pharmacologically active substances owe their effects to irreversible inhibition. Both tight-binding and irreversible inhibition manifest themselves in ways that allow them to be confused with noncompetitive inhibition, as in equation 15 with the two inhibition constants equal. This is because the practical effect of irreversible inhibition is not on any of the kinetic parameters in the Michaelis–Menten equation, but on e_0, the total enzyme concentration. However, a decrease in e_0 can be confused with a decrease in the apparent value of k_0, as they occur as a product in equation 1. Although uncompetitive inhibition affects k_0, it does so without affecting k_A, and thus also changes K_m. Decreasing k_0 without affecting K_m, similar to what one would observe if e_0 decreases, is a definition of noncompetitive inhibition.

Inhibitory Effects on Physiological Activities

Competitive and uncompetitive inhibition are sufficiently similar in their effects in artificial experiments on isolated enzymes that they are often not distinguished, and an uncompetitive component in mixed inhibition often passes unnoticed. Many inhibitors are described in the literature as competitive inhibitors in the absence of any real evidence of the type of inhibition. This sort of confusion can easily lead to the entirely false idea that they are similar in their effects in systems where the inhibited enzyme is mixed with other enzymes and catalyzes a step in the middle of a pathway.

In a typical experiment in vitro, one decides the concentrations of the various components in advance and measures the rate that results; however, this is very different from what happens in the cell. To a first approximation, an enzyme catalyzing a step in the middle of a pathway must transform its substrate at the rate at which it arrives, that is, within certain limits it has little or no effect on the rate of its reaction, but instead determines the concentrations of the metabolites around it. (This is an oversimplification, but is useful for discussion.)

It is useful therefore to transform equations 13 and 14 into expressions for *a* in terms of *i*:

$$a = \frac{\nu K_m(1+i/K_{ic})}{k_0 e_0 - \nu} \qquad \text{...(20)}$$

$$a = \frac{\nu K_m}{k_0 e_0 - \nu(1+i/K_{iu})} \qquad \text{...(21)}$$

However similar equations 13 and 14 may seem, their transformed versions are drastically different. Equation 20 shows a linear dependence of *a* on *i*, which means that increasing *i* can never result in uncontrolled increases in *a*. Even at an inhibition concentration equal to the inhibition constant, the substrate concentration is only doubled. By contrast, the curve defined by equation 21 is a rectangular hyperbola that produces a steep and uncontrolled rise in substrate concentration at quite moderate inhibitor concentrations.

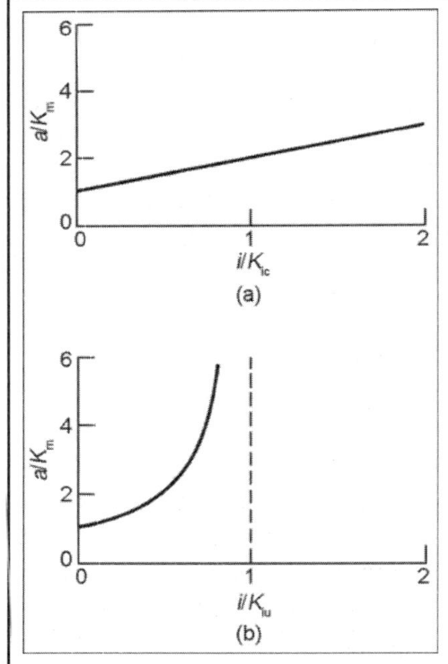

Figure 17.5 Effects of (a) competitive and (b) uncompetitive inhibition on the concentration of substrate in a constant-rate system.

The point is that in competitive inhibition, rises in substrate and inhibitor concentrations oppose one another—not only does the inhibitor compete with the substrate, but equally, the substrate competes with the inhibitor. In uncompetitive inhibition, however, these effects potentiate one another.

It follows that although it is relatively easy to find molecules that will act as competitive inhibitors, it is also largely useless as a strategy for designing pesticides or drugs because it is correspondingly easy for the organism to counteract the effect of the inhibition. To produce major metabolic effects one needs uncompetitive inhibitors, irreversible inhibitors, or tight-binding inhibitors: None of these are as easy to produce as weakly binding competitive inhibitors, but they are far more effective.

NON-MICHAELIS-MENTEN BEHAVIOR

All of the cases considered so far can be regarded as generalizations of the Michaelis–Menten equation (equation 1). However, although many enzymes do behave in this way, at least as a first approximation, there are some important exceptions. It is simple to calculate from equation 1 that if $a = K_m/9$ then $v = 0.1V$ and if $a = 9K_m$ then $v = 0.9V$; in other words, spanning the 10–90% range of available rates requires an 81-fold range of substrate concentrations, almost two orders of magnitude. Similar calculations may be done with any of the equations of the Michaelis–Menten type for additional substrates, inhibitors, or activators. Their implication is that as long as enzymes follow Michaelis–Menten kinetics, their rates cannot be adequately varied by manipulating concentrations of substrates, for example, because effective regulation will often require sensitivity to small changes in signals— certainly changes much smaller than two orders of magnitude. A second difficulty arises from the fact that inhibition of the types considered commonly derives from structural similarities between inhibitors and substrates or products, whereas there is no reason to expect the molecules needed for metabolic signals to resemble the substrates or products of the enzymes that need to respond to the signals.

In reality, the concentration of the end product of a pathway often serves as such a signal: too low, and the pathway needs to be activated; too high, and it needs to be inhibited. It is often found, therefore, that the enzyme that catalyzes the first committed step of a pathway, that is, the first step after a branch point, in the branch that leads to the end product in question, is inhibited by that end product. For inhibition of this kind to be possible, the enzyme must have a specific binding site for the end product, independent of the binding sites for substrates and products. Such a site is called an allosteric site, and the phenomenon is called allosteric inhibition. Because the need for it often coincides with the need for higher sensitivity than is provided by Michaelis–Menten kinetics and the common kinds of inhibition, allosteric inhibition is often cooperative: This means that the equations that define it are more complicated than those considered above, allowing, for example, a change from 10% to 90% inhibited over a concentration range much smaller than 81-fold, and typically less than 10-fold (though rarely, if ever, less than 3-fold). The fact that allosteric and cooperative behavior are often found together has led many authors to treat the two terms as synonymous, a tendency encouraged by the fact that in one of the most widely accepted models of non-Michaelis–Menten behavior, that of Monod, Wyman, and Changeux, both result from the same structural properties attributed to the enzyme. Nonetheless, most careful authors consider the two properties to be conceptually distinct and not necessarily occurring together, so the two terms should be considered distinct as well.

KINETICS OF MULTIENZYME SYSTEMS

As noted in the introduction, nearly all kinetic studies of enzymes have been carried out using isolated enzymes, and although this has been very valuable for arriving at a good understanding of the nature of enzyme catalysis, it is quite inadequate as a guide to how systems of enzymes will behave. One cannot assume that the flux through a metabolic pathway is a property of a unique enzyme catalyzing the rate-limiting step, and that the properties of the pathway as a whole can be deduced from studies of the kinetics of this one enzyme in isolation. Space does not permit a full analysis of this subject, which has developed from a classic paper by Kacser and Burns, but it should suffice to examine an example of a pathway in which biotechnologists have attempted to increase the flux by identifying the rate-limiting enzyme and using genetic manipulation to increase its activity.

Tryptophan biosynthesis in yeast provides such an example, tryptophan being a commercially valuable metabolite for which increased production would be very desirable. The tryptophan pathway consists of five enzymes, and in the classical model any one of these could be the "key" enzyme catalyzing the rate-limiting step. However, when the activity of each of these enzymes was increased in turn, either singly or at the same time as others in the pathway, the results were trivial: Increases of enzyme activity of 20-fold or greater produced flux increases of perhaps 30%. Only when all five enzyme activities were increased (or all but one, apparently unimportant, activity) was there a substantial increase in flux, which even then was much smaller than the increase in enzyme activity. The object here is not to analyze in detail why manipulation of tryptophan biosynthesis did not produce the desired results, but to use it to illustrate the point that the whole approach is misconceived. One cannot treat the kinetic behavior of systems of enzymes as if it were determined by the properties of a single component. Moreover, abundant evidence exists to show that all organisms have evolved regulatory mechanisms to control metabolic fluxes and concentrations to satisfy their own requirements.

Artificially trying to force more activity in a pathway by increasing the activities of certain enzymes simply stimulates the regulatory mechanisms to resist. Tryptophan biosynthesis is just one example of a general result, and similar efforts in other pathways—for example, increased alcohol production in yeast and increased starch production in potatoes—have produced similar results. The essential point is that flux control is not a property of a single enzyme in a pathway, but is shared among all the enzymes in the pathway. Strictly speaking, it is shared among all the enzymes in the cell, but it is usually safe to assume that enzymes catalyzing reactions remote from the pathway of interest have very little effect on the flux through it, so to a first approximation one can consider flux control to be shared among the enzymes of the pathway.

To express this idea in quantitative terms, one can define a flux control coefficient for any enzyme by the following equation:

$$C_i^J = \frac{\partial \ln J}{\partial p} \Big/ \frac{\partial \ln v_i}{\partial p} \qquad \ldots(22)$$

This definition compares the effect on the flux J through a pathway of some perturbation of the activity of the ith enzyme, represented by a change in the parameter p, with the effect the same perturbation would have on the rate v_i of the same reaction if it were considered in isolation. The identity of the parameter p does not have to be specified, because as long as it affects only one enzyme, the control coefficient defined by equation 22 is independent of the manner in which the flux and isolated rate are perturbed. However, to make the definition more concrete, consider the case where p is the logarithm of the enzyme concentration, e_i. As most reactions are considered under conditions where the rate is proportional to the enzyme concentration, it will often be true to write $\partial \ln v_i = \partial \ln e_i$, so that the denominator in equation 22 has a value of unity and the whole equation simplifies to the following:

$$C_i^J = \frac{\partial \ln J}{\partial \ln e_i} \qquad ...(23)$$

This equation is less abstract and simpler to understand than equation 22, and in the past was often used as a primary definition of a control coefficient. However, this is not recommended, first because it is not always true that the isolated rate is proportional to the enzyme concentration, and second because equation 23 can give the false impression that control coefficients are concerned only with effects brought about by changes in enzyme concentration.

Flux control coefficients are measures of how much the flux through a pathway is dependent on the activities of the individual enzymes. Mathematical analysis shows that they satisfy a property called the summation relationship:

$$\Sigma C_i^J = 1 \qquad ...(24)$$

The interpretation of this relationship would be straightforward if all flux control coefficients were always positive, but in reality negative flux control coefficients also occur; for example, if a pathway contains a branchpoint, the enzymes in one branch normally have negative flux control coefficients for flux through the other. Nonetheless, in practice, negative flux control coefficients are nearly always small in magnitude, so although the sum of positive flux control coefficients may be greater than 1, it is rarely much greater than 1, so that the idea of sharing flux control among all the enzymes of a system is reasonably accurate. It follows that we should not expect any enzyme to have complete control over flux, and the closer to complete control any enzyme approaches, the less control all of the others have, taken together. Moreover, a flux control coefficient is not a constant, but tends to decrease when the activity of an enzyme is increased. In other words, even if one can identify one enzyme that has a large proportion of the total flux control, increasing its activity will tend to decrease that proportion, so that the amount the flux can be increased by increasing the activity of one enzyme will always be small. Once this is understood, the failure to increase the flux to tryptophan (in the example mentioned) or to achieve significant flux increases (in other similar examples), ceases to be a mysterious result, but is simply what was to be expected.

CONCLUDING REMARKS

The topic of this article is referred to in all textbooks of biochemistry, but unfortunately the treatment is nearly always very superficial and potentially misleading. Since Mahler and Cordes, there has been no serious attempt to cover enzyme kinetics in a general biochemistry text. More specialized textbooks were abundant in the past, but have become much less so. There are, however, two recent ones that may be consulted for more information about most of the points discussed here, as well as topics such as the pH- and temperature-dependence of enzyme-catalyzed reactions, which require more detail than can be covered adequately in a short article. Fast reactions are not dealt with in this article, and as they require both techniques and methods of analysis different from those of steady-state kinetics, it is best to refer to a specialized source, such as Hiromi, who provides a thorough treatment. Two recent books deal with the kinetics of multienzyme systems, one at a very readable level, the other much more advanced. This is also covered much more briefly in two of the books mentioned eariier.

18

IMMOBILIZED ENZYMES

In 1916, Nelson and Griffin found that an enzyme in water-insoluble form showed catalytic activity. They reported that invertase extracted from yeast was adsorbed on charcoal, and the adsorbed enzyme showed the same activity as native enzyme. In 1948, Sumner found that urease from jack bean became water insoluble on standing in 30% alcohol and sodium chloride for 1 to 2 days at room temperature and that the water-insoluble urease was active. Thus, it has been known for a long time that enzymes in water-insoluble form show catalytic activity. However, the first attempt to immobilize an enzyme to improve its properties for a particular application was not made until 1953, when Grubhofer and Schleith immobilized enzymes such as carboxypeptidase, diastase, pepsin, and ribonuclease by using diazotized polyaminopolystyrene resin. On the other hand, before this (1949), immobilization of physiologically active protein was carried out by Micheel and Ewers.

Further, in 1951, Campbell et al. prepared immobilized antigen by binding albumin to diazotized *p*-aminobenzylcellulose. Subsequently, a number of reports on the preparation and application of immobilized antigens and antibodies appeared. Such studies may also be applicable to immobilized enzymes. In the 1960s, many papers appeared; in particular, Professor Katzir-Katchalski and his co-workers at the Weizmann Institute of Science, Israel, carried out extensive studies on new immobilization techniques, and on the enzymatic, physical, and chemical properties of immobilized enzymes. In 1969, Dr. Chibata et al. at Tanabe Seiyaku Co., Ltd., in Japan succeeded in the industrialization of a continuous process for the optical resolution of DL-amino acids using immobilized aminoacylase. This was the first industrial application of immobilized enzymes in the world. In 1971, the first Enzyme Engineering Conference was held at Henniker, New Hampshire, and the predominant theme of this conference was immobilized enzymes.

A definition and classification of immobilized enzymes were proposed at the conference. This conference is biannual, and the main topics have continued to be immobilized enzymes. Immobilized enzymes are defined as "enzymes physically confined or localized in a certain defined region of space with retention of their catalytic activities, and which can be used repeatedly and continuously."

Accordingly, enzymes modified to a water-insoluble form by suitable techniques satisfy this definition of immobilized enzymes. In addition, when an enzyme reaction using a substrate of high molecular weight is carried out in a reactor equipped with a semipermeable ultrafiltration membrane without leakage of the enzyme from the reactor, this can also be considered as a kind of immobilized enzyme system.

The term *immobilized enzyme* was recommended at the First Enzyme Engineering Conference in 1971. Before that time, various terms such as water-insoluble enzyme, trapped enzyme, fixed enzyme, and matrix-supported enzyme had been used. As functional groups involved in enzyme immobilization, free, amino and carboxyl groups, the sulfhydryl group of cystein, the imidazole group of histidine, phenolic groups, and hydroxyl groups of serine and threonine may be considered. For immobilization of an enzyme, it is necessary that functional groups in the active center should not be involved in the reaction leading to immobilization of the enzyme. Further, because the tertiary structure of enzyme protein is maintained by relatively weak binding forces, such as hydrogen, hydrophobic, and ionic bonds, it is necessary to carry out the immobilization reaction under mild conditions. Methods for enzyme immobilization can be classified into three basic categories as follows.

1. Carrier-binding method: the binding of enzymes to water-insoluble carriers such as polysaccharide derivatives, synthetic polymers, and porous glass, etc.
2. Cross-linking method: intermolecular cross-linking of enzymes by means of bifunctional or multifunctional reagents such as glutaraldehyde, bisdiazobenzidine, and hexamethylene diisocyanate, etc.
3. Entrapping method: incorporating enzymes into the lattice of a semipermeable gel on enclosing the enzymes in a semipermeable polymer membrane, such as collagen, gelatin cellulose triacetate polyacrylamide, and κ-carrageenan, etc.

The carrier-binding method is the oldest immobilization method for enzymes, and many papers have been published on this method. When enzymes are immobilized in this way, care is required regarding the selection of carriers as well as in binding techniques. The carrier-binding method can be further divided into three categories according to the binding mode of the enzyme, that is, physical adsorption, ionic binding, and covalent binding. The cross-linking method is based on the formation of chemical bonds, but water-insoluble carriers are not used in this method. The immobilization of enzymes is performed by the formation of intermolecular cross-linkages between the enzyme molecules by means of bifunctional or multifunctional reagents. The entrapping method is based on confining enzymes in the lattice of a polymer matrix or enclosing enzymes in semipermeable membranes, and it can be classified into the lattice and microcapsule types. This method differs from the covalent binding and cross-linking method in that the enzyme itself does not bind to the gel matrix or membrane.

Thus, this method may have wide applicability. However, if a chemical polymerization reaction is used for entrapping, relatively severe conditions are required and lose of enzyme activity occurs in some cases. Therefore, it is necessary to select the most suitable conditions for the immobilization of various enzymes. Immobilization of enzymes by covalent binding or cross-linking is carried out under relatively severe conditions in comparison with those of physical adsorption or ionic binding. Thus, in the former cases, conformational change of the enzyme structure and partial destruction of the active center may occur. Accordingly, unless immobilization of an enzyme by covalent binding is carried out under well- controlled conditions, immobilized enzyme having high activity cannot be obtained. However, the binding forces between the enzyme and carrier are strong, and the enzyme cannot easily be lost from carriers even in the presence of substrates or salts at high concentrations. However, when the activity of enzymes immobilized by covalent binding decreases on long-term operation, regeneration is impossible.

Table 18.1 Preparation and characteristics of immobilized enzyme

	Carrier-binding method				
Characteristic	*Physical adsorption*	*Ionic binding*	*Covalent binding*	*Cross-linking method*	*Entrapping method*
Preparation	Easy	Easy	Difficult	Difficult	Difficult
Enzyme activity	Low	High	High	Moderate	High
Substrate specificity	Unchangeable	Unchangeable	Changeable	Changeable	Unchangeable
Binding force	Weak	Moderate	Strong	Strong	Strong
Regeneration	Possible	Possible	Impossible	Impossible	Impossible
General applicability	Low	Moderate	Moderate	Low	High
Cost of immobilization	Low	Low	High	Moderate	Low

On the other hand, immobilization of enzymes by the ionic-binding method can be achieved simply under mild conditions. Accordingly, in this case, preparations having relatively high activity are obtained. However, the binding forces between enzyme and carrier are weak in comparison with those in the covalent-binding methods. Therefore, leakage of the enzyme from the carrier may occur after changes in the ionic strength, pH of the substrate, or product solution. This type of immobilized enzyme can be regenerated when the enzyme activity decreases after prolonged operation. Thus, the ionic-binding method is advantageous in comparison with the covalent-binding method, particularly when expensive carriers or enzymes are used. Unlike the ionic-binding, covalent-binding, and cross-linking methods in the entrapping method, no binding between enzyme and carrier should occur in theory. Therefore, in many cases, preparations having high activity are obtained. However, in this method, regeneration of activity losses is impossible, as in the case of the covalent-binding method.

A major disadvantage of the entrapping method is that its application is limited to small molecular substrate and product molecules; the entrapped enzyme shows little or no activity toward macromolecular substrates. Although a number of enzyme immobilization methods have been studied, no ideal general methods applicable for the immobilization of many enzymes have yet been developed. Each method has specific disadvantages. Therefore, in practice, it is necessary to find a suitable method and conditions for the immobilization of a particular enzyme in the light of the intended application. Because of the special characteristics of immobilized enzymes, a considerable amount of work has been carried out on application of immobilized enzymes as solid catalysts having high substrate specificity. Immobilized enzymes are used in various fields, such as chemical processes, analysis, medical treatment, food processing, affinity chromatography, and so forth. For the efficient utilization of such immobilized systems, chemical engineering studies are necessary.

IMMOBILIZED REACTORS

The use of enzymes in food as well as in pharmaceutical and chemical industries has increased steadily since the early 1970s. Applications range from synthesis in the milligram scale for natural or nonnatural oligosaccharides up to industrial synthesis for the production of bulk chemicals such as acrylamide. This article highlights some of the most important aspects for the application of immobilized enzymes and presents selected examples. The most common definition for immobilized enzymes is that proposed by Katchalski-Katzir in the 1960s: "Enzymes physically confined or localized in a certain defined region of space with retention of their catalytic activities, which can be used repeatedly and continuously." According to this definition, three types of immobilized enzymes can be distinguished:

1. Heterogenization of the soluble enzyme by coupling to an insoluble support by adsorption or covalent binding, by cross-linking of the enzyme or entrapment in a lattice or in microcapsules such as alginate beads
2. Retention of the enzyme by means of ultrafiltration membranes
3. Use of whole cells for biotransformations using their enzyme apparatus

Table 18.2 Examples of the industrial use of immobilized enzymes

Enzyme	*Substrate*	*Product*	*Scale (ton a^{-1})*
Penicillin amidase	Penicillin G	6-Amino penicillanic acid	6,000
Cephalosporin amidase	Glutaryl-7-ACA	7-ACA	?
Amino acylase	Acyl-D-L-amino acid	L-amino acid	approx. 300
Aspartase	Fumaric acid	L-aspartic acid	1,200
Fumarase	Fumaric acid	L-malic acid	360
Lactase (galactosidase)	Lactose GalGlc	Low lactose milk Gal + Glc	?
Aspartase β-decarboxylase	Aspartic acid	L-alanine	120
Nitrile hydratase	Acrylonitrile	Acrylamide	30,000
Glucose isomerase	Glucose	High fructose corn syrup	8,000,000
Lipase (expected)	Rac-1-phenylethyl-amine	(*S*)-1-phenylethyl-amine	200

Because of the scope of this article, only the first type is considered. Membrane reactors have been reviewed quite recently, as have applications of whole cell processes. Retention of the cells within the reactor may be achieved by membrane separation or by the same immobilization methods that are used for isolated enzymes. In principle, the cell itself can be regarded as a form of native immobilization of enzymes. Biosensors are a very special form of carrier-fixed biocatalysts. The major goal behind immobilization (in its most common definition) is the recovery of the biocatalysts, separation from products and reactants, and subsequent reuse in batch or continuous processes. The latter is especially important for a reduction of the catalyst costs. Enzymes immobilized on a support often show enhanced stability when compared with the soluble form. When considering the use of soluble or carrier-fixed enzymes, the following topics have to be addressed:

1. Additional costs for support and chemicals performing the immobilization have to be balanced against the increase of stability.
2. Loss of activity during the immobilization step.
3. When the catalyst is immobilized only by adsorption or entrapment without covalent attachment, its leakage from the carrier support has to be examined and compared with the overall deactivation rate.
4. Mass transfer limitations for enzymes on a support may cause problems when adjusting of the pH is necessary during the reaction.
5. With soluble enzymes, higher volumetric activities at high catalyst concentrations are possible, enabling conversion of poor substrates at reasonable rates.
6. Whereas membrane reactors can be easily sterilized before use, this is not possible for reactors with carrier-fixed enzymes. To prevent microbial contamination, these processes are quite often operated at higher temperatures. These are discussed in more detail later.
7. When enzymes are to be used together with organic solvents to increase reactant or product solubility or to alter their kinetics, it may become necessary to immobilize them on a support. The support will at the same time act as a water pool to maintain the enzymatic activity. In such systems, water-insoluble organic solvents have less effect on the enzyme stability than water-soluble solvents.

Table 18.3 summarizes some arguments in favor or against immobilization of biocatalysts, namely enzymes. The final decision for a certain reactor design should be based on an optimization process covering all relevant factors contributing to the overall costs, including investment, catalyst consumption, or productivity.

Table 18.3 Comparison of processes using soluble or carrier-fixed enzymes

Enzyme fixed on a carrier or membrane	*Soluble enzyme retained by a membrane*
Continuous or batch process	Continuous or batch process
Improved stability	Lower enzyme stability
Loss of activity during immobilization	No loss of activity during immobilization
Immobilization costs	No costs for immobilization
Mass transfer limitations (heterogeneous reaction)	No mass transfer limitation (homogeneous reaction)
Established technology	New technology
	Sterile process possible
Higher investment costs	Easy supply of new enzyme

Heterogenization of Enzymes

The methods for the heterogenisation (or localization) of enzymes by coupling them to insoluble supports or by entrapment have been discussed in the literature. The principles developed for general heterogeneous catalysis in synthetic chemistry are valid, resulting in well-known reactor configurations. Differences between enzyme catalysis and other systems result from the nature of the biocatalyst and reaction medium. For example, soft particles containing the biocatalyst, such as alginate beads, may limit the pressure drop in fixed-bed reactors. The decision as to specific reactor design will be based on a careful analysis of the kinetic properties of the reaction system. For example, if the enzyme shows a strong substrate-surplus inhibition, a continuously operated reactor with complete back-mixing working at high conversion is advantageous. A reaction with strong product inhibition may utilize a batch reactor or a plug flow reactor to achieve higher volume and catalyst specific productivities. An extractive bioreactor may be used if substrates and products show different solubilities.

By using this reactor configuration, the destabilizing effect of organic solvents may also be overcome, because the enzyme is separated from the organic phase, which is used to extract the insoluble product. The aqueous phase containing the enzyme will be saturated until the maximum solubility

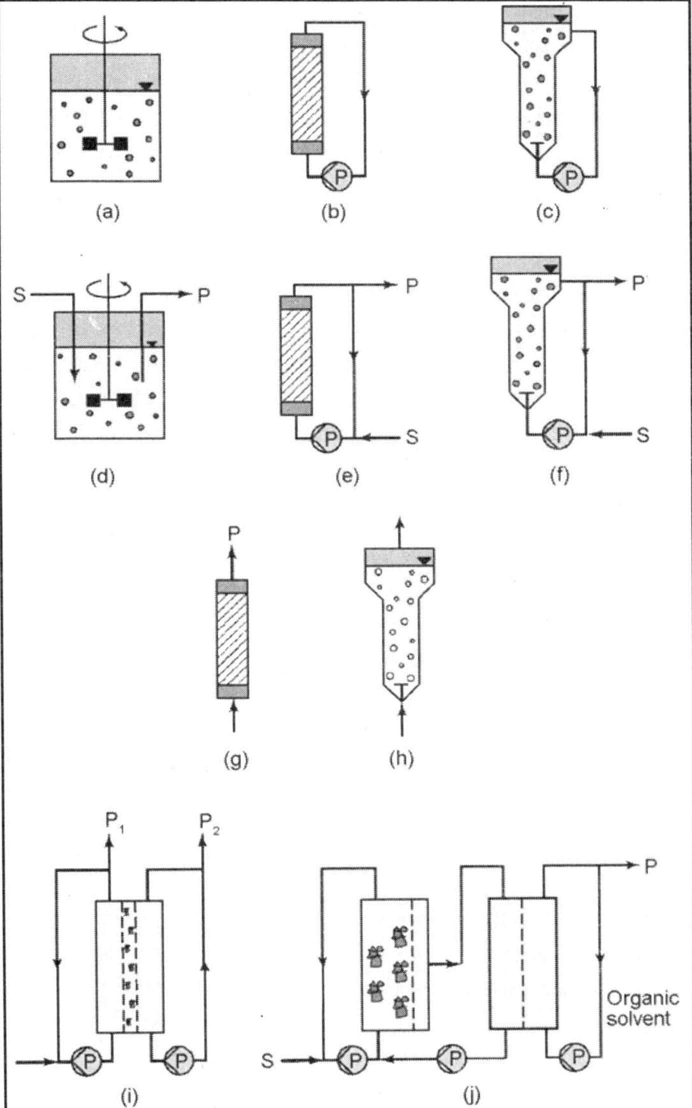

Figure 18.1 Reactors for immobilized enzymes. (a-c) Batch reactors with complete backmixing; (a) Stirred-tank reactor; (b) Fixed-bed reactor; (c) Fluidized-bed reactor. (d-f) Continuously operated reactors with complete backmixing. (g-h) Continuously operated reactors with plug-flow behavior. (i) Reactor with the enzyme immobilized in or on a membrane that may at the same time separated two phases such as water and organic solvent. (j) Reactor with physical separated enzyme and organic solvent in order to prevent denaturation of the protein.

of with the substrate is reached Reactions using biocatalysts are normally performed in aqueous solution at temperatures between 10 and 80°C and at ambient pressure. Due to the inhibition of some enzymes by heavy metals, the materials of construction must not release these elements. Reactors are operated under conditions that prevent microbial contamination. The reactor itself as well as the substrate may be sterilized prior to reaction by using chemical agents (ethanol, formaldehyde, ethylenoxide, Velcorin) or steam.

Ultraviolet rays or γ-rays may be used to sterilize the immobilized enzyme on its support. Alternatively, the immobilization may be performed under sterile conditions. Antibacterial agents may be added to the reaction mixture to prevent microbial growth while the reactor is running. In some cases, the reactants may act as sterilants or inhibitors of microbial growth, such as ketones or alcohols. At higher concentrations (more than 500 mmol/L), solutions may become autosterile because of osmotic pressure effects. Industrial processes are often performed at elevated temperatures, above 55°C, reducing the danger of microbial contamination. For a constant product quality and reproducibility of downstream processing, the reactor should be operated at constant conversion. To overcome the deactivation per unit of time that shows all biocatalysts as a result of denaturation processes, either the residence time has to be increased or fresh enzyme has to be supplied. The latter is especially easy for soluble enzymes. For carrier-fixed enzymes, a combination of both methods is used, as discussed later.

IMPORTANT FEATURES OF IMMOBILIZED ENZYMES

An exhaustive treatment of all aspects of immobilized enzymes, including a complete mathematical description, is beyond the scope of this article. This information may be found elsewhere. The well-known Michaelis-Menten equation (equation 1) describes the enzymatic activity as function of the substrate concentration for an irreversible one-substrate reaction without inhibition. This idealized situation is rarely found in nature. Equation 2 gives the rate for a reversible reaction A⇔B. Using this form, product inhibition is considered as a competition of A and B for the active site as expressed by the two K_M values. This form might be extended to cover substrate surplus inhibition as well as multisubstrate reactions by adding additional terms in the numerator. The form of equation 2 resembles the kinetics of heterogeneously catalyzed reactions.

$$\nu = c_E \times \frac{A_{max} \times c_A}{K_M + c_A} \qquad \ldots(1)$$

$$\nu = c_E \times \frac{\dfrac{A_{max,A} \times c_A}{K_{M,A}} - \dfrac{A_{max,B} \times c_B}{K_{M,B}}}{1 + \dfrac{c_A}{K_{M,A}} + \dfrac{c_B}{K_{M,B}}} \qquad \ldots(2)$$

Upon immobilization of the enzyme on a support, the intrinsic properties expressed as kinetic constants A_{max} and K_M may be altered. This may be because of to conformational changes of the protein by the attachment resulting in a variation of the active site. More often, however, there might be a direct influence of the support or the micro-environment. For example, the polymer matrix may cause a pH difference between the microenvironment of the immobilized enzyme and the bulk fluid. Thus, the activity as a function of pH is often different for the soluble and the immobilized enzyme. In a similar way, the apparent K_M value of an immobilized enzyme may be altered compared to that of the soluble form.

Additional attracting or repulsing interactions of the support matrix with the substrates or products may lead to different values when compared with the native enzyme. A similar problem arises when lipases are used in a heterogeneous reaction system. Because of their natural properties, these enzymes enrich at hydrophopic/hydrophilic interfaces as found in oil-in-water emulsions. If the inhibiting product enriches there, too, the enzyme will experience a very high local product concentration that will drastically alter its activity.

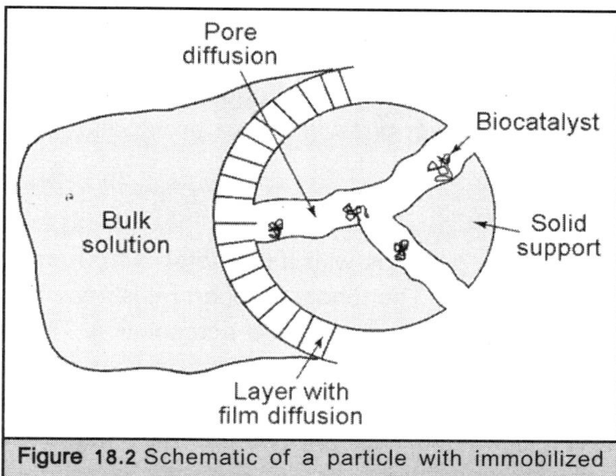

Figure 18.2 Schematic of a particle with immobilized biocatalyst.

Scheme 1

In heterogeneous catalysis, mass transport affects control kinetics at the solid–liquid interphase. The external mass transfer or film diffusion is determined by the velocity difference between the particle and the solution. The influence of this type of mass transfer resistance can be minimized by rapid flow through a fixed bed or fluidized bed reactor or vigorous mixing in a stirred tank reactor. The limiting property will be the mechanical stability of the support used. The example given in Scheme 1 illustrates the influence of film diffusion. The enzyme fructose-1,6-bisphosphate aldolase from rabbit muscle catalyzes the reaction between dihydroxyacetone phosphate (**1**) and various aldehydes (**2**). The enzyme has been immobilized on an insoluble support, thereby enhancing its stability and enabling the reaction in a continuously operated slurry reactor. If the mass transport is only diffusion controlled (relative velocity between particle and medium = 0), the Sherwood number describing the mass transport with spheres is defined as

$$Sh = \frac{k \times d}{D} = 2 \qquad \ldots(3)$$

The molar flux of mass transport is given by

$$v_{mt} = k \times (c_0 - c) \qquad \ldots(4)$$

Assuming Michaelis-Menten-type kinetics, the enzymatic reaction will be rate limiting as long as

$$k > \frac{A_{max,A}}{K_M} \qquad \ldots(5)$$

For rabbit muscle aldolase bound to Eupergit C1Z (nonporous carrier, diameter 1 μm), typical values for $A_{max,A}$ are 50 μmol m^{-2} s^{-1} corresponding to 405.9 nkat g^{-1} (24.3 U g^{-1}). K_M values range between 0.1 and 10 mmol L^{-1}. From equation 3, typical k values of 2×10^{-3} m s^{-1} result. Thus, the mass transport is much faster than the enzymatic reaction, since

$$5 \cdot 10^{-6} ms^{-1} < \frac{A_{max,A}}{K_M} < 500 \cdot 10^{-6} ms^{-1}$$

If, however, particles with a diameter of 100 μm were used for immobilization, film diffusion would be rate limiting. The mass-transport resistance might be reduced to a certain extent by establishing a high velocity between particle and solution. The influence of internal mass transfer or pore diffusion becomes more dominant with increasing particle size. This may reduce the effectiveness by substrate limitation or product inhibition in the inner part of the particle, and when the reaction causes a strong pH- shift, local high values may deactivate the enzyme. This may be avoided by higher buffer concentrations. In a very simple way the effect of internal mass transfer may be considered as an efficiency coefficient η, which is a function of size, shape, and porosity of the support matrix and the diffusion of the compounds involved:

$$\eta = \frac{\text{activity of immobilized catalyst}}{\text{activity of catalyst without mass transfer}} \quad \text{...(6)}$$

$$V_{app} = \eta \times V \quad \text{...(7)}$$

Pore diffusional resistance is indicated when apparent activity rises after grinding the catalyst particles (homogeneous catalyst distribution). If no increase of activity is observed, film diffusion is already limiting. These effects, together with the kinetics of the immobilized catalyst, can be easily investigated in a differential recirculation reactor. Only a small amount of catalyst is used to minimize conversion per pass through the bed, as measured in the bypass. By adjusting the flow rate, film diffusion and pore diffusion may be easily distinguished. For a more elaborate description of pore diffusion, the Thiele modulus, interpreted as the reaction rate relative to the diffusion rate within the particle, has to be used. The number of catalytically active reaction sites per unit mass of catalyst can be easily determined for immobilized enzymes. For this active-site test, a suitable substrate, such as an ester, is converted by the enzyme and the product is determined analytically. This test is independent from mass transport phenomena because only the additional amount of liberated product after addition of the catalyst is used for calculation. According to the Michaelis-Menten kinetics of an enzymatic reaction with varying reaction order, the efficiency coefficient g is a function of conversion. This has been demonstrated for immobilized α-chymotrypsin in the cleavage of amino acid esters. The use of immobilized enzymes in processes with precipitated substrates or products has been addressed recently.

SUITABLE EXAMPLES

One of the first enzymes shown to be active after immobilization by adsorption on charcoal was invertase. This enzyme is still used for the large-scale production of invert sugar by hydrolysis of sucrose. One kilogram of immobilized enzyme supports production of approximately 10,000 kg of invert sugar.

Formation of L-Amino Acids

Enantiomerically pure L-amino acids are required for use in parenteral nutrition, as food additives or as chiral synthons. In addition to isolation from natural sources and direct fermentation, racemic mixtures of L- and D-amino acid can be resolved by enzymatic reaction with one of the two enantiomers. Scheme 2 illustrates this for a racemic amino acid using amino acylase I from *Aspergillus*. (*Note*: the EC number, in parentheses, is the internationally recognized Enzyme Commission number. The six main classes are oxidoreductases, transferases, hydrolases, lyases, isomerases, and lipases). The unreacted D-enantiomer can be racemized easily and recycled to increase the overall yield.

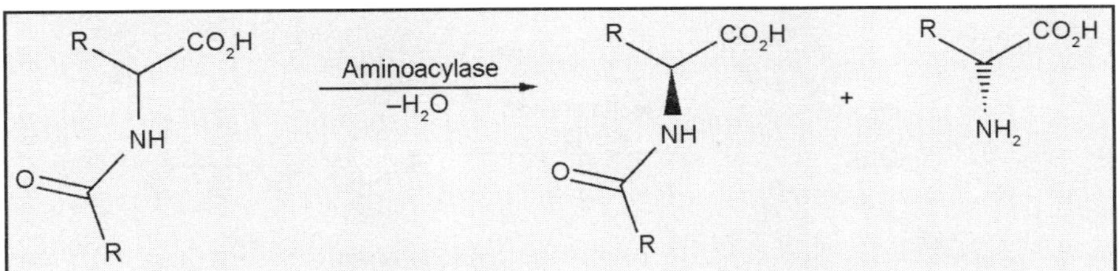

4 —Aminoacylase, $-H_2O$→ 5 + 6

Scheme 2

The first industrial process of this type was established by Tanabe Seiyaku Co. in 1969 using enzyme immobilized on an insoluble support. One of the best methods to obtain a stable catalyst in their example has been to immobilize by ionic binding to DEAE-Sephadex. To compensate for the loss of activity during operation (20% after 1 month), fresh enzyme has been passed through the column periodically to establish the former activity again. The main products of this process are L-methionine, L-phenylalanine, and L-valine. Typically, a 0.2 mol L^{-1} solution of D, L-acetyl amino acids is feed into a 1-m^3 fixed-bed reactor with various flow rates. For L-methionine, the space-time yield reported is about 715 g L^{-1} d^{-1}.

Formation of 6-Amino Penicillanic Acid and 7-Amino Cephalosporanic Acid

6-Amino penicillanic acid (6-APA) (**8**) and 7-amino cephalosporanic acid (7-ACA) (**11**) are important intermediates for semisynthetic β-lactam antibiotics. Both compounds can be obtained readily by amidase-catalyzed cleavage of the corresponding amides (**7** and **10**) that are accessible by fermentation (Scheme 3). Whereas penicillin amidase has been used for production of 6-APA since the early 1970s, cephalosporin C amidase has been used just recently in a chemoenzymatic synthesis of 7-APA. Amidases have been prepared from several different sources, such as *Pseudomonas, Bacillus,* and (recombinant) *Escherichia coli*. In general, the enzyme is immobilized on and in various supports such as polyacrylonitrile fibers, polyacrylamid beads, or Sephadex. For 6-APA batch production, typically a 10% solution of penicillin G is circulated through the bioreactor or processed in a stirred-tank reactor. One cycle lasts about 3 h. The life time for the enzyme is more than 360 cycles. The space-time yield for this process is about 800 g L^{-1} d^{-1}. One of the main problems for both processes is the liberation of a strong acid. To avoid large pH shift, an effective method to control and maintain the desired pH is needed. High recirculation rates are used for the enzyme immobilized on hollow fibers. For the enzyme immobilized in beads, both a small bead diameter (typically around 100 to 600 /m) and good control of film and pore diffusion by high fluid velocities are needed.

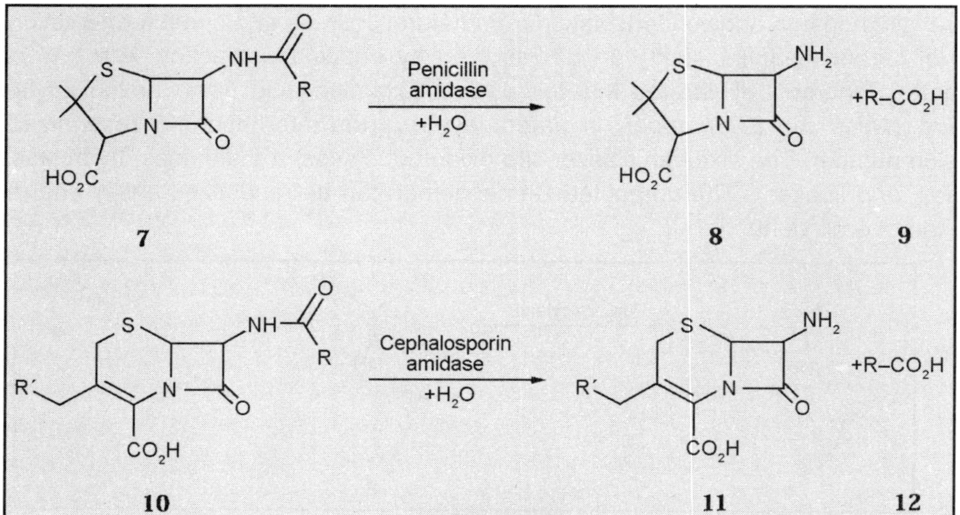

Scheme 3

Formation of High-Fructose Corn Syrup

The production of high-fructose corn syrup, a mixture of glucose (**13**) and fructose (**14**), is the largest enzymatic process in commercial practice so far utilizing immobilized biocatalysts (Scheme 4). The process using glucose isomerase was first commercialized in 1970. The enzyme is available from different bacterial sources. By site-directed mutagenesis, more thermostable mutants have been obtained. Most of the enzyme preparations commercially available are cross-linked with glutaraldehyde. The enzyme is used in fixed-bed reactors, and the feed flow is adjusted to overcome enzyme deactivation and ensure a constant product quality (conversion). Different columns of different enzyme age are operated in parallel, resulting in an average conversion of 45% in a process of uniform throughput. The equilibrium conversion is about 50%. If the residual activity has dropped to about 10% of the initial activity, the enzyme is replaced. For the most-stable commercially available enzyme preparations, the half-life is more than 100 days. Typical temperatures range from 50 to 60°C with 50% substrate solution.

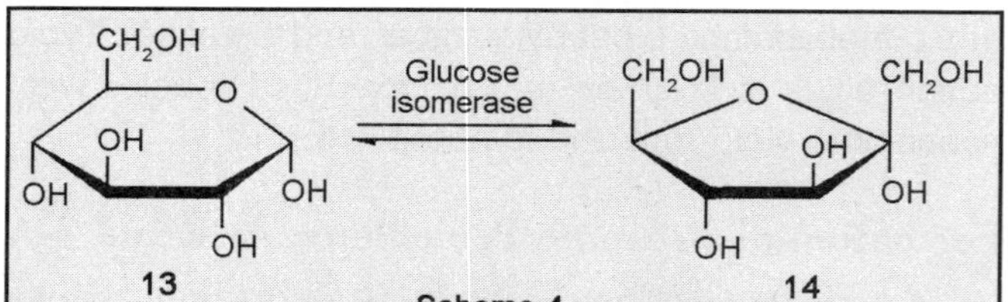

Scheme 4

Formation of Acrylamide

Biotransformations have been also established for the production of bulk chemicals. The enzymatic production of acrylamide (**16**) from acrylonitrile (**15**) using nitrile hydratase from *Rhodococcus rhodochrous* has replaced the copper-catalyzed process in Japan. For this process, whole cells were immobilized in a cationic acrylamide-based polymer gel. Typical reaction conditions range from 0 to 15°C and pH 6.5 to 8.0.

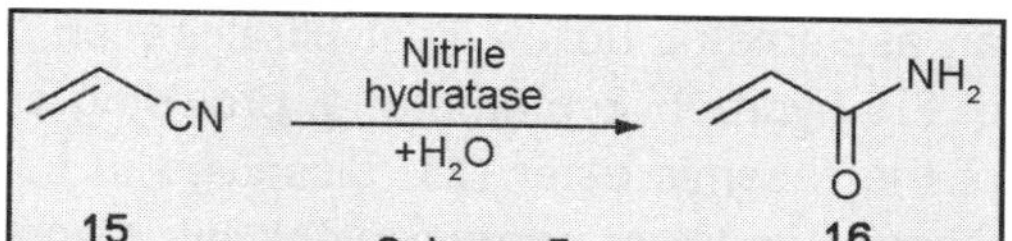

Scheme 5

Acrylonitrile is added gradually in order to achieve a final concentration of acrylamide of 400 g L^{-1}. The conversion of acrylonitrile is more than 99%, with a selectivity for acrylamide of more than 99%. Because of its broad substrate specificity, the nitrile hydratase may be also used for production of nicotinamide, isonicotinamide, or pyrazinamide.

Formation of Glycerol-3-Phosphate

Glycerol-3-phosphate (G3P) (**18**) is a useful starting material for the synthesis of phospholipids. These compounds are of importance for numerous medicinal, diagnostic, and research applications. Their potential applications include microemulsion and liposomal drug delivery systems. The enzymic synthesis of G3P using glycerol kinase has been commercialized by Genzyme. Adenosine triphosphate (ATP) is used as the phosphorylating agent, and regeneration of ATP is achieved by pyruvate kinase using phosphoenol pyruvate (PEP) (**22**). The enzymes used are immobilized on a solid support. Especially for the glycerol kinase the stability was increased dramatically when compared with the native form. Its half-life in aqueous solution is about 15 min, because of dissociation of the subunits.

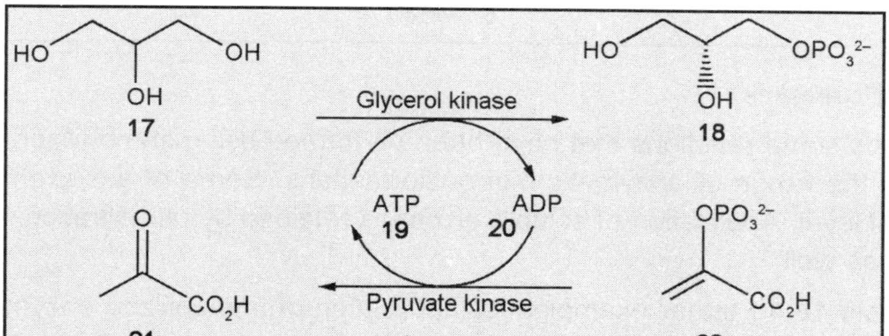

Scheme 6

The reaction is typically performed in a batch reactor containing the immobilized enzyme in a fixed bed. Multikilogram-scale production typically uses 1 mol L^{-1} glycerol and 1 mol L^{-1} phosphoenol pyruvate. For this process, space-time yield is about 81 g L^{-1} d^{-1}. Catalytic amounts of ATP (1 to 10 mol%) are used, resulting in total turnover numbers up to 100. These and even higher values are attainable without requiring coimmobilization of the cofactors. NADH values of about 1,000 may be reached easily. These processes are economically feasible because of the low cost of the cofactors.

Formation of (2R,3S)-Methyl-*p*-Methoxyphenylglycidate

Optically pure (2R,3S)-methyl-*p*-methoxyphenylglycidate (**24**) is used as chiral building block for the synthesis of diltiazem. Sepracor has developed a process for enantioselective hydrolysis of the racemic ester using a lipase, which has reached a 50 t a^{-1} scale. The reaction is performed in a multiphase membrane reactor with the enzyme reversibly entraped within the porous matrix of an asymmetric hollow fiber ultrafiltration membrane. An aqueous buffer is circulated on one side of the hydrophilic membrane, providing an aqueous environment for the enzyme. A 17% solution of the racemic ester (**23**) dissolved in toluene is circulated on the other side of the membrane. Despite its lower enantioselectivity, *Candida cylindracea* lipase is being used; lipases from other sources were not retained sufficiently within the membrane. The reactor is used in a batch mode. The space-time yield, based on the membrane area, is about 125 g m^{-2} d^{-1}. This type of reactor has also been used for the hydrolysis of butter oil or the production of peroxycarboxylic acids.

MeO OME 23 Lipase $+H_2O$ MeO OMe 24 + MeO OME 25 +MeOH 26

Scheme 7

Miscellaneous Processes

Table 18.4 lists some reactions that have been performed using immobilized enzymes to give an impression of the range of enzymatic biotransformations. Some of the examples are already used on industrial level. Application of soluble enzymes retained by ultrafiltration membranes have been successful as well.

Table 18.4 Further examples for application of immobilized enzymes

Enzyme	*Substrate*	*Product*
Glucose kinase	Glc, ATP	Glucose-6-phosphate
Oxynitrilase	Aldehydes, HCN	Cyanohydrines
Transaminase	Ketone	Amine
Pyranose oxidase	Sugar	Keto-sugars
Hydantoinase	Hydantoin	Amino acid
Oxidoreductase	Benzoylformate	Mandelate
Sialidase	Neu5Ac, Gal	Neu5Ac-Gal
Glucosidase	Glc, octanol	Octylglycoside
Chymotrypsin	Amino acids	Oligopeptides
Glycosyltransferase	Carbohydrate	Oligosaccharide

19

BYPASS REGULATION IN *E. COLI*

The glyoxylate bypass of the bacterium *Escherichia coli* has been used as a model system for the study of metabolic regulation. *E. coli* is an ideal organism for these studies because of the powerful genetic tools that can be used to achieve experimental manipulations that are not possible in many other systems. The long-term goal of these projects is to extract fundamental principles of metabolic control from a detailed analysis of how each regulatory mechanism functions and how these mechanisms interact to control the entire process.

GLYOXYLATE BYPASS

Adaptation of *E. coli* to growth on acetate or fatty acids requires the induction of the enzymes of the glyoxylate bypass: isocitrate lyase and malate synthase. This pathway is essential for growth under these conditions because it prevents the net loss of the entering carbon as CO_2 in the Krebs' cycle. The glyoxylate bypass is regulated at two levels. Coarse control of the glyoxylate bypass is achieved by regulating the transcription of *aceBAK*. The *aceBAK* operon encodes malate synthase, isocitrate lyase, and isocitrate dehydrogenase (IDH) kinase/phosphatase, respectively. Once induced, the flow of isocitrate through the

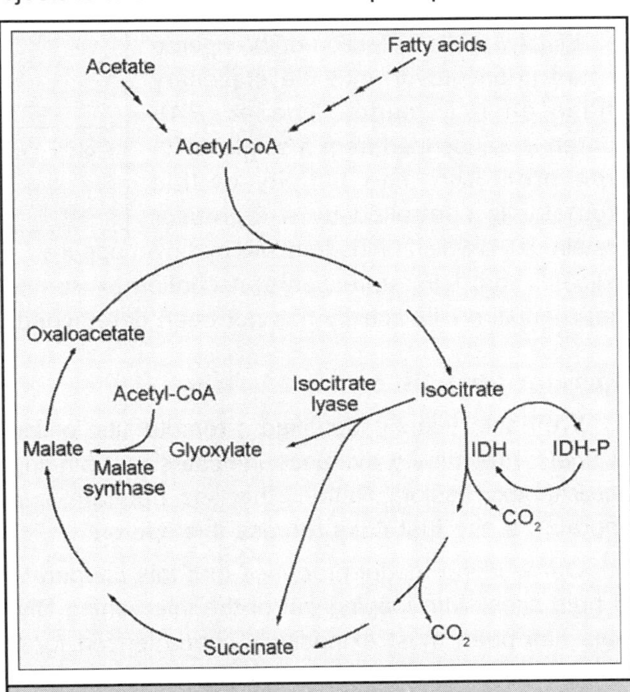

Figure 19.1 The Kreb's cycle and the glyoxylate bypass.

bypass is regulated by the phosphorylation of IDH, the Krebs' cycle enzyme that competes with isocitrate lyase. This phosphorylation cycle is catalyzed by a bifunctional kinase/phosphatase.

CONTROL OF TRANSCRIPTION

Glyoxylate Bypass Operon

The *aceBAK* operon appears to respond to metabolic stress. Full expression is observed only during growth on acetate or fatty acids. However, the presence of these substrates is not sufficient for induction because the operon is repressed by the simultaneous presence of preferred carbon sources such as glucose or pyruvate. Poor carbon sources produce a partial induction of *aceBAK* even though the glyoxylate bypass is not required for growth on these substrates. This partial induction appears to be a response to general metabolic stress because there is a roughly linear relationship between the culture doubling time and the extent of induction when the different growth conditions are compared.

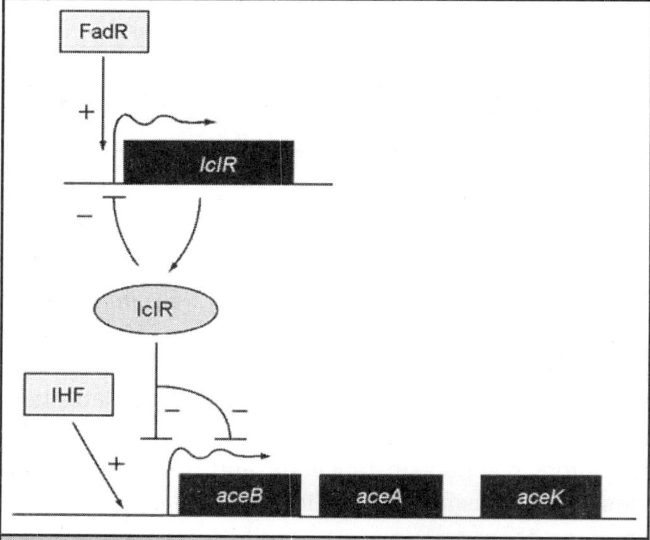

Figure 19.2 Regulation of the glyoxylate bypass operon, aceBAK encodes malate synthase, isocitrate lyase, and IDH kinase/phosphatase, respectively.

Finally, partial induction of the operon occurs upon growth into the stationary phase on rich media. The *aceBAK* operon is transcribed from a single, $\sigma70$-type promoter. The activity of this promoter is controlled by a repressor protein, IclR, which binds to a site that overlaps the "–35" box of the promoter. Preliminary evidence indicates that IclR represses transcription by competing with RNA polymerase. The mechanism that regulates IclR activity has not been determined.

Operator for IclR

We have recently identified a remote site for IclR. This operator lies within the coding region of *aceB*. Preliminary evidence indicates that binding of IclR to this site does not directly repress *aceBAK* expression. Rather, this site appears to act cooperatively with the promoter-proximal operator, a site that does repress this operon.

Our working model proposes that this cooperativity results from simultaneous binding of IclR at both sites, with looping out of the intervening DNA. This type of cooperativity between remote sites has been most extensively characterized for the *lac* operon but appears to occur in other operons as well.

Integration Host Factor (IHF)

Integration host factor (IHF) is a sequence-specific DNA binding protein that binds to many sites in the *E. coli* chromosome. This histone-like protein has been implicated in a wide variety of processes including plasmid replication, lambda phage recombination, and transcriptional activation and repression. We have identified a binding site for IHF upstream of the *aceBAK* promoter. IHF activates *aceBAK* expression under inducing conditions by preventing IclR repression.

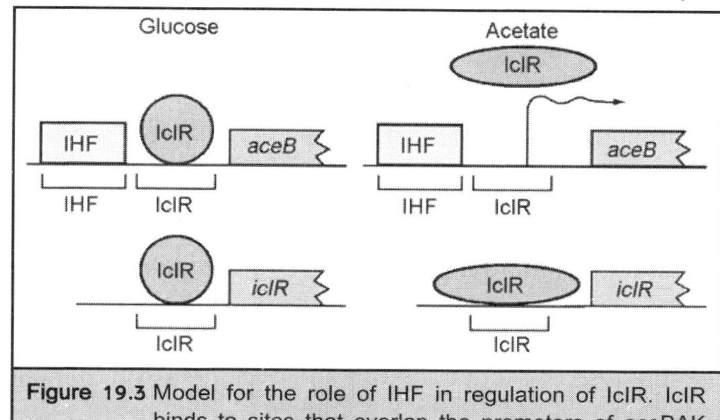

Figure 19.3 Model for the role of IHF in regulation of IclR. IclR binds to sites that overlap the promoters of aceBAK and iclR.

In contrast, IHF does not appear to affect IclR activity under repressing conditions. The carbon-source specificity of the IHF effect amplifies regulation of *aceBAK*. IHF can affect transcription by a variety of mechanisms. The most clearly established mechanism is one in which IHF binds to a site well removed from the promoter, bending the DNA at an angle of about 140° and so bringing a remote activator site near the promoter. Similar mechanisms have been proposed for IHF in processes as diverse as plasmid replication and recombination.

IHF may also act by direct interactions with other proteins. For example, IHF has been suggested to activate transcription at three promoters by direct contact with RNA polymerase, although this remains to be established. IHF can also repress transcription by binding to sites that overlap a promoter. In these systems, IHF probably acts by competing with RNA polymerase. IHF plays a novel role in the regulation of *aceBAK* expression.

In their recent review, Goosen and Van de Putte have suggested that IHF may not participate in transcriptional regulation. This is clearly not true for *aceBAK*, which requires IHF for full carbon-source sensitivity. However, the mechanism by which IHF plays this role must still be determined. Our working model proposes that IHF and IclR bind independently near the *aceBAK* promoter under repressing conditions. Under inducing conditions, we suspect that IclR undergoes a structural change that causes the binding of IHF and IclR to become competitive. IclR repression is relieved under these conditions because the IHF site is occupied.

Control of *iclR* Expression

The gene encoding IclR is subject to repression by its own product. IclR achieves this autorepression by binding to an operator that overlaps the promoter of *iclR*. This provides a feedback control mechanism that helps maintain a constant level of IclR. The expression of *iclR* is also regulated by FadR. FadR was initially identified because it represses the genes encoding the enzymes of fatty acid degradation. It was subsequently shown to activate the transcription of *fabA*, a gene whose product participates in unsaturated fatty acid biosynthesis. FadR activates *iclR* expression and may provide a link between *aceBAK* and fatty acids, a carbon source that also requires the glyoxylate bypass.

Regulation by IclR

Perhaps the most unusual feature of this system is the ability of IclR to differentially regulate *aceBAK* and *iclR*. Whereas IclR-mediated repression of *aceBAK* undergoes a striking response to the carbon source, repression of *iclR* is nearly constant. This difference is probably an advantage to the cell because induction of *iclR* expression on acetate would oppose induction of *aceBAK*. IHF, which is required for the carbon-source sensitivity of IclR, appears to be responsible for differential control by this repressor. *aceBAK* responds to the carbon source because it has an associated site for IHF. In contrast, *iclR* does not respond to the carbon source because it does not have a site for IHF. In addition to the regulation of *aceBAK* and *iclR*, IclR also appears to regulate *acs*, the gene encoding acetyl coenzyme A synthetase in *E. coli*. Mutations in *iclR* reduced consumption of acetate mediated by Acs. However, the precise role of IclR in regulating *acs* remains uncertain.

IclR has lent its name to a family of transcriptional regulatory proteins. This family is characterized by about 25% identity throughout their sequences, with particularly high homology in a region near the N-terminus. This is the region where we identified sequences that appear to match the consensus for helix-turn-helix DNA-binding motifs. An "IclR signature" in the C-terminal regions of these proteins has also been defined in the PROSITE database. The IclR family is distinct from other families of transcriptional regulators in bacteria. It is broadly distributed, currently including 19 members found in 12 different Gram-positive and Gram-negative bacterial species; in some cases, the function of the protein is known. Others were identified during genomic sequencing projects, and their functions remain to be identified. The proteins of known function regulate a wide variety of catabolic genes. IclR also fits this pattern because the glyoxylate bypass is essential for the catabolism of acetate.

FINE CONTROL OF PROTEIN PHOSPHORYLATION

IDH Phosphorylation

Fine control of the glyoxylate bypass is achieved by the phosphorylation of IDH, the Krebs' cycle enzyme that competes with isocitrate lyase. This phosphorylation cycle is catalyzed by a bifunctional protein, IDH kinase/phosphatase.

Regulation of IDH

IDH is a member of a large family of dehydrogenases. This family, comprising at least 17 proteins, includes members from prokaryotes, yeast, and humans. Although some of these enzymes act on isocitrate, others act on isopropyl malate or tartarate. Members of this family exhibit extensive sequence homology, particularly among residues that have been implicated in substrate binding and catalysis for *E. coli* IDH. These enzymes lack the Rossman fold, which is an important structural feature of many other dehydrogenases. The phosphorylated form of IDH is completely inactive.

Inactivation results from the phosphorylation of Ser-113, an active site residue that forms a hydrogen bond with isocitrate in dephospho-IDH. Phosphorylation has little effect on the three-dimensional structure of IDH. Rather, it blocks isocitrate binding primarily by electrostatic and steric repulsion. The mechanism by which phosphorylation controls IDH is a striking contrast with regulatory phosphorylation of glycogen phosphorylase, the only other phosphoprotein whose structure has been extensively characterized. The phosphorylation site of glycogen phosphorylase is remote from the active site. Phosphorylation of this enzyme activates catalysis by inducing a conformational change.

Mechanism of IDH Kinase/Phosphatase

IDH kinase/phosphatase is expressed from a single gene and appears to have a single active site that catalyzes the kinase and phosphatase reactions. This active site also appears to catalyze an ATPase activity that is more efficient then either the kinase or phosphatase reaction. Our working model proposes that the IDH phosphatase reaction results from the kinase back reaction tightly coupled to ATP hydrolysis.

E•ATP ⇌ E•ATP•IDH

E ⇌ E•ATP; E ⇌ E•ADP

E•ATP ⇌ E•ADP•P_i ⇌ E•ADP

E•ADP•P_i ⇌ E•ADP•P_i•IDH

E•ADP ⇌ E•ADP•IDH(P)

E•ATP•IDH ⇌ E•ADP•IDH(P)

Figure 19.4 Model for the catalytic mechanism of IDH kinase/ phosphatase.

Control

The IDH phosphorylation cycle is controlled by a wide variety of metabolites. Examples include isocitrate, NADP, 3-phosphoglycerate, pyruvate, and AMP. There appear to be two distinct mechanisms by which these ligands regulate IDH kinase/phosphatase. Access to the phosphorylation site of IDH is controlled by the substrates of this enzyme. When these substrates bind to IDH, they induce a conformational change, narrowing the active site cleft and burying the site of phosphorylation. Because isocitrate binds only to dephospho-IDH, this ligand inhibits only IDH kinase. Isocitrate activates IDH phosphatase by preventing product inhibition by

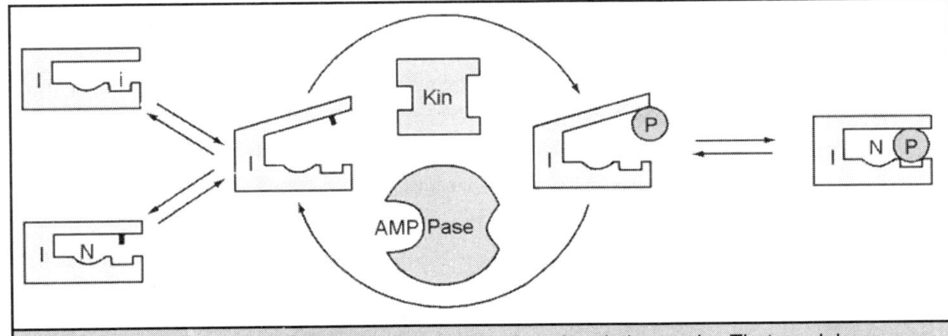

Figure 19.5 Model for the regulation of the IDH phosphorylation cycle. That model proposes that IDH kinase/phosphatase can exist in two conformations, one favoring the kinase reaction (Kin) and the other favoring the phosphatase (Pase).

isocitrate. A second mechanism for regulating this cycle results from a conformational equilibrium in IDH kinase/ phosphatase. The equilibrium between the kinase and phosphatase conformations is controlled by metabolites that bind preferentially to the phosphatase conformation. Preferential binding activates this activity and inhibits IDH kinase.

PHYSIOLOGICAL CONTROL

Glucose as a Medium

The glyoxylate bypass serves no useful purpose during growth on carbon sources such as glucose or pyruvate. Under these conditions, *aceBAK* is repressed by IclR. Although a basal level of IDH kinase/phosphatase is present, the conditions favor the dephosphorylation of IDH. Pyruvate, which activates IDH phosphatase and inhibits IDH kinase, appears to be the primary regulator of this phosphorylation cycle under these conditions.

Acetate as a Medium

Adaptation to growth on acetate requires the induction of the glyoxylate bypass operon. The mechanisms responsible for transcriptional regulation have already been discussed. During steady-state growth on acetate, about 75% of the IDH is converted to the phosphorylated form. Because phospho-IDH is completely inactive, partial phosphorylation of IDH inhibits the Krebs' cycle and so forces isocitrate through the glyoxylate bypass. Under these conditions, isocitrate must be partitioned between the Krebs' cycle for the production of energy and the glyoxylate bypass for the production of metabolic intermediates. The primary regulators of IDH kinase/phosphatase during growth on acetate appear to be isocitrate, 3-phosphoglycerate, and AMP. Isocitrate and 3-phosphoglycerate act as indicators of the levels of metabolites, whereas AMP reflects the energy needs of the cell.

Culture Conditions

The flux through the glyoxylate bypass responds dramatically to the culture conditions. For example, addition of glucose to a culture growing on acetate virtually eliminates flux through this pathway. Inhibition of the bypass is probably advantageous because, when glucose is available, the glyoxylate bypass is no longer required to replenish metabolites used for biosynthesis. Inhibition of the bypass results, in part, from the rapid dephosphorylation of IDH, increasing this enzyme's activity and thus diverting isocitrate through the Krebs' cycle. IDH kinase/phosphatase catalyzes the rate-limiting step in this transition. Long-term adaptation to the carbon source results from control of *aceBAK* transcription. The response of the IDH phosphorylation cycle during these metabolic transitions is amplified by two mechanisms. Zero-order ultrasensitivity results when one of the modification enzymes (e.g., kinase or phosphatase) is saturated by its protein substrate. Multistep effects, which can either enhance or reduce sensitivity, result when an effector controls more than one step in a protein modification cascade.

The glyoxylate bypass provided the first experimental demonstrations of these effects. Both phenomena have since been found in other systems. Study of the glyoxylate bypass led to the discovery of another mechanism for sensitivity enhancement: the branch point effect. This effect results from the competition between enzymes for a limiting amount of a common substrate. During growth on acetate, isocitrate levels are nearly saturating for IDH but are within the pseudo-first-order range of isocitrate lyase. A consequence of this difference in saturation is that the flux carried by isocitrate lyase is ultrasensitive to the phosphorylation state of IDH. Both the theoretical and experimental studies responsible for identifying this phenomena were the products of studies on the glyoxylate bypass.

CONCLUDING REMARKS

The glyoxylate bypass has proven to be a very useful model for metabolic control. The basic control mechanisms have all been identified, with the important exception of the mechanism that controls the IclR repressor. We also have considerable understanding of metabolite levels, flux rates, and how these factors are controlled. These studies have led to the identification of several fundamental principles of metabolic control. Although the basic control processes have been identified, much remains to be learned concerning their mechanisms of action. How, for example, does IHF control the carbon-source sensitivity of IclR? Are the models that we proposed for the catalytic mechanism and regulation of IDH kinase/phosphatase correct and, if so, what are their structural bases? As so often happens in biology, the more we know, the more we realize how much we do not know.

20

GLUCOSIDASES

β-Glucosidase has the potential for practical use in the saccharification of cellulosic materials and the synthesis of aryl or alkyl glucosides in a nonaqueous reaction medium. The latter topic, especially synthesis in an aqueous–organic two-phase system, has been described in detail.

β-Glucosidase (β-D-glucoside glucohydrolase) catalyzes the cleavage of the β-glucosidic ether bond of aryl and alkyl glucosides. Aryl glucosides are, in general, better substrates than alkyl glucoside. β-Glucosidase from some origins can also hydrolyze β-galactosides and β-fucosides. Its hydrolytic potential in practical use would be as a component of cellulases for saccharification of cellulose. The enzyme also has a synthetic potential for aryl or alkyl glucosides through reverse hydrolysis (condensation) or transfer reactions. Research concerning β-glucosidase-catalyzed synthesis of glucosides in a nonaqueous phase has recently been active, although its industrial application has not been reported yet.

DENATURATION

Synergistic Actions

β-Glucosidase is one of the key enzymes for saccharification of cellulose. The enzymatic saccharification proceeds through synergistic actions of *endo*-glycanase, *exo*-glycanase, and β-glucosides, and β-glucosidase catalyzes the last step of the saccarification, hydrolysis of cellobiose to glucose. β-Glucosidase favors shorter cellooligosaccharides, especially cellobiose.

β-Glucosidase activity of cellulase from *Trichoderma reesei*, which is one of the most promising microorganisms for cellulose saccharification on an industrial scale, is still weak, so addition of β-glucosidase from other origins for the cellulose saccharification process has been attempted, resulting in a significant increase in cellulose conversion. Reinforcement of β-glucosidase activity of *T. reesei* cellulase by gene technology has also been investigated.

Substrates for β-Glucosidase

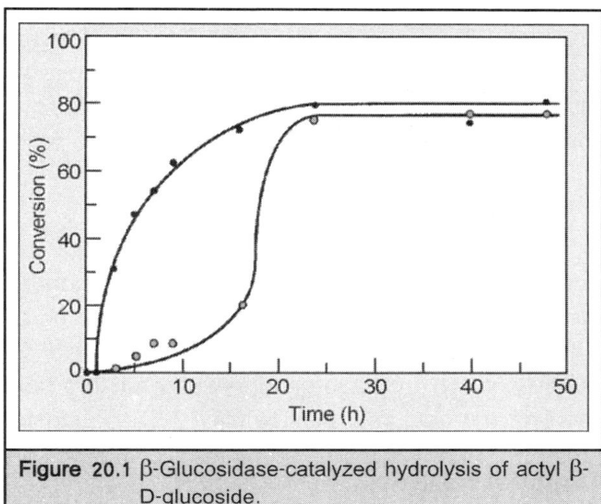

Figure 20.1 β-Glucosidase-catalyzed hydrolysis of actyl β-D-glucoside.

Alkyl glucosides, which are substrates for β-glucosidase, have an amphiphilic structure with hydrophilic and hydrophobic moieties and are a group of nonionic surfactants. The phase diagram for a ternary mixture of octyl β-D-glucoside–buffer(water)–octanol was prepared, and it included the regions of micellar solution, microemulsion, liquid crystalline phase, and an undefined structure. In the aqueous monophasic system, octyl β-D-glucoside exists as monomers at concentrations below its critical micellar concentration (CMC), and it forms aggregates at concentrations above the CMC, while the monomer concentration remains at the CMC. For the lower substrate concentrations (< ca. 20 mmol/L), the relation between the reaction rate and substrate concentration followed a classical Michaelis–Menten scheme, but the reaction rate remained constant at the higher substrate concentrations. The relation could be divided into two regions, and two lines intersected at 19 mmol/L, which corresponded to the CMC of the system. This indicates that octyl β-D-glucoside monomer is the actual substrate for the enzyme.

PRODUCTION OF ALKYL GLUCOSIDES

β-Glucosidase-catalyzed synthesis of aryl or alkyl glucosides in a nonaqueous system has been an active area of research, as have lipase-catalyzed acylation and proteinase-catalyzed peptide synthesis. Chemical synthesis of alkyl glucosides is well developed and used at the commercial level. However, the processes are complicated by the presence of identically reactive hydroxyl groups and by the toxicity of some of the catalysts used.

β-Glucosidase is effective for synthesis through transglycosylation or condensation. Although the reaction rate of condensation is slower and hence a longer time is required to reach equilibrium compared with transglycosylation, condensation has some advantages, such as cheaper substrates and easier purification of the product.

Alkyl glucoside synthesis through transglycosylation is kinetically controlled, but the yield of alkyl glucoside through transglycosylation was almost equal to that through condensation. On the basis of substrate costs, condensation would be a choice for alkyl glucoside synthesis. Equation 1 shows the alkyl glucoside synthesis through β-glucosidase-catalyzed condensation between glucose and alcohol.

$$\text{Glucose (HO, HO, OH, OH, OH)} + ROH \xrightarrow{\beta\text{-glucosidase}} \text{Alkyl glucoside (HO, HO, OH, OH, OR)} + H_2O \quad \ldots(1)$$

where ROH indicates aliphatic alcohol. Thermodynamically, hydrolysis predominates over synthesis in an aqueous reaction medium. One approach to shift the equilibrium is to add a water-miscible solvent, which decreases the thermodynamic water activity and thereby favors synthesis. An alternative approach is to add a water-immiscible solvent. In this case, a hydrophobic product is extracted to the organic phase so that more product is synthesized.

Single-Phase System

Aryl and alkyl glucosides were synthesized in a mixture of acetonitrile and water through condensation using β-glucosidase from almonds. The synthetic reaction was entirely regio- and stereoselective. The enzyme had a pronounced affinity for the alcohols containing a phenyl group. The ratio of water to acetonitrile was a crucial factor for both the product yield and the enzyme stability, a 9:1 (v/v) acetonitrile–water mixture being found to be the best. It was also reported that the enzyme could synthesize alkyl β-D-galactosides.

Double-Phase System

In the synthesis of alkyl glucosides in an aqueous–organic two-phase system, one of the substrates, alcohol, itself is used as an organic phase. The aqueous phase includes glucose and free or immobilized β-glucosidase. This synthesis has mostly been used for alkyl glucoside synthesis. The water activity a_w or the water content in the system plays an important role in both the reaction rate and the equilibrium yield in alkyl glucoside synthesis. An a_w of at least 0.67 was needed for synthesis of octyl β-D-glucoside. The reaction rate increased, but the equilibrium yield decreased with increasing a_w. A high yield of octyl glucoside in a short time was realized by controlling the a_w during the condensation reaction. The optimum water content was higher for the synthesis of glucoside with a shorter alkyl residue. Physicochemical understanding for these results have not been established yet. Figure 20.2 shows the transient changes in the yield of alkyl β-D-glucosides, Y_P, through β-glucosidase-catalyzed condensation between glucose and alcohols at 60°C in the two-phase system. The volume ratio of organic to aqueous phase was fixed at 16.7. The length of the alkyl chain significantly affected both the equilibrium yield and the reaction rate. The longer the alkyl chain, the lower were both the yield and the rate. The figure also shows that the enzyme could catalyze the condensation between glucose and secondary

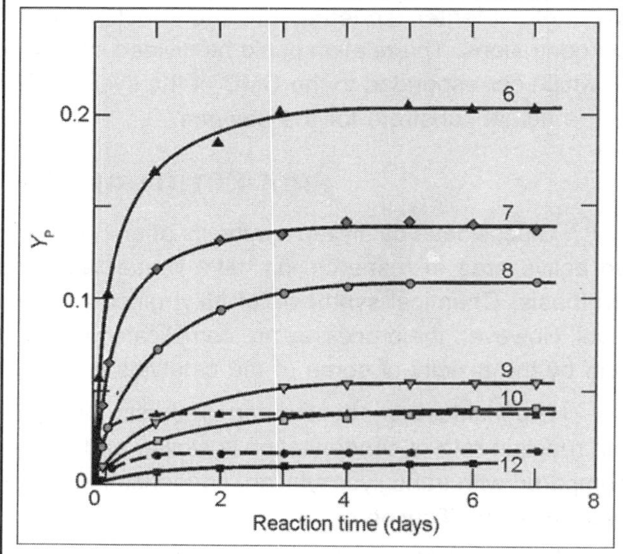

Figure 20.2 Kinetics for the synthesis of alkyl β-D-glucosides through β-glucosidase-catalyzed condensation of glucose and various alcohols in a two-phase system.

alcohols, although the yield of 2-alkyl glucoside was lower than that of 1-alkyl glucoside. In the two-phase system, the product partitioned to the organic phase.

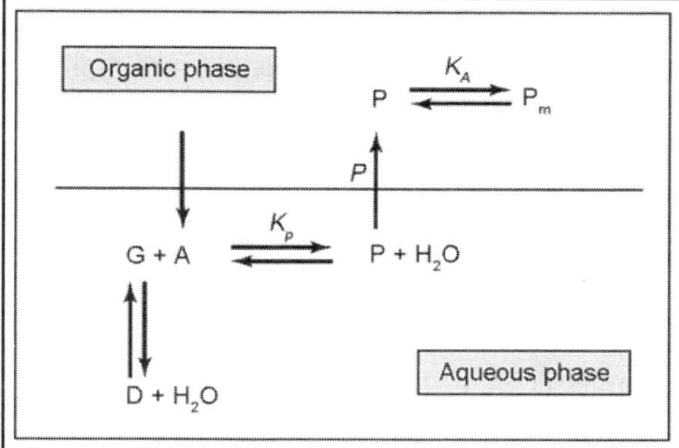

Figure 20.3 Reaction scheme supposed to occur in the synthesis of alkyl β-D-glucoside through β-glucosidase-catalyzed condensation in an aqueous-organic two-phase system.

Because it is known that β-glucosidase catalyzes the condensation of glucose molecules to form oligosaccharides, mainly β-glucobioses, at higher glucose concentrations, the undesirable condensation would also occur in the system as well as the intended condensation between glucose and alcohol in the system. Under the following assumptions, (i) the condensations occur only in the aqueous phase containing the enzyme; (ii) the product P, alkyl glucoside, can partition to the organic phase with a partition coefficient P_{app}; (iii) the glucose, G, and produced β-glucobioses, D, locate only in the aqueous phase due to their high hydrophilicity; (iv) the aqueous phase is always saturated with an alcohol, A, that is, the concentration of the alcohol in the aqueous phase, $C_{A,aq}$, is given by its solubility in water, and (v) the water concentration, C_W, is kept constant during the reaction, equations for estimation of the equilibrium yields of alkyl glucoside, Y_P, and β-glucobioses, Y_D, through β-glucosidase-catalyzed condensation in the two-phase system have been derived:

$$Y_P = \frac{X}{4K_D C_{G0}/C_W}\left\{\left[(1+X)^2 + \frac{8K_D C_{G0}}{C_W}\right]^{1/2} - (1+X)\right\} \quad \ldots(2)$$

$$Y_D = 1 - \frac{C_W(1+X)}{4K_D C_{G0}} \cdot \left\{\left[(1+X)^2 + \frac{8K_D C_{G0}}{C_W}\right]^{1/2} - (1+X)\right\} \quad \ldots(3)$$

where

$$X = (1+\alpha P_{app})K_P C_{P,aq}/C_W \quad \ldots(4)$$

C_{G0} is the initial glucose concentration in the aqueous phase, $C_{P,aq}$ is the alkyl glucoside concentration in the aqueous phase, α is the volume ratio of aqueous to organic phase, and K_P and K_D are the equilibrium constants for the alkyl glucoside and β-glucobiose formation, respectively. The water concentration can be calculated by $C_W = -6.43\ C_{G0} + 54.58$, where both C_W and C_{G0} are in units of moles per liter.

The equilibrium constant for formation of β-glucobioses, K_D, was determined to be 1.17. The equilibrium constant for formation of 1-alkyl β-D-glucosides with various alkyl chains, K_P, was evaluated for the experimentally observed Y_P values according to equations 2 and 4. The K_P value was independent of alkyl chain length and was 1.90. It was also shown that alkyl glucoside with any alkyl chain formed a dimeric aggregate in the alcohol phase with an association constant K_A of 4.2 L/mol, which was independent of alkyl chain length, and that the partition coefficient of the monomeric alkyl glucoside, P, increased exponentially with the alkyl chain length, n ($\ln P = 1.08\,n - 5.17$). The apparent partition coefficient, P_{app}, can be calculated by

$$P_{app} = P(1 + 2K_A P C_{P,aq}) \qquad ...(5)$$

Using these equations and parameters, we can calculate the equilibrium yields, Y_P and Y_D, under any conditions. The volume ratio of aqueous to organic phase, α, is an important factor affecting the yields. At higher α values, the equilibrium yields of alkyl glucosides are higher, while the equilibrium yields of β-glucobioses become lower. These results indicate that higher α values are preferable for alkyl glucoside synthesis. However, the concentrations of alkyl glucosides in their corresponding alcohol phase, $C_{P,org}$, decrease with an increase in α. The initial glucose concentration, C_{G0}, is another important factor. The yield, Y_D, of β-glucobioses, undesirable products in alkyl glucoside synthesis, largely increases at higher C_{G0}. Although the yield of alkyl glucoside, Y_P, also increases with increasing C_{G0}, the extent is not as great as that for the yield of β-glucobioses.

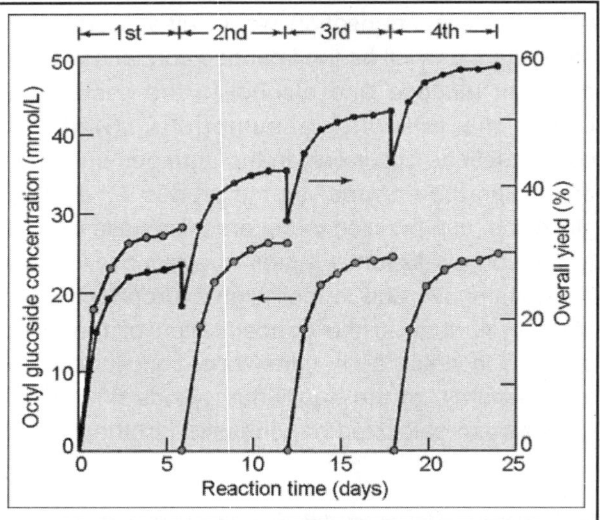

Figure 20.4 Repeated use of β-glucosidase for the synthesis of 1-octyl β-D-glucoside in the heterogeneous system.

Synthesis in a Heterogeneous System

Alkyl β-D-glucoside could be synthesized in a heterogeneous system where solid glucose and β-glucosidase are added in a buffer-saturated alcohol. In such a system, a very viscous phase appeared, and it could be easily separated from the alcohol phase after the condensation reaction. Because the enzyme in the viscous phase was expected to be still active, the phase was repeatedly used for the synthesis of alkyl glucoside. The enzyme was added only in the first run, and no further addition was made in subsequent batches. In each batch, glucose was added in an amount equal to that of the product in the previous batch. In this system, the overall yield of alkyl glucoside could be significantly improved without immobilization of the enzyme. The reason for the increase in the overall yield by the consecutive synthesis can be explained by

$$Y_{P,n} = \frac{\sum_{i=1}^{n}(VC_{P,i})}{VC_{G0} + \sum_{i=1}^{n}(VC_{P,i})} = \frac{\sum_{i=1}^{n}C_{P,i}}{C_{G0} + \sum_{i=1}^{n}C_{P,i}} \qquad ...(6)$$

where $Y_{P,n}$ is the overall yield of alkyl glucoside at the end of nth batch, $C_{P,i}$ is the product concentration at the end of ith batch, C_{G0} is the initial glucose concentration in the first batch, based on the total volume of reaction medium, V, which is nearly equal to the volume of the alcohol phase, and n is the number of cycles. Equation 6 indicates that the overall yield, $Y_{P,n}$, asymptotically approaches unity when n increases.

21

ERYTHROPOIETIN

Erythropoietin (EPO) stimulates red blood cell production by promoting both the growth of late erythroid precursor cells and their maturation into proerythroblasts, in which globin synthesis starts. It is widely accepted that EPO is a major physiological regulator of the process of erythroid differentiation and that it controls the maintenance of physiological levels of the circulating erythrocyte mass. EPO is a heavily *glycosylated protein* and produced mainly by the kidney in human adults. Anemia associated with renal failure often results from a decreased level of EPO, and EPO administration results in a dramatic improvement with few side reactions in these anemic patients. Carbohydrates attached to EPO play a critical role in expression of the in vivo activity. In this paper, we mainly focus on the production of recombinant human EPO by mammalian cells and structure and function of carbohydrate chains, comparing with urinary EPO and recombinant product. Basic aspects of EPO, including a novel function of EPO in brain, are also described.

CHEMICAL NATURE

It has taken a long time to isolate and characterize EPO. More than 100 years ago, it had been suggested that hypoxia stimulated red blood cell formation. In 1977, human EPO was purified from patients with aplastic anemia. The human gene for EPO was cloned in 1985, and recombinant EPO (rEPO) is now available. One unit (U) of EPO activity was defined as an erythropoiesis-stimulating activity elicited when 5 μmol of $CoCl_2$ are injected into a starved Sprague Dawler male rat. Cobaltous ion promotes erythropoiesis through stimulation of EPO production, because this ion mimics hypoxia in terms of EPO production. The latest standard human rEPO gives 130,000 U/mg protein when it is fully glycosylated. The values of the serum concentration in healthy adult humane range 10 to 30 mU/mL. In 1989, EPO receptor was cloned from erythroleukemia cell line, and the signal transduction pathway of EPO has been analyzed in recent decades. Human EPO gene has been mapped on chromosome 7q11-q22. The mouse EPO gene is localized on chromosome 5.

There is a single copy with a length of 4 kb that covers a minimum promoter, five exons, four introns, and hypoxia-inducible enhancer. The promoter, first intron, and hypoxia-inducible enhancer are highly homologous between human and mouse EPO. Deduction from cDNA indicated 193 amino acid residues in human EPO. The mature EPO isolated from human urine consists of 165 residues and is formed by removal of N-terminal 27 signal peptide and C-terminal Arg residue through the post-translational processing. Two disulfide bond are formed between positions 7 and 161 and 29 and 33. The disulfide linkage between 7 and 161 is essential for EPO bioactivity. Because there is 80% protein homology between mouse and human EPO, human EPO crossbinds with mouse EPO receptor and vice versa. EPO protein has a significant homology to mature thrombopoietin, which is a primary regulator for megakaryopoiesis. A tertiary conformation for EPO has been predicted based on the protein structure modeling.

Despite little homology in the amino acid sequence, EPO has a similarity to the three-dimensional structures of growth hormone, prolactin, interleukin-6, and granulocyte-colony stimulating factor. By analogy with growth hormone that has been shown to contain four antiparallel α-helices, EPO is also thought to contain four antiparallel α-helices and three loops, which is supported by the analysis of site-directed mutation within EPO molecule. Mutations of buried portion of the helix A (first helix from the N-terminal of EPO) and D (last helix) show little effect in vitro activity, whereas those of surface portion of the helix A and D reduced the activity. The mutational analysis suggests that EPO has two separate receptor binding sites, which is consistent with the ligand-induced receptor homodimerization, an initial step for the signal transmission.

Production of Erythropoietin

As far as kidney cells responsible for EPO production are normal, the relationship between concentrations of EPO and hemoglobin in blood is normally inverse. A decrease in the hemoglobin content, for instance from blood loss, is detected by the EPO-producing cells in kidney as a reduction of oxygen supply. Low oxygen stimulates transcription of EPO gene, and the resulting increase in blood EPO promotes the proliferation and differentiation of colony-forming unit erythroids (CFU-Es), erythroid precursor cells on which EPO acts, into matured red blood cells. Recovery of the red blood cell mass and the concomitant oxygen supply suppresses EPO production. The molecular mechanism of hypoxic activation of EPO gene transcription has been extensively studied, although the molecule that senses changes of oxygen concentration is poorly understood. An important *cis*-acting sequence required for the hypoxic induction of Epo gene was defined in the 3′-flanking region.

This enhancer is a 50-bp element consisting of three important segments. The highly conserved sequence near the 5′ end of the enhancer is a hypoxia-inducible factor-1 (HIF-1) binding site. HIF-1 is a transcription factor consisting of two basic helix-loop-helix-PAS proteins (HIF-1α and HIF-1β). The middle segment containing CA repeats in human gene is not well conserved between human and mouse, but the mutation of this region abolished the inducible behavior of both the human and murine enhancers. Thus far the specific protein that binds to this region has not been demonstrated. The third element is the 3′ segment, which is again highly conserved and amplifies the hypoxic signal. Binding of hepatocyte nuclear factor-4 (HNF-4), which belongs a member of the steroid hormone receptor family, to this element appears to increase the hypoxic inducibility.

Organ Involved

The kidney is the primary site of EPO production in adults, whereas the liver is the production site in the fetal stage. *Hypoxia* activates expression of the EPO gene in kidney and liver. The-EPO producing cells in the kidney have been controversial but recent studies indicate that EPO is produced by peritubular interstitial fibroblasts. The hepatic EPO production is essential for fetal erythropoiesis, because homozygous mice carrying null mutations of EPO gene die around embryonic day 13 due to deficiency in erythropoiesis. Although EPO production in liver is hypoxia inducible, the failure of renal EPO production cannot be substituted by the liver, because uremic patients with severe anemia show low levels of serum EPO. EPO has been believed to act exclusively on erythroid precursor cells, but several recent findings indicate that EPO acts on neurons in the central nervous system (CNS). Astrocytes produce brain EPO, and low oxygen tension stimulates the production of EPO as it does in liver and kidney.

Physiological Functions

Hematopoiesis is governed by several hematopoietic growth factors. EPO is a glycoprotein that plays a essential role in the regulation of erythropoiesis by stimulating the proliferation and differentiation of *erythroid precursor cells*. All the terminally differentiated blood cells have been derived from pluripotent stem cells with self-renewing activity. Hematopoiesis occurs in bone marrow, and the spleen is also an active hematopoietic site in murine. Pluripotent stem cells are committed to differentiate to multipotent stem cells with differentiation activity to several lineage-specific progenitors that have lost the self-renewality. Each lineage-specific progenitors differentiate to mature blood cells such as erythrocytes, monocytes, granulocytes, and megakaryocytes. The most primitive erythroid-specific progenitor cells derived from multipotent stem cells are burst-forming unit-erythroids (BFU-Es).

Although BFU-Es express EPO receptor, their response to EPO is low, and they require additional cytokines such as interleukin 3 or granulocyte-macrophage–colony-stimulating factor for their proliferation and differentiation into CFU-Es that show the highest sensitivity to EPO. Homozygous mice carrying null mutation in EPO receptor gene die around embryonic day 13, because mature erythrocytes are not formed, although CFU-Es abundantly exist. Thus, EPO is essential for differentiation and proliferation of CFU-E. Recently, a novel function in the CNS has been proposed. Neuronal cell lines such as PC12 and SN6 express EPO receptor, and binding of EPO to PC12 cells increases the intracellular concentration of monoamines. EPO protects primary cultured hippocampal and cerebral cortical neurons from NMDA receptor-mediated glutamate toxicity, which is believed to be a major cause of neuron death by ischemia. These findings suggest that EPO plays a neuroprotective role when brain is exposed to hypoxia or ischemia.

ISOLATION OF ERYTHROPOIETIN

In some cases, carbohydrates attached to the proteins play a key role in expression of their biological activities. The carbohydrates may be required for the in vitro activity as well as the in vivo activity. Even when the carbohydrates have nothing to do with the in vitro activity, they may be very important in vivo; carbohydrates protect glycoproteins from heat inactivation and proteolytic degradation, resulting in an increased lifetime in circulating blood. Eukaryotic cells are used for production of recombinant glycoproteins.

For clinical applications, not only glycosylation but also the structure of attached carbohydrates are critical; a glycoprotein whose terminal carbohydrate residues are recognized by hepatic receptors would be rapidly eliminated from the circulation. The kind of carbohydrate structure may also affect the antigenicity of the recombinant product. When recombinant glycoproteins are produced in heterologous animal cells, their amino acid sequences govern the positions onto which carbohydrate chains are attached, but the maturation of the chains may entirely depend on the host cells. It is important, therefore, to compare the properties of recombinant glycoproteins produced in different cell lines with those of naturally occurring ones. The glycoprotein may also be dependent on the culture conditions.

Gene Involved in the Production

Peptides derived from urinary EPO (u-EPO) were sequenced to prepare synthetic DNA probes for screening a human fetal liver genomic library for EPO gene clones. Restriction endonuclease mapping and Southern analysis showed the gene fragments contained in three phage clones to be identical. One, designated λEPO41, was used for sequence analysis. The entire amino acid sequence of EPO was determined. The clone λEPO41 contained a whole EPO coding region, although half of 3′-untranslated region was missing. The EPO expression vector, pZIP-NeoSV(X) 1-EPO, contained the EPO gene preceded by a long terminal repeat (LTR) functioning as transcriptional promoter and followed by DNA sequences derived from the transposon Tn5 that confers G418 resistance upon mammalian cells. The expression vector was introduced into two mammalian cells, BHK cells established from baby hamster kidney, and ψ2 cells derived from NIH/3T3. Both cells were transfected with pZIP-NeoSV(X) 1-EPO, using calcium phosphate method. EPO-producing cells were isolated by selection in G418 media and by two rounds of limiting dilution. The EPO levels in culture supernatants of EPO-producing cells assayed by radioimmunoassay (RIA) were 150 U/mL for BHK cells and 300 U/mL for ψ2 cells.

Cultural Activities

BHK cells were cultured in Eagle basal medium supplemented with 10% calf serum and 10% bacto-tryptose phosphate broth. The ψ2 cells were cultured in Dulbecco's modified Eagle's medium supplemented with 10% calf serum. Cell cultures for EPO production were started at a cell density of 2×10^5 cells/mL in plastic culture trays containing 2 L of the medium. After 4 days of culture, the cells grew to be nearly confluent. The culture medium was replaced with the low-serum medium (1.5% calf serum) and culture supernatants were harvested every 3 days.

Human Renal Erythropoietin

Human urinary EPO directed MAb was effective in the purification of EPO from human urine and culture supernatants of BHK cells that produced rEPO, but pretreatment of the EPO with SDS was needed for EPO to bind with the immobilized MAb. To develop a more rapid purification procedure, the new MAbs, R2, were used for rEPO purification. The culture supernatant (100 L) was filtered with suction and concentrated to 2 L by ultrafiltration on a hollow-fiber device with a nominal molecular cutoff of 10,000. The concentrate was put on an immunosorbent column (4.4 x 6.6 cm) containing EPO-directed MAb R2 fixed on Affi-Gel 10 equilibrated with phosphate buffered saline (PBS). The column was extensively washed with PBS, 10 mM NaPi, pH 7.4, containing 0.5 M NaCl and 0.15 M, in this order. EPO was eluted by 0.2 M acetate. pH 2.5, containing 0.15 M NaCl. After neutralization of the eluted fraction, the protein was concentrated by ultrafiltration with a hollow-fiber device and was precipitated with 90% ethanol.

The precipitate was dissolved in a small amount of PBS and put on a Sephadex G-100 column (2.2 x 94 cm) equilibrated with PBS. The EPO in the fractions was concentrated and precipitated with 90% ethanol. The precipitate was dissolved in 10 mM NaPi, pH 6.8, containing 10 μM $CaCl_2$ and put on a hydroxyapatite column (100 x 7.8 mm: Bio-Gel HPHT, Bio-Rad) equilibrated with the

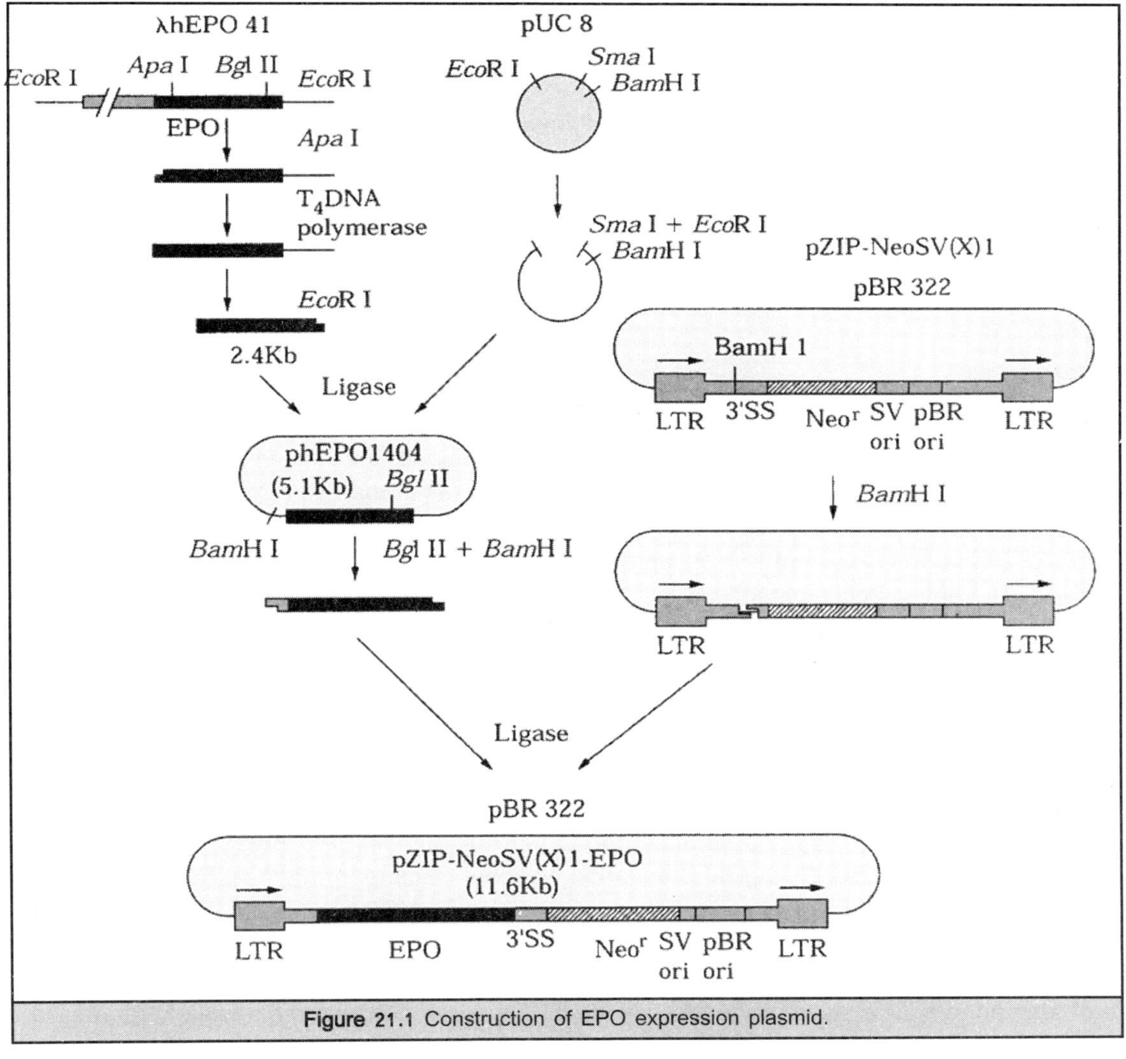

Figure 21.1 Construction of EPO expression plasmid.

same buffer. The adsorbed EPOs were eluted by increasing concentration of NaPi. The rEPO-B appeared in the flow-through of a column equilibrated with 10 mM NaPi, whereas rEPO-ψ was eluted by 50 mM NaPi. The large-scale purification of rEPO-B in the supernatant of BHK cells was attempted. Purification with the immunosorbent column was effective; most of the protein and little EPO activity in the concentrate of culture supernatants emerged in the flow-through fractions without being adsorbed, and EPO was eluted sharply at pH 2.5.

About 2,800-fold purification was achieved by this single step, and 84% of the activity was recovered. Mab R2 was suitable for purification of EPO with a high yield. Subsequent chromatography with Sephadex G-100 and then hydroxyapatite yielded purified EPO with 3,260-fold purification and 52% recovery of the starting activity. The purified EPO had the specific activity of 137,000 U/A_{280nm}, which was higher than that of the EPO purified with an earlier preparation of an MAb. The final EPO preparation was pure on SDS-polyacrylamide gel electrophoresis and high-pressure liquid chromatography. The immobilized R2 could be used 30 times without any deterioration.

Table 21.1 Purification of recombinant human EPO from BHK cells

Procedures	*Protein ($A_{280\ nm}$)*	*Activity ($U \times 10^{-4}$)*	*Specific activity ($U/_{A280\ nm}$)*	*Yield (%)*	*Purification (fold)*
Concentrated culture supernatant	473,000	1,990	42	100	1
Immunosorbent column	141	1,680	119,000	84	2,830
Sephadex G-100	97	1,310	135,000	66	3,210
Hydroxyapatite	75	1,030	137,000	52	3,260

Purified uEPO showed two bands. The minor one with the lower molecular weight is the asialylated form of uEPO. Both rEPOs migrated as a single band, but they differed in electrophoretic mobility. Recombinant EPO from ψ2 cells migrated slightly faster than uEPO, and the rEPO-B band was broader than those of rEPO-ψ and uEPO. Thirty amino acids in the N-terminal portion of rEPO-ψ and rEPO-B were consistent with those of uEPO and also with those deduced from the cDNA sequence. The amino acid sequence in the C-terminal region of both rEPOs and uEPO determined by the carboxypeptidase method was identical to that of uEPO, indicating that the recombinant products had the same primary structure as uEPO. Therefore, differences between purified rEPOs in their electrophoretic mobility on SDS-polyacrylamide gel, therefore, are attributable to the difference in the carbohydrate chains.

Biology of Erythropoietin

The Second International Reference Preparation for human EPO obtained from the WHO International Laboratory for Biological Standards, National Institute for Medical Research, London, was used as the standard. The in vivo and in vitro activities of three purified EPOs were assayed. The stimulatory effect of EPO on the incorporation of ^{3}H-thymidine into DNA of cultured mouse fetal liver cells was used to assay the in vitro activity, and the in vivo activity was assayed with the use of starved rats. The stimulatory effect of the EPO preparations were completely repressed by incubation with antiserum against EPO, indicating that the mitogenic effect of a contaminant did not contribute to the in vitro activity. The in vivo activity of rEPO-B was equivalent to that of uEPO, and its in vitro activity was slightly higher than that of uEPO. The in vitro activity of rEPO-ψ was 1.6-fold that of uEPO and of rEPO-B, whereas, the in vivo activity was only 25% that of other EPOs.

The in vitro activities of the three EPO preparations slightly increased after sialidase treatment, but their in vivo activities were abolished. It seems to be that the lower the amount of sialic acid, the higher the in vitro activity. The high activity of the native rEPO-ψ as compared with other native EPOs may be due to the low content of sialic acid, because the asialo EPOs have an increased affinity to EPO receptor. However, the low in vivo activity of rEPO-ψ is probably due to the small amount of sialic acid, because the removal of sialic acid from EPO by sialidase treatment abolished in vivo activity. Loss of in vivo activity is caused by the removal of asialoglycoproteins from the circulation by the liver. Thus, it is likely that the in vivo lifetime of rEPO-B is longer than that of rEPO-ψ.

Table 21.2 Biological activities of human EPO

EPO	*Treatment*	*EPO activity ($U/_{A280}$)* *In vitro*	*In vivo*
u-EPO	Untreated	77,000	72,000
	Sialidase treated	99,000	0
r-EPO-B	Untreated	96,000	72,000
	Sialidase treated	109,000	0
r-EPO-ψ	Untreated	153,000	18,000
	Sialidase treated	173,000	0

IMMUNOLOGY

We reported an EPO-directed monoclonal antibody (MAb) that binds to the antigen after it is treated with SDS. It appears that this antibody binds to the epitope exposed by a conformation change of EPO caused by the SDS. This antibody worked efficiently in the purification of EPO from human urine and human recombinant EPO from the supernatant of cultured cells, but treatment of preparations containing EPO with SDS is time consuming. We have prepared MAbs capable of binding with the native EPO and thereby to achieve the following purposes: (i) establishment of a more rapid procedure for the purification of EPO by immobilized MAb; (ii) development of a sandwich-type ELISA, which is easier to handle and more rapid and sensitive than radioimmunoassay (RIA), by use of MAbs that bind to EPO at different antigenic determinants.

Hybridomas Cells

Female BALB-c mice 8 weeks old were immunized intraperitoneally with 90 μg of pure rEPO emulsified 1:1 in Freund's complete adjuvant. After three times of immunization, hybridomas were prepared by procedures described previously. Hybridoma cells were maintained in RPMI 1640 supplemented with 10% fetal calf serum. Two hybridoma clones (R2 and R6) that were stable in their rapid growth and their production of MAbs to EPO were isolated. Both antibodies were produced in the ascitic fluid of BALB/c mice and purified by the ammonium sulfate precipitation and diethylaminoethyl (DEAE)-cellulose column chromatography. The subclass of both antibodies was IgG1. The apparent dissociation constant, k_d was calculated by the binding to ^{125}I-labeled EPO. Both MAbs bound to EPO with high affinities, with k_d = 1–2 nM, respectively. The fully deglycosylated rEPO bound to the MAbs, indicating that they recognized peptide sequences of the antigen but not the carbohydrates attached to the antigen. Moreover, each antibody recognizes a different epitope.

ELISA

The finding that the antigen site recognized by MAb R2 differed from the sites recognized by R6 prompted us to develop a sandwich-type ELISA for the measurement of EPO levels. EPO was sandwiched between R6 fixed in microtiter wells and the enzyme-linked R2. R2 was linked covalently with alkarine phosphatase with 0.2% glutaraldehyde. The MAb R6 solution (10 μg/mL) was added to the wells of microtiter plates with 96 flat-bottomed wells and incubated for 2 h. After washing, 100 μL/well of the sample solution containing 0.05% Tween 20 was added to the wells, which were incubated for 2 h to form the fixed R6 and EPO complexes. After washing, 100 μL/well of the alkaline phosphatase-linked R2 solution was added and the wells were incubated for 2 h more to form ternary complexes (fixed R6, EPO, and enzyme-linked R2). The wells were washed, and then 100 μL of 2 mg/mL *p*-nitrophenyl phosphate disodium was added. After incubation for 30 min, the reaction was stopped by the addition of 3 M NaOH.

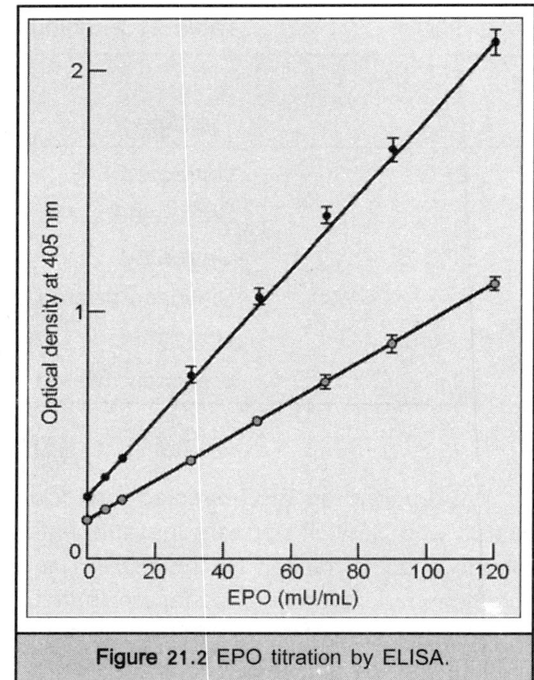

Figure 21.2 EPO titration by ELISA.

The optical density at 405 nm was measured with a microtiter reader. This procedure is referred to as a two-step reaction. In the two-step reaction, EPO in samples was bound to the fixed R6 and then alkaline phosphatase-linked R2 was added to the wells to form ternary complexes of R6, EPO, and the enzyme-linked R2. In the alternative procedure, the one-step reaction mixtures of EPO samples and enzyme-linked MAb R2 were added to the wells coated with MAb R6. The relationship of the amounts of rEPO and the enzyme activity was linear regardless of different procedures; use of the one-step reaction shortened the time needed for completion of the process to 7 h. Once the microtiter wells in which the first antibody, R6, have been fixed, results can be obtained within 3 h. To check the reproducibility of the ELISA, intra-assay and interassay studies were done. In both studies, the coefficients of variation were satisfactory (<20%). The assay measured plasma levels of EPO as low as 2 mU/mL within several hours. We have developed RIA for EPO by the use of antisera against human u-EPO and measured EPO levels in plasma samples from normal subjects and anemic or polycythemic patients.

A sample diluted to about 1 U/mL EPO was mixed with 10,000 cpm of ^{125}I-rEPO, rabbit anti-rEPO antiserum. The mixture was incubated at 4°C overnight and sheep antirabbit γ-globulin (Cappel) was added. The immunocomplexes were precipitated by centrifugation. The radioactivity in the precipitate was counted with a γ-counter. The amounts of EPO in the sample were estimated from a standard curve (2 to 128 mU/assay). From total 38 samples, EPO levels of normal subjects and patients were measured with ELISA to compare with RIA. The results indicate a high degree of correlation between these two methods, with a correlation coefficient of 0.965. The ELISA was comparable with the RIAs in sensitivity. One can assay EPO for a short time without using any radioactive compounds. Moreover, the assay conditions, once established, can be used semipermanently, because MAbs with the constant properties can be supplied continually. The sandwich-type ELISA developed by the use of MAbs thus had advantages over the RIAs in practicality. The findings that the MAbs R6 and R2 bound to EPO with high affinities and recognized different antigenic determinants made it possible to develop a sensitive and easier-to-handle method for EPO assay. The sandwich-type ELISA should be useful to assay EPO levels in human plasma for investigating an etiology of defective erythropoiesis and also to monitor rEPO given to anemic patients.

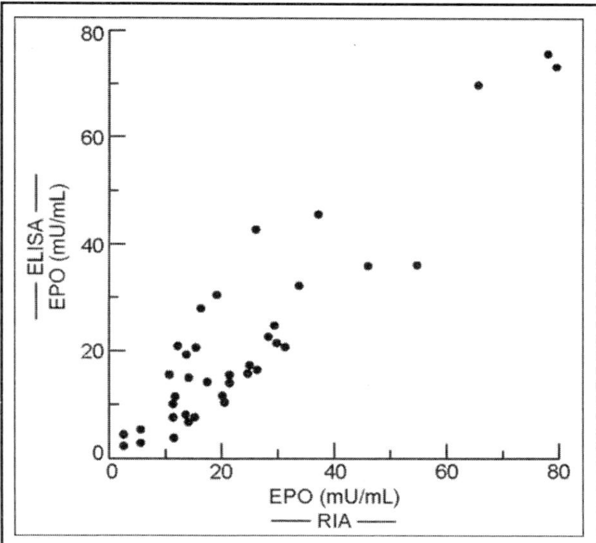

Figure 21.3 Comparison of RIA and ELISA in human plasma. The ordinate shows EPO in the plasma samples assayed with ELISA by two-step reaction, and the abscissa indicates EPO estimated with RIA by using polyclonal antibody raised in a rabbit.

ROLE OF CARBOHYDRATES IN ERYTHROPOIESIS

The following biological functions can be assigned to carbohydrates attached to proteins such as EPO that act in the endocrine system. First, the carbohydrates of glycoproteins are needed for their biosynthesis and secretion, including folding of the protein during translation and protection of the protein from intracellular degradation. Second, carbohydrates confer stability to the protein so that it is not eliminated from the circulation during its travel from the production site to its target cells. Mammalian cells, particularly liver cells, express proteins (lectins) that bind to carbohydrates in glycoproteins. Glycosylation of proteins thus can be a signal determining their lifetime in the circulation. Third, carbohydrates are involved in the expression of biological activity, including the steps of the binding of the protein to its target cells and intracellular signal transduction.

To examine these possible functions of carbohydrates, the following strategies can be used. First, there are some mammalian cell lines lacking an enzyme involved in glycosylation. One can use these cells to produce a recombinant glycoprotein and compare with the natural product or the fully glycosylated recombinant product. Second, one may produce the protein in procaryotes incapable of attaching carbohydrates, and if the product has the biological activity, one can get some information of functions of carbohydrates. Third, one can examine the effects of chemical or enzymatic removal of carbohydrate residues from a glycoprotein on its biological properties. Discovery of enzymes that remove carbohydrate residues from a glycoprotein is particularly useful, because the enzymatic treatment is milder by far than the chemical treatment. Fourth, glycosylation sites can be removed by site-directed mutagenesis and the biological properties of the glycosylation defective mutant glycoproteins are compared with those of wild type.

Removal of Carbohydrates from Erythropoietin

Figure 21.4 shows representative structures of N-linked and O-linked oligosaccharides of recombinant human EPO produced by BHK cells and cleavage sites of the enzymes that we used for removal of carbohydrate residues. The native EPO is represented as EPO-0, and deglycosylated forms are as EPO-1~5 in Table 21.3. Sialidase removes terminal sialic acids, producing asialo-EPO (EPO-1). *β*-Galactosidase cleaves off the exposed Gal residues of N-linked oligosaccharides in EPO-1 but does not split the $b_1 \rightarrow 3$ linkage between Gal and GalNAc in the O-linked oligosaccharide, giving EPO-2. Double digestion of EPO-1 with *β*-galactosidase and *β-N*-acetylhexosaminidase removes all the exposed Gal and GlcNAc residues from N-linked oligosaccharides, and the core structure is exposed (EPO-3). N-Glycanase completely eliminates N-linked oligosaccharides from EPO-0, yielding EPO-4. *Endo-α-N-* acetyl-galactosaminidase removes

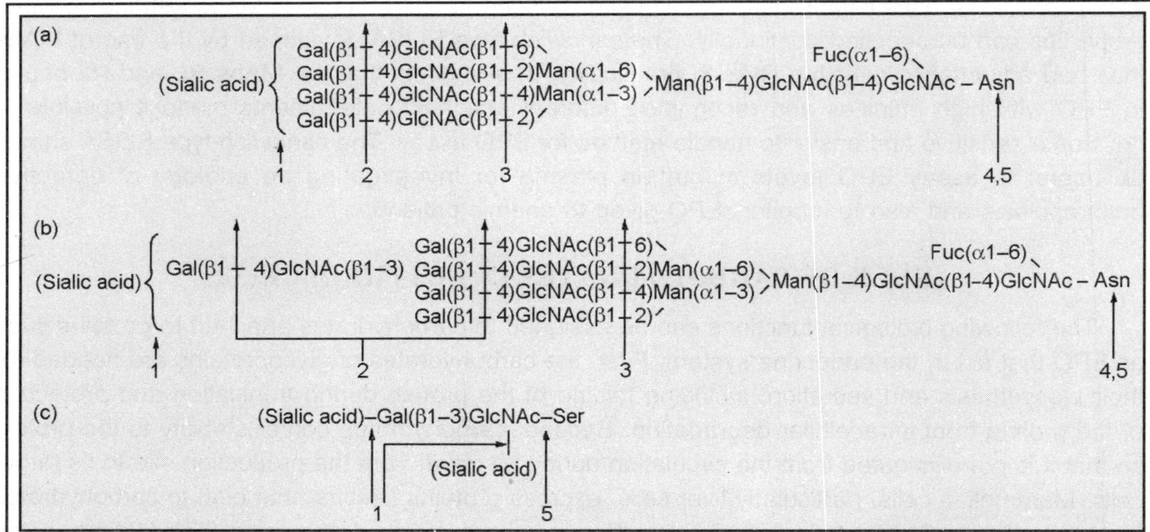

Figure 21.4 Enzyme removal of sugars from recombinant human EPO. Cleavage site of sugars in EPO by glycosidases are shown. (a and b) Structures of representative complex-type N-linked sugars (c) O-linked sugar.

O-linked oligosaccharides from asialo-EPO. Therefore, successive digestion of EPO with sialidase, *endo*-a-*N*-acetylgalactosaminidase, and *N*-glycanase yields the fully deglycosylated EPO, EPO-5. Completion of the enzyme reactions were confirmed by SDS-polyacrylamide gel electrophoresis of the products.

Table 21.3 Biological activities of recombinant EPO deglycosylated by glycosidases

EPO	*In vitro activity (%)*	*Affinity to EPO-R [I_{50} (nM)]*	*In vivo activity (%)*
EPO-	100	1.2	100
EPO-1	252	0.3	0
EPO-2	252	0.3	0
EPO-3	266	0.3	0
EPO-4	81	0.6	0
EPO-5	106	0.3	0

Affinities to EPO receptors do not represent the true dissociation constants but the relative values; the lower value, the higher affinity. Carbohydrates with terminal sialic acids are essential for expression of the in vivo activity. The carbohydrates are not important for expression of the in vitro activity. Deglycosylation increased the in vitro activity by increasing the affinity to EPO receptor. The presence of terminal sialic acids is responsible for an decreased affinity. EPO-4 and 5 have in vitro activity similar to that of the native EPO (EPO-0), although their affinity to EPO receptor is higher than EPO-0. They may be partially inactivated during assay thorough aggregate formation. The removal of terminal sialic acids from EPO has been shown to destroy its in vivo activity.

Desialylated EPO is cleared from the circulation faster by far than the native EPO. In contrast, the in vitro activity of EPO increases with desialylation. A possible function of carbohydrates in the in vitro activity of EPO was controversial. It has been reported that *N*-glycanase digestion of EPO produced in Chinese hamster ovary (CHO) cells caused almost complete loss of activity, which was contradictory to our conclusion that carbohydrates were not important in expression of the in vitro activity. However, all the results published later support our conclusion. We have to be very careful to assay the activity of EPO that has lost N-linked sugars, because the deglycosylated EPO is temperature-sensitive and sticks to experimental vessels.

Production of Mutant EPO

We constructed EPO genes mutated by site-directed mutagenesis so that EPOs lacking N-linked carbohydrates in every possible combination of the three N-glycosylation sites were produced. Because the codon of Asn (N) at the N-glycosylation sites is replaced with that of Gln (Q) in these mutants, we referred to them as NQ0, NQ1, and so on. In this nomenclature, for example, NQ13 means the mutant at positions 24 and 83. NQ0 corresponds to the wild type and NQ123 is the mutant with no N-glycosylation sites. Transient production of mutant EPOs by BHK cells was examined to find roles of sugars in biosynthesis of EPO. The mutant EPOs produced by stably transfected BHK cells were partially purified and their properties, including the biological activity both in vivo and in vitro, were investigated.

In some mutants, the in vitro activity was higher than that of wild-type, and the high activity is probably due to an increased affinity to EPO receptor. The in vitro activity of EPO with no glycosylation sites was comparable with that of the wild-type EPO, whereas the in vivo activity severely decreased. The elimination of all three N-glycosylation sites decreased EPO production to 10% of that of the wild-type EPO. These results indicate that N-linked sugars of EPO have two major functions: N-glycosylation is important for (i) proper biosynthesis and/or secretion and (ii) expression of the in vivo activity. N-Linked sugars of EPO affect binding affinity of the ligand to the receptor but do not play a key role in expression of the in vitro activity.

It has been reported that replacement of Asn to Gln at position 24 or 38 gives EPOs with the specific activity in vitro of about 20% that of the wild type, which is contradictory to our results. Preventing O-linked glycosylation at position 126 by site-directed mutagenesis severely reduced the production into the culture medium, suggesting that O-linked sugar is important for proper biosynthesis and/or secretion. But the result from other laboratories showed the normal production of the mutant. The latter was supported by the result obtained using the mutant cells that are deficient in an enzyme involved in the carbohydrate synthesis. Involvement of O-linked sugar in expression of the biological activity is unlikely. O-linked sugar is present only in primate EPOs but not in rodents.

Genetic Engineering

Interestingly, fractionation of recombinant EPO containing biantennary and tetraantennary N-linked sugars revealed that EPO containing more extensively branched chains has greater in vivo activity. This suggests a possible unique role of the N-linked sugar in vivo. Attempt to produce EPO in *Escherichia coli* has been made. They were unable to compare the biological activity of EPO produced in *E. coli* with that of EPO produced in mammalian cells because of the poor stability of *E. coli* EPO, but they confirmed that *E. coli* EPO had the in vitro activity. *Bacillus brevis* cells secrete recombinant human EPO, which has in vitro activity similar to that of the fully glycosylated EPO. These results support our conclusion that the N-linked sugars in EPO are not required for expression of the in vitro activity. When human wild-type EPO was expressed in the polarized Madin-Darby canine kidney (MDCK) epithelial cells, EPO was preferentially secreted from the apical domain. This polarized secretion was perturbed by the treatment of the cells with tunicamycin, suggesting the involvement of N-linked sugars in the apical sorting mechanism in MDCK cells. N-Linked sugar chain at position 38 is critical for the polarized secretion, but the biological significance of this finding remains unknown.

22

BIOCHEMICAL CONVERSION PROCESS

Cultivation media in biotechnology are basically multiphase fluid systems with a continuous liquid phase in which cells and gas bubbles are dispersed. In such systems, the gas holdup ε is defined as the fraction of the gas phase of the total volume V_D of the dispersion considered.

$$\varepsilon = \frac{V_G}{V_D} \qquad \ldots(1)$$

V_G is the total gas volume within V_D (i.e., the gas holdup is a volume-related quantity). When considering the liquid phase, the gas bubbles are void areas. Thus, the gas holdup is sometimes called the void fraction.

When the entire culture within a bioreactor is considered, we then more precisely refer to the corresponding holdup as the total or global gas holdup. However, we may also speak about spatial distributions of the gas phase, particularly in systems where the gas phase is not quasi-homogeneously distributed. We then prefer to consider local gas holdups ε_l, which are defined by the differential quotient

$$\varepsilon_1 = \frac{dV_G}{dV_D} \qquad \ldots(2)$$

Usually we assume that we can consider the dispersion to be a quasi-homogeneous fluid. This is a reasonable assumption when discussing technical reactor sizes, where the bubbles are usually very small as compared to the reactor dimensions.

However, when we discuss effects on small scales, we should keep in mind that in gas–liquid dispersions the gas phase is composed of individual bubbles of volume V_j. It does not make sense to consider gas holdups when looking for the effects on scales in the order of the mean bubble size. On such small scales, the gas–liquid two-phase flow system must be treated in a discontinuous way. Hence, discussions about the gas holdup in bioreactors are more or less restricted to large scales as compared to the bubble size. In most practical cases we think of global gas holdups when discussing dispersions within bioreactors. When we consider the individual bubbles of volume V_j, we may rewrite the total gas holdup as

$$\varepsilon = \frac{V_G}{V_D} = \frac{\Sigma V_j}{V_D} \qquad \ldots(3)$$

where the sum is drawn over all bubbles in the entire volume V_D of the dispersion. The volume V_j characterizes the bubble size. However, it is often more convenient to use diameter instead, since the diameter can most easily be estimated by visual inspection of the dispersion. Unfortunately, the bubble diameter is not well defined, since different bubbles of the same volume might have different forms. This problem has been circumvented by defining the so-called equivalent diameter d_j, which is the diameter of a sphere with the bubble's volume V_j. We can then write the gas volume V_G as

$$V_G = \Sigma V_j = \frac{\pi}{6} \Sigma d_j^3 \qquad \ldots(4)$$

Bubbles have several functions in bioreactors, the most important being the transport of a gaseous reactant from the dispersed gas phase across the bubble surfaces into the liquid phase. In this respect, the transport cross section *A* available for this mass transfer is of primary importance. *A* is the total interfacial area between the gas and the liquid phase. This is, apart from the top surface of the dispersion, the sum of all bubble surfaces

$$A = \Sigma A_j \approx \pi \Sigma d_j^2 \qquad \ldots(5)$$

where A_j is the surface area of the *j*th bubble. In biochemical reaction engineering, one prefers to discuss the transport cross section as a specific, and in this particular case a volume-related, quantity. This specific interfacial area *a* is defined as

$$a = \frac{A}{V_D} = \frac{\Sigma A_j}{V_D} \approx \frac{\pi \Sigma d_j^2}{V_D} \qquad \ldots(6)$$

With these definitions we are able to relate the gas holdup to the quantities of primary interest in biochemical reaction engineering. Starting from equation 1, we obtain

$$\varepsilon = \frac{V_G}{V_D} = \frac{AV_G}{V_D A} = \frac{A}{V_D} \frac{\frac{\pi}{6} \Sigma d_j^3}{\pi \Sigma d_j^2} = a \frac{\frac{\pi}{6} \Sigma d_j^3}{\pi \Sigma d_j^2} \qquad \ldots(7)$$

Because the bubbles in biotechnologically relevant dispersions are not of uniform size, such expressions are of little practical use. One is more interested in characterizing the bubbles by some mean bubble diameter. In light of the preceding discussion it is convenient, as proposed by Sauter, to take the quantity

$$d_s = \frac{\Sigma d_j^3}{\Sigma d_j^2} \qquad \ldots(8)$$

as such a mean bubble diameter. The sums are drawn over all bubbles *j*. In the literature, d_s is referred to as the Sauter diameter. With the Sauter diameter d_s, we obtain a simple very important practical relation between the relevant gas-phase characteristics of a bubble dispersion:

$$\varepsilon = a\frac{d_s}{6} \text{ or } a = \frac{6\varepsilon}{d_s} \qquad \ldots(9)$$

which essentially states that the important specific interfacial area *a* is proportional to the gas holdup ε, the proportionality constant being dependent in a most simple way from the Sauter mean bubble diameter d_s. In other words, the specific interfacial area can increase by increasing the gas holdup, and is more effective, when the mean bubble size is kept smaller.

Assuming that the volume V_L of the liquid phase of a culture within a bioreactor is given, the higher the gas holdup ε, the larger the total volume V_D of the dispersion. Usually, in discussions about gas-phase behavior within bioreactors, the solid phase, in particular the cells, can be attributed to the liquid phase.

In reactors where the reactor horizontal cross-sectional area A_{RCS} is uniform with respect to the height H_D of the dispersion, we can write the gas holdup as a function of the height H_D of the fully aerated liquid dispersion and the height H_L of the liquid without aeration, i.e., when there are no gas bubbles present.

$$\varepsilon = \frac{V_G}{V_D} = \frac{V_D - V_L}{V_D} = \frac{(H_D - H_L)A_{RCS}}{H_D A_{RCS}} = \frac{(H_D - H_L)}{H_D} \qquad \ldots(10)$$

This relative enhancement of the culture level in the reactor by aeration, or simply by the presence of gas bubbles, is of practical importance, since, as discussed later, one must monitor this level during the operation of the bioreactor. In cases where it is possible to interrupt the gas throughput through the reactor and to remove the bubbles, this equation may be used to measure the total gas holdup. If such a degassing is possible, this is the most simple measurement of the total gas holdup in dispersions.

From a global view, a higher gas holdup of the dispersion leads to a lower mean density ρ_D of the dispersion. When the gas phases can be assumed to be quasi-homogeneously distributed, the density ρ_D can be represented as a linear function of the gas holdup ε:

$$\rho_D = (1 - \varepsilon)\rho_L \qquad \ldots(11)$$

where ρ_L is the density of the liquid phase. This has an immediate consequence for the static pressure at a particular height *H* within the reactor. Since the static pressure drop Δp over a height interval ΔH is dependent on the density of the dispersion, it is possible to directly relate the static pressure drop and the gas holdup

$$\Delta p = \rho_D g \Delta H \qquad \ldots(12)$$

$$\varepsilon = 1 - \frac{\Delta p}{\rho_L g \Delta H} \qquad \ldots(13)$$

which is a relationship that is often used to measure the gas holdup in multiphase reactors. Here, *g* is the gravitational acceleration.

The gas holdup can also be considered in terms of the aeration rate. The aeration rate is often represented by the gas throughput Q_G relative to the cross-sectional area A_{RCS} of the reactor. This quantity is referred to as the superficial gas velocity w_{sg}, since it formally has the dimension of a velocity

$$w_{sg} = \frac{Q_G}{A_{RCS}} \qquad \ldots(14)$$

When the mean bubble velocity in the laboratory coordinate system is $w_{bl,}$ then with

$$w_{b1} = \frac{H_D}{\tau_B} \qquad \ldots(15)$$

where τ_B is the mean residence time of the bubbles in the reactor, and

$$Q_G = \frac{V_G}{\tau_B} \qquad \ldots(16)$$

we obtain

$$\varepsilon = \frac{V_G}{V_D} = \frac{V_G}{A_{RCS}H_D} = \frac{V_G\tau_B}{\tau_B A_{RCS}H_D} = \frac{w_{sg}\tau_B}{V_D} = \frac{w_{sg}}{w_{b1}} \qquad \ldots(17)$$

In other words, the gas holdup is proportional to the superficial gas velocity, the proportionality constant being the reciprocal value of the mean bubble rise velocity in the laboratory coordinate system. The superficial gas velocity is the variable that can be manipulated. The effective bubble velocity $w_{bl,}$ however, depends on the fluid-dynamic properties within the reactor, as well of the broth rheology. Equation 17 has some practical consequences, as discussed later.

GAS HOLDUP

Of all the properties that determine the performance of a bioreactor with respect to a biochemical conversion process, it is the gas-liquid mass transfer capacity or capability that is the most important. This mass transfer capacity is primarily determined by the specific gas-liquid interfacial area *a* that the bioreactor is able to provide. As discussed, this quantity is primarily related to the gas hold-up. When there is no gas holdup, then there is no mass transfer.

The mass transfer rate is usually described by simple transport equation analogues to Fick's first or Ohm's law, where it is assumed that the mass current through a region that represents a resistance for the mass transfer is drawn by some driving force. In the gas–liquid mass transfer between gas bubbles and the bulk liquid, the essential mass transfer resistance is assumed to be the liquid-side boundary layer at the physical bubble surface, and the driving force is the difference between the dissolved gas concentrations at both sides of this boundary layer. In this simple picture, the mass transfer resistance is characterized by the product $k_L a$, where *a* is the specific interfacial area, which by equation 9 is proportional to the gas holdup.

The mass transfer coefficient k_L is defined by this simple transport law. In the particular case ofthe oxygen transfer from a bubble into its surrounding liquid, the transport resistance is essentially the liquid-side boundary layer around the bubble. At the physical interfacial area the liquid is assumed to be saturated with oxygen (oxygen concentration O^*), while on the bulk liquid side we assume the mean dissolved oxygen concentration *O* present. The oxygen transfer rate *OUR* can then be written as

$$OUR = k_L a \Delta O \propto \varepsilon \qquad \ldots(18)$$

where $\Delta Q = O^* - O$ is the mean driving force.

Although practically most important, the mass transfer effects that are of interest in biotechnology are not restricted to oxygen gas–liquid mass transfer. The transfer of CO_2 is also of importance, particularly in anaerobic systems. Beer production may serve as an important example of this transfer. Here, however, the mass transfer goes in reverse (i.e., CO_2 is mainly transferred from the liquid to the gas bubbles).

The straightforward way to increase the gas holdup ε, and thus the mass transfer rate, is to increase the volume of gas sparged per time unit ($Q_G = dV_G/dt$) into the dispersion. Considering the specific interfacial area *a*, it would be preferable to use a sparger that produces smaller bubbles or to apply strong mechanical agitation ofthe dispersion to redisperse larger bubbles. By equation 9, a smaller mean bubble diameter d_s leads to a higher specific mass transfer area *a*. These simple arguments, however, have their clear practical limits, as the following discussion will show.

The quality ofthe gas holdup may be defined by the size of the bubbles, characterized by the mean (Sauter) diameter d_s and by some surface properties ofthe bubbles. The size of a bubble determines its rise velocity with respect to the surrounding bulk liquid. When this relative velocity decreases, the bubble more closely follows the liquid motion. Since the liquid phase is operated in a batch mode within most of the practically used bioreactors, the residence time of the bubbles within the bioreactor becomes higher with smaller bubble size.

Very small bubbles may even stay within the bioreactor for a very long time before leaving the dispersion at its top surface. Additionally, the escape of the bubbles from the dispersion may be delayed or even suppressed by surface active molecules dissolved in the liquid phase, which stabilize the boundary layer. Both effects increase the gas holdup, often in an undesirable way. Long residence times of very small bubbles are undesirable when they exceed the mean time spans during which the bubbles are able to positively contribute to the oxygen transfer.

To get an idea of the time constant of the mass transfer, consider that single bubbles with an equivalent diameter of 5 mm can be assumed to have a mass transfer time constant of roughly 5 min. Consequently, when such bubbles stay within the reactor for periods much longer than 5 min, they no longer actively contribute to mass transfer. They must then be regarded as dead volume, which essentially reduces the effective reactor volume (i.e., it reduces the productivity).

When we are dealing with very long gas residence times, the dispersion accumulates a great amount of gas, and we consider foamy behavior of the dispersion. Such situations are extremely undesirable, since with continuing gas supply, the gas holdup becomes larger and larger until the headspace of the reactor becomes full and the dispersion enters the gas vent line. There it is usually stopped at the off-gas filter. This filter quickly becomes clogged by the material dissolved or dispersed in the liquid (e.g., the cells or dissolved proteins).

Since the gas throughput is usually controlled, the reactor pressure will be enhanced to hold the throughput. The pressure may quickly rise, and eventually some part of the reactor, most often the tubing by which substrate or base is supplied, bursts. To prevent such accidents, in general, the gas holdup in a bioreactor must be kept under control. This is most often done by measuring the height of the dispersion with some level sensor and by reducing the gas holdup with chemical additives (antifoam agents) when the maximal level is reached.

Antifoam agents increase the coalescence behavior of the gas bubbles, thus leading to a larger bubble size. Larger bubbles are much more easily disengaging from the dispersion, so that the gas holdup drops. Soya oil or silicone-based solutions most often serve as antifoam agents. However, their addition to biological cultures must be well dosed. One argument for keeping the amount of antifoam agents within the medium low is that these molecules, as they are surface active, most often negatively influence many downstream processing procedures. Also, the addition of too much antifoam agent may drastically reduce the gas holdup in a bioreactor.

This discussion shows that, with respect to mass transfer, there are competing requirements. Thus, the addition of antifoam must be optimized to obtain a maximum performance of the mass transfer in the reactor while avoiding the problems with foam. There is another problem associated with high gas holdups or foam formation. It is well known that microbial cells are in some sense surface active as well; that is, they easily attach themselves to bubble surfaces and are then transported upward toward the surface of the dispersion. This flotation effect is a negative influence on reactor performance when a stable foam layer develops on top of the dispersion. Then, a significant amount of the biomass might become captured in the foam layer.

In particular, in the beginning of a cultivation (after inoculation) when the biomass concentration is low, the biomass can become nearly quantitatively trapped in the foam layer. Then, consequently, one does not find a significant biomass concentration in the bulk. After addition of an antifoam agent, the biomass appears back in the bulk. Hence, it is necessary to avoid these kinds of gas holdup effects by initially adding a sufficient, but not too large, amount of antifoam agent to the medium before inoculation. While the small-scale shear flow that appears in the surroundings of the bubbles when they rise in the liquid does not have any negative effect on microbes, there are some problems with aerated cultures of shear-sensitive mammalian cells. It was observed that some cell lines die when their cultures are aerated with air bubbles, while they grow without problems when they are indirectly supplied with oxygen (e.g., by means of oxygen diffusion through silicone tubes submerged in the culture medium). The current explanation of this phenomenon is that it is not the shear flow around the rising bubbles that damages the cells but the sudden disruption of the liquid film that appears immediately before the gas bubble is disengaged into the reactor's headspace.

When a cell happens to be within this bursting film, then it is subject to extreme shear rates that will most probably destroy the cell membrane. Similar effects are also to be expected during bubble coalescence or redispersion effects, since in such situations the new bubble surfaces are formed very quickly. Thus, very high local velocity gradients appear on the scale ofthe diameters of these cells. Another important function of the dispersed gas phase in a bioreactor is mixing. In bubble column bioreactors, the bubbles are the only agitators. The individual bubble agitates the liquid medium locally; however, when the total gas holdup is high enough, eddies with diameters up to the reactor diameter are induced within the dispersion. With higher gas holdups, excellent mixing can be obtained, at least in dispersions with a lower liquid-phase viscosity. Mixing by means of the rising gas bubbles is not only important in bubble columns; even in industrial-scale stirred tank bioreactors, a considerable part of mixing is performed by the gas phase, and this effect is proportional to the gas holdup. This point will be discussed in more detail in the next section.

BIOTECHNOLOGICAL CULTIVATION

Practically, the gas holdup within a biotechnical cultivation broth is first of all a function of the aeration rate Q_G. Bioreactors are usually aerated with compressed air, which is dispersed into the culture medium by means of a sparger. The mass transfer via the top surface of the culture is most often comparably small and can thus be neglected. In stirred tank reactors, gas can be drawn into the dispersion by an impeller. This is the case when the impeller is operated in an unbaffled vessel where whirling occurs. However, such operational modes are generally avoided in technical bioreactors. Hence, when there is no aeration, the gas holdup is negligible.

In mechanically stirred bioreactors, which are used in most practical cases in biochemical engineering, a second influence parameter gains importance. This is the mechanical power P_D transferred into the dispersion. For this case of practical importance, we obtain

$$\varepsilon \propto \left(\frac{P_D}{V_D}\right)^{0.47} w_{sg}^{0.53} \qquad \text{...(19)}$$

as an experimental or engineering correlation, where both (P_D/V_D) and w_{sg} are power-related quantities. The quantity w_{sg} is related to the power P_c released by the rising bubbles into the dispersion. This power was originally supplied by the air compressor against the static pressure ofthe liquid at the sparger location.

$$P_c = Q_G \Delta p = w_{sg} W \qquad \text{...(20)}$$

where Δp is the pressure difference between the surface of the dispersion and the position ofthe sparger and W is the weight ofthe dispersion above the sparger (i.e., essentially the weight of the liquid phase). The point to note is that the mechanical power pneumatically introduced into the dispersion by the gas is proportional to the superficial gas velocity $w_{sg.}$ In bubble column bioreactors this is the only form of mechanical power input.

What is the reason for this power dependency ofthe gas holdup? With the mechanical power, two fluid–mechanical effects relevant to the gas holdup appear:

1. The first effect is that with higher energy inputs, the velocities of the circulatory fluid flows within the reactors become more intense. Hence, a larger number of the bubbles within the flow are recirculated. Thus, the mean residence time increases and, consequently, by equation 17, so does the gas holdup. In particular, intensive eddy motions with high local flow velocities may trap some smaller bubbles within the dispersion for some time.
2. The second effect supports the first one. At higher power input levels, the chaotic eddy flow motions lead to higher shear rates in the liquid phase on the scale of the bubble size. This reduces the mean bubble diameter, as characterized by the Sauter diameter, due to redispersion. Consequently, even more bubbles closely follow the liquid motions and thus the gas holdup increases.

In stirred tank reactors, one also induces a global circulation of the dispersion by means of the agitator to keep the mean gas residence time high and thus to increase the gas holdup and the specific interfacial area a or mass transfer coefficient $k_L a$.

To obtain a high enough gas holdup or specific interfacial area *a* in dispersions that are not foaming, one might come to the conclusion that the increase of the superficial gas velocity will help to solve the problem. It is true that with higher gas throughputs we usually increase the gas holdup; however, this has its limits, at least with respect to its quality characterized by the mean bubble diameter d_s or the specific interfacial area *a*.

Two competing processes come into play:

1. With higher gas throughputs we increase the energy flow into the liquid-phase motion, thus the dispersion of gas bubbles or, better, the redispersion by shear rates increases.
2. On the other hand, with higher holdups the number density of the bubbles increases and consequently the number of collisions between bubbles. This results in an increased coalescence probability and hence an increase in the mean bubble size. It is simple to understand that the coalescence probability is directly proportional to the probability that the bubbles meet each other, and this is proportional to the gas holdup. Larger bubbles most probably disengage from the dispersion. In this way the gas holdup is affected.

Consequently, the mean bubble diameter depends on the relative rates of coalescence and redispersion. Both are critically dependent on the composition of the medium, particularly, the concentrations of surface active agents. It is unlikely that this problem can be treated quantitatively in the near future, at least not for a real biochemical cultivation broth, where the composition with respect to the surface active components varies with time in a complex, virtually unpredictable way. The current knowledge is essentially of qualitative nature.

According to equation 17 we can increase the gas holdup with the superficial gas velocity, provided the residence time of the gas phase or the corresponding mean bubble velocity w_{bl} relative to the reactor coordinates remains practically constant. At higher gas throughput, the circulation flows in the reactor are intensified and, hence, so is the mean residence time of the bubbles. This leads to a lower mean bubble-rise velocity and an increased holdup. When the holdup increases too much, the bubble coalescence probability exceeds the redispersion probability. Consequently, the bubble size increases.

At higher viscosity the mean bubble-rise velocity is lower, and, according to equation 17, large gas holdups and larger bubbles appear at even lower superficial velocities. On the other hand, in dispersions with liquid phases of higher viscosity, one observes drift of the gas holdup toward higher values with a long time constant. This effect can be understood from the fact that due to the smaller slip velocity between bubble and bulk liquid, the probability of small bubbles escaping from the dispersion at its top surface becomes lower. Thus, very small bubbles accumulate in the dispersion, leading to an increased holdup. The bubble size distributions tail toward small equivalent bubble diameter values. The effective bubble rise velocity w_{bl} does not sensitively depend on the superficial gas velocity w_{sg} for bubbles in the relevant Sauter diameter region of some millimeters.

Hence, with larger superficial gas velocity w_{sg} the gas holdup will rise according to equation 17. Then, coalescence appears to a considerable extent. Big bubbles appear, and the dispersion becomes quite inhomogeneous with respect to bubble size. At very high superficial gas velocities the reactor will become flooded with gas. At small gas holdups, the interfacial area that can be produced in a stirred tank reactor is essentially independent of the agitator type, and practically only dependent on the effectively introduced mechanical power. Higher mechanical power inputs by means of the agitator lead to a more intense redispersion of coalesced bubbles; the mean bubble size becomes smaller and the gas holdup higher. However, it is not possible to increase the gas holdup to arbitrarily high values, since agitators are restricted in the amount of gas they are able to disperse. When a critical gas throughput is exceeded, they become flooded. An essential criterion for an agitation system is the maximal gas throughput it can disperse. In other words, with respect to gas holdup and mass transfer considerations, it is straightforward to primarily distinguish different agitators by their ability to disperse more or less gas.

BIOLOGICAL VALUES

To control the gas holdup, at least in the sense of keeping it below a critical value, it must be measured. Many techniques have been proposed in the literature; however, only a few of them can routinely be applied in biotechnical production practice. The most convenient method to estimate the gas holdup in large reactors is to measure static pressure differences Δp between sites in the reactor at different heights and to relate Δp to the gas holdup ε with equation 12. The liquid-phase density ρ_L can be estimated reliably in simple density measurements. Some pressure sensors on the market are sensitive and also robust enough for measurements in production reactors. The most straightforward way to proceed is to use equation 10 and register the height H_D of the surface of the dispersion. In any case, it is necessary to realize when it exceeds a limit H_{Dmax}. This can be done by various level meters or sensors. Since in most real bioprocesses the oxygen supply cannot be switched off and it is not possible to remove the bubbles, the volume of liquid phase must be determined indirectly.

The most simple way is to make use of the culture weight, which is often measured even at production reactors and some estimate on the density of the nongas phase of the dispersion. The most simple level monitor is a conductivity probe, which makes use of the fact that biotechnical culture media are electrically conductive. Hence, a single electrode placed at the critical height in the headspace of the bioreactor, connected with some counter electrode in the dispersion (e.g., the wall of the reactor) via a voltage source, would suffice. Only when the dispersion reaches the electrode does the current loop close. The flowing electric current then indicates that the gas holdup reached its maximal value. The same measurement principle can be used to measure local gas holdups. Then, however, the probe must be point-like (i.e., small as compared to the bubble size). Whenever the point is within a bubble, the current flow between the electrodes is interrupted. The ratio ofthe sum of the time intervals Δt_j when the current is interrupted to the total measurement time t_{tm} is exactly the local gas holdup when the total measurement time t_{tm} was not too short. All other point-like sensors that can distinguish between gas and liquid phases can be used in the same way. One example is the total light reflection probe.

There are other probes for local gas hold measurements with larger measurement volumes, such as capacitance sensors, which measure the local gas hold-up by introducing a small capacitor into the dispersion, so that the dispersion forms the dielectric. The dielectric constant is dependent on the gas holdup within this capacitor. After some calibration, this sensor can be used to monitor the local gas holdup, provided it can be made certain that the holdup between the plates of the capacitor is representative of the holdup of its surroundings.

CONCLUDING REMARKS

In bioreactors, air is dispersed primarily to provide a sufficiently large specific interfacial area a for oxygen mass transfer into the liquid phase of the culture medium. Although a, or its product with the mass transfer coefficient k_L, the k_La value, is the key quantity from the point of view of allowing a high oxygen transfer rate, the gas holdup is a very important quantity that must be kept under control in practically all bioreactors. Too little gas holdup does not allow for a sufficient mass transfer; too much gas holdup leads to problems with the reactor operation, in particular when due to foam formation. The gas holdup can be controlled by the gas throughput and, in stirred tanks, also by the stirrer speed. However, when foam is formed, it might become necessary to destroy it. Mechanical foam destroyers are applied in a few cases; however, in most bioreactors, foam is combated by chemical antifoam agents. Since the gas holdup and hence the total volume of the dispersion may become too high when foam is being formed, most reactors are equipped with a level control. This impressively underlines the importance attributed to the gas holdup in biotechnology.

23

NUTRITIVE SWEETENERS

Starch is the main storage carbohydrate of plants and a primary source of energy in the human diet. It can be found in cereals such as wheat or barley, in the seeds of the maize plant, and in the tuberous roots of potato and cassava. Industrially, starch and starch-derived sweeteners are mainly produced from maize, wheat, potato, and tapioca. The use of carbohydrate hydrolyzing enzymes to convert starch to nutritive sweeteners is not a new technology. For centuries malt has been used to produce a sweet syrup from starch, although it was not known at the time that the enzymes in malt were responsible for the reaction. It was the development of amylases from fungal and bacterial sources at the turn of the twentieth century that paved the way for the use of enzymes in starch hydrolysis.

Table 23.1 Starch content of basic raw materials

Raw material	*Starch content (%)*
Maize	72
Wheat	64
Potato	15
Tapioca	20–30

Even so, it was not until the 1960s that enzymes began to replace mineral acids in the production of dextrose from starch. The term *glucose syrup* is often used to describe the refined, concentrated, aqueous solution of saccharides obtained by the controlled partial hydrolysis of edible starch. This conversion may be catalyzed by food-grade acids and enzymes. Depending on the degree of hydrolysis, the resultant syrup contains varying amounts of D-glucose. Syrup produced from maize starch is often called corn syrup rather than glucose syrup. The term *glucose syrup* is ambiguous. In the glucose syrup industry, glucose is used to describe any starch-derived syrup; however, from a chemical viewpoint, it is the monomeric saccharide α-D-glucose. To distinguish the monomer from the syrup, the term *dextrose* is used to describe the monomer. Because starch is composed of D-glucose units, dextrose is the ultimate product of nondestructive starch hydrolysis. The extent to which the starch molecules are broken down is described by the term *dextrose equivalent* (DE). Starch has a DE of 0, whereas the ultimate product, dextrose, has a DE of 100.

In general, sweetness will increase with increasing DE value, although the nutritive value will remain the same. It is the DE value that is important in describing a glucose syrup, since the characteristics of the syrup will vary with the DE. Products with a DE of less than 20 are referred to as maltodextrins, those with a DE of between 20 and 80 as glucose syrups, and those with a DE greater than 80 as hydrolysates. At a DE lower than 20, the functionality of the maltodextrin is often related to the botanical source of the starch. This is related to the ratio of amylose and amylopectin within the particular starch structure.

Nature of Starch

In its native state, starch is present in plant cells as microscopic granules arranged in layers. These granules vary in size and shape—from the relatively large granules found in potato (50–150 μm) to the smaller granules of maize and wheat (2–40 μm). Starch is made up of two types of high molecular weight polymers of glucose, amylose and amylopectin. Amylose and amylopectin are found concomitantly within the starch granule, and the ratio varies depending on the source of the starch. Starches from tubers, such as potato and tapioca, tend to have lower levels of amylose compared to starches from grain.

Table 23.2 Level of amylose and amylopectin in common starches

	Amylose (%)	*Amylopectin (%)*
Maize	28	72
Wheat	28	72
Potato	20	80
Tapioca	17	83

The ratio of amylose to amylopectin also varies with the variety of the crop. For example, some maize varieties yield up to 99% amylopectin (waxy maize), whereas other starches have a high amylose content (high amylose starches). These are specialist starches, and unlikely to be used in the production of glucose syrup. The basic building block of both amylose and amylopectin is α-D-glucose monomer linked through an α-1,4 glucosidic bond. In amylose, the glucose monomers are linked solely through the α-1,4 glucosidic bonds to yield linear chains of several hundred glucose units. In solution, the amylose chains may be coiled in the form of a helix. In amylopectin, the basic structure is still linear chains of glucose monomers linked through α-1,4 glucosidic bonds, but these are linked to other chains through α-1,6 glucosidic bonds. These branch points occur every 10–12 glucose units. This gives amylopectin a branched structure that may contain many thousands of glucose units per molecule.

In a starch granule, amylose and amylopectin are linked together through hydrogen bonding to form structures that contain both crystalline and amorphous regions. It is the intra- and intermolecular hydrogen bonding that gives starch its insolubility in water. On heating an aqueous suspension of starch, the hydrogen bonds are disrupted and the starch disperses. On cooling, the hydrogen bonds reform and the starch precipitates. The rate at which these hydrogen bonds form is related to the concentration of starch in the dispersion and the temperature. This precipitation is called retrogradation and can cause low yields and processing difficulties if allowed to occur in the production of glucose syrups.

PRODUCTION OF NUTRITIVE SWEETENERS

The manufacture of glucose syrup and nutritive sweeteners from starch-containing raw materials can be divided into a number of individual processes:

1. Method of isolation
2. Gelatinization and liquefaction
3. Dextrinization
4. Saccharification
5. Isomerization

Within each of these defined stages, the use of microbial enzymes facilitates the process to give a product that in turn becomes the raw material for the subsequent stage. The first of the stages that makes use of enzymes is the separation of the starch from the starch-containing raw material.

Method of Isolation

The separation of starch is primarily based on physical processes. With tubers such as potatoes and tapioca, the starch is separated using a process of chopping, disintegration (rasping), and sieving. It is unusual to use enzymes to facilitate the separation of starch from tuberous raw materials. The processes of starch separation for maize and wheat are much more complex than those for potato and tapioca. In recent years enzymes have come to be used extensively to improve the starch separation from wheat. Enzymes can be used to improve the yield of starch from maize. In both processes the enzymes employed act on hemicellulose compounds such as arabinoxylans and β-glucans.

Table 23.3 Percentage of β-glucan and arabinoxylan in maize and wheat

	β-glucan (%)	*Arabinoxylan (%)*
Maize	0.3	3.5
Wheat	0.4	8.5

The particular enzyme depends on the type of hemicellulose present in the raw material. In the case of maize, whole maize kernels are steeped in an aqueous environment under controlled conditions of temperature, time, and concentrations of sulfur dioxide and lactic acid. The main purpose of the steeping process is to soften the maize kernel for milling while inhibiting microbial growth to enable optimum starch recovery. It has been suggested that the addition of enzymes to the steep tank improves the separation of the starch from the fiber by degrading the arabinoxylans that bind the starch to the fiber. Studies have shown that the use of an enzyme preparation in the steeping of maize can result in increased starch yield, increased grind capacity, and decreased steep times.

The effectiveness of enzyme preparations when added to the steeping of maize is a subject of some debate due to the increase in starch yield versus the amount of enzyme added. It has also been suggested that enzymes are not absorbed along with the water into the maize kernel because the enzyme is a large molecule that could not penetrate the kernel through the tip cap. This problem could be overcome by adding the enzyme during the first grind, after the initial cracking of the maize, allowing the enzyme access to the arabinoxylan gums that bind the starch. Many maize starch processors do not have the facilities to have a holding time between the first and second grind to allow enzyme action. It is up to the individual maize starch processor, therefore, to determine the feasibility of using enzymes. In contrast to the separation of corn starch, wheat starch is produced by at least five different commercial processes. In all of these, wheat is ground and combined with water to form either a dough or a batter.

The fractionation is achieved by physical means, and vital wheat gluten and starch are products of high value. Wheat flour contains a high level of arabinoxylans, a large proportion of which are soluble in water. These high molecular weight gums can bind significant amounts of water and hence give rise to high viscosities in doughs or batters. Addition of an enzyme (xylanase) to the dough or batter can reduce the viscosity by degrading the arabinoxylan and, hence, improve the separation process. The enzymes used not only improve the separation of gluten and starch but also enhance gluten coagulation. In all enzymes used to aid starch separation, the presence of proteinase or amylase side activities is detrimental to the process, since they will break down protein and starch, cause material loss, and cause deterioration in the quality of the end products, gluten and starch. Once separated, the starch may be dried as a product in itself or further processed to nutritive sweeteners. The conversion of starch to nutritive sweeteners is the main use of enzymes in the starch industry.

Gelatinization

At ambient temperatures starch granules are insoluble in water, however, if an aqueous starch suspension is heated to above 60°C the granules begin to swell and disrupt. Eventually the starch molecules are dispersed into solution. This process is known as gelatinization. The temperature required for complete dispersion or gelatinization depends on the source of the starch. For most starches, a temperature of 105–110°C is sufficient. At 25–40% dry solids, the viscosity of gelatinized starch is extremely high, and so a thinning agent is required.

Viscosity reduction may be achieved by a combination of the shear forces exerted on the slurry as it is continuously pumped through a jet cooker (mechanical thinning) and the action of acid or enzyme. During the process of gelatinization and thinning of a starch slurry, proteins, lipids, and hemicellulases that complex with the amylose and amylopectin are released and coagulate to some extent. The lipids that are complexed mainly with the amylose are dissociated at higher temperatures. Subsequent cooling may again lead to the precipitation of the amylose–lipid complexes. These precipitates cause hazes and flocculation later in the process and are difficult to remove. They are often the cause of a positive iodine reaction to starch in saccharification reactors. To prevent retrogradation and reprecipitation of starch–amylose complexes, it is necessary to continue the thinning process until partial hydrolysis occurs. This process is called liquefaction.

Table 23.4 Gelatinization temperature for some common starches

Starch type	*Gelatinization temperature (°C)*
Maize	72–76
Wheat	60–64
Potato	65–70
Tapioca	70–75

Liquefaction

Liquefaction is the first primary action in hydrolysis or thinning of the starch slurry whereby viscosity is reduced and first-stage degradation of starch molecules takes place.

Acid liquefaction process

Traditionally, the process of thinning, liquefaction, and dextrinization of starch was carried out using acid. In acid liquefaction, a starch slurry of between 30 and 40% dry solids is adjusted to a pH of 1.5–2 with addition of an acid, usually hydrochloric acid. The acidified slurry is then heated to 140–150°C by either direct steam injection in a jet cooker or indirect heating in a reactor for approximately 5–8 min. The hydrolysate is then flash-cooled to atmospheric pressure and neutralized.

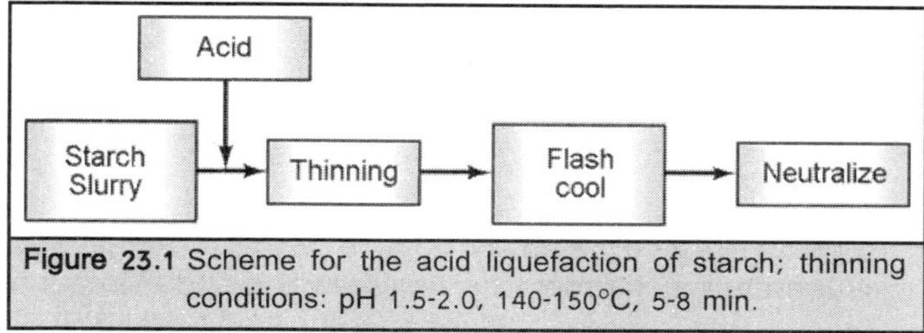

Figure 23.1 Scheme for the acid liquefaction of starch; thinning conditions: pH 1.5-2.0, 140-150°C, 5-8 min.

Acid liquefaction results in complete gelatinization, thinning, and a liquefact that can be easily filtered. However, the process also results in a product with a high color and ash content, and significant purification costs. The action of the acid on the starch molecules is nonspecific, and by-products: such as furfuraldehyde derivatives are formed, resulting in a lower yield of desired products. The use of a heat-stable endo-α-amylase allows for less rigorous conditions to be employed in the liquefaction process. By-product formation becomes less of a problem and refining costs are reduced.

Two-stage enzyme liquefaction process

The first enzyme to be employed in starch liquefaction was the endo-α-amylase produced by *Bacillus subtilis* var. *amyloliquefaciens*. The enzyme is utilized in a two-stage enzyme liquefaction process. A starch slurry at 30-40% dry solids is thinned at a temperature of 85°C for 1 h at pH 6.0–6.5. This is followed by a short heat treatment at 140°C, flash-cooled to 85°C (at which point a second enzyme addition is made), and the reaction continued until liquefaction is complete. For optimum enzyme stability a calcium concentration of 150 ppm is required; this is usually added as calcium hydroxide or calcium chloride into the starch slurry. If a thermostable α-amylase is employed, such as that from *Bacillus licheniformis* or *Bacillus stearothermophilus*, the heat treatment at 140°C may be eliminated. The use of thermostable α-amylases has largely replaced acid liquefaction and processing using the less heat-stable α-amylase from *B. subtilis*.

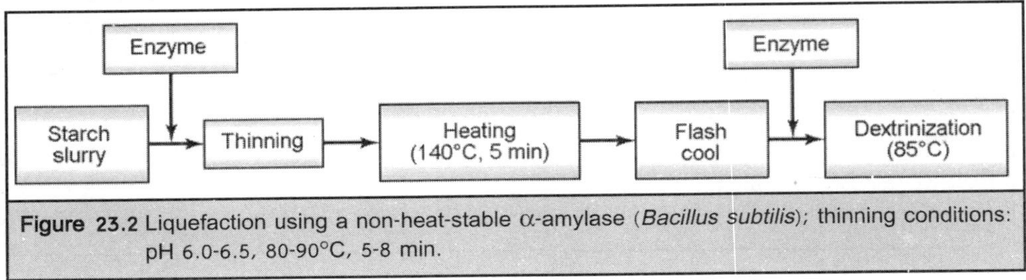

Figure 23.2 Liquefaction using a non-heat-stable α-amylase (*Bacillus subtilis*); thinning conditions: pH 6.0-6.5, 80-90°C, 5-8 min.

Single-stage enzyme liquefaction process

A starch slurry at 30–40% dry solids is prepared, and the pH is adjusted to 5.8–6.2 by the addition of sodium of calcium hydroxide. Calcium salts are added, if necessary, to give a level of 75–100 ppm of free calcium ions. The slurry is pumped continuously through a jet cooker operating at 103–107°C, enzyme having already been added to the slurry stream. The higher shear produced by the action of pumping through the jet cooker gives a degree of mechanical thinning, and this, in addition to the action of the α-amylase, means that peak viscosities are avoided. The slurry is then transferred to a holding coil, which maintains the temperature for 5–10 min before the liquefied starch is flash-cooled to 95°C, where dextrinization continues. In the single-stage liquefaction process, the control of the jet cooker is critical. Control of the temperature of the slurry through the jet cooker and into the holding coil must be accurate. If the temperature is allowed to drop much below 105°C, incomplete gelatinization will occur and filtration problems may be encountered later in the process. However, at temperatures much above 105°C, the enzyme will be inactivated and the viscosity of the slurry is not effectively reduced.

Operating pH should be between 5.8 and 6.2, although lower pH values (down to 5.5) have been reported with the use of *B. stearothermophilus* heat-stable α-amylase. At lower pH values the enzymes are less stable and, in combination with high temperatures, lead to rapid inactivation; on the other hand, higher pH levels may cause color formation and production of by-products, resulting in lower end product yields. Other factors that affect the performance of the liquefaction process are dry substance and calcium ion concentration. Higher levels of dry substance have a stabilizing effect on the enzyme but may also lead to incomplete gelatinization and subsequent filtration problems. A dry substance level of 30–35% gives adequate enzyme stability, and at the same time ensures complete gelatinization. Calcium is required to give the heat-stable α-amylases stability at higher

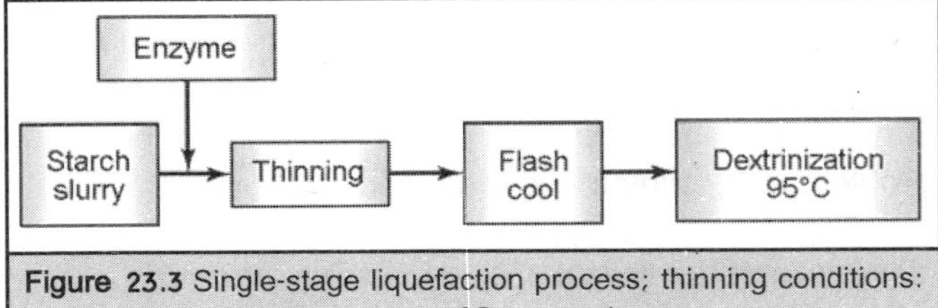

Figure 23.3 Single-stage liquefaction process; thinning conditions: pH 5.8-6.2, 105-117°C, 5-11 min.

temperatures; for *B. licheniformis* enzyme the requirement is 75–100 ppm free calcium ions, and for *B. stearothermophilus* it is 100–150 ppm. The level of calcium ions is usually adjusted by addition of calcium hydroxide or calcium chloride to adjust the pH of the slurry. The final parameter that requires consideration is enzyme dose rate: Too little enzyme leads to incomplete liquefaction and starch positive reaction after dextrinization; too much enzyme results in a large enzyme carryover into dextrinization and a dextrose equivalent higher than desired.

The DE is used to monitor the degree of starch breakdown. It is calculated by expressing the reducing sugars, such as dextrose, as a percentage of the dry substance. Typical enzyme dose rates for the single-stage liquefaction process range from 0.6 to 1.0 L of a heat-stable α-amylase per ton of dry substance. Enzymes produced from *B. licheniformis* give different starch breakdown products to heat-stable α-amylases produced using *B. stearothermophilus*. The difference shows in the measurement of DE for the same degree of starch degradation. The DE given by *B. stearothermophilus* enzyme can be as much as two percentage points higher than *B. licheniformis* enzyme. Because of their different breakdown products, mixtures of heat-stable α-amylases from *B. licheniformis* and *B. stearothermophilus* give a more efficient breakdown of starch during liquefaction. A further potential use of the *B. stearothermophilus* heat-stable α-amylase is using lower pH liquefaction; the enzyme is more stable than *B. licheniformis* enzymes at pH values down to 5.5. The advantage of lower-pH liquefaction is reduced by-products and color formation.

Acid or enzyme liquefaction

A process that takes advantage of acid thinning (which can improve filtration downstream due to better separation of fat and protein) and enzyme liquefaction is the acid or enzyme liquefaction process. A starch slurry containing between 30 and 40% dry substance is adjusted to give a pH of 2–5. The slurry is jet cooked at 140°C for about 5 min, where mechanical thinning due to both shear and acid thinning takes place. The pH is very important, if it is too low, color and by-product formation occurs, but if it is too high, insufficient thinning results. After this initial cooling, the thinned starch is cooled to 95–100°C, the pH is adjusted to 5.8–6.2, and enzyme is added. The major disadvantages of this process are increased energy consumption and potential by-product formation.

Dextrinization

After liquefaction of the starch slurry, the action of the heat-stable α-amylase on the starch is allowed to continue in a process called dextrinization. This process involves further starch degradation until a syrup with the required dextrose equivalent is obtained. Dextrinization may involve the further addition of α-amylase, as in the case of maltodextrin production, or there may be sufficient enzyme remaining after the cooling process. The degree to which dextrinization is allowed to continue is determined by the type of product manufactured. In the production of high-dextrose syrups (i.e., syrups containing 94–97% dextrose), dextrinization is allowed to continue to a DE of 15–18. However, in the production of very-high-maltose syrups (68–75% maltose) the DE must be controlled to a maximum of 12. Addition of a further quantity of bacterial α-amylase allows maltodextrin with intermediate DEs up to 30 to be prepared. Once the desired DE has been reached during dextrinization, the α-amylase may require inactivation. This can be achieved by a second jet-cooking step at temperatures of 120°C and above with a low pH, preferably between 3.8 and 4.5. The resulting hydrolysate may then be filtered, clarified with carbon, concentrated, and dried or spray-dried to give a maltodextrin. The process of liquefaction and dextrinization produces a hydrolyzed starch suspension of low viscosity that is stable to retrogradation and suitable for further enzyme processing to high-dextrose, glucose, or high-maltose syrups.

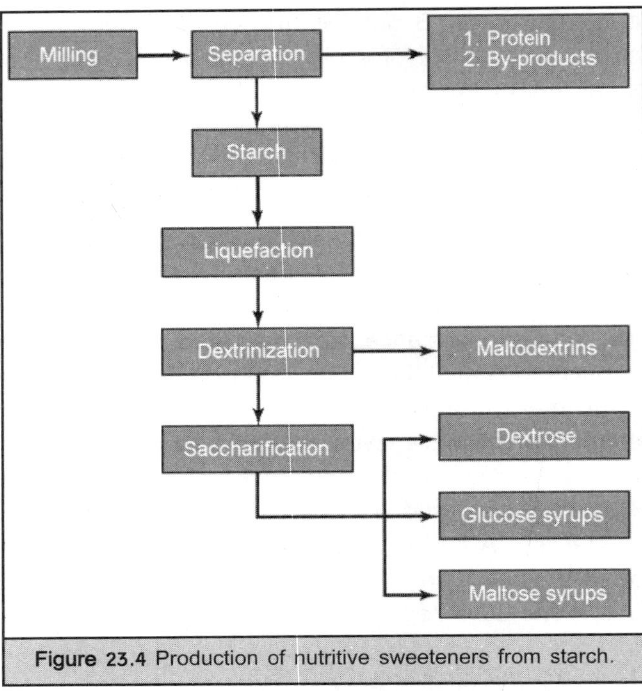

Figure 23.4 Production of nutritive sweeteners from starch.

Saccharification

The process in which the dextrinized starch is converted to nutritive sweeteners is called saccharification. Further hydrolysis of the oligosaccharides produced by the process of liquefaction and dextrinization may be carried out by a number of different exoamylases and debranching enzymes, depending on the required product.

Dextrose (glucose) production

During the production of glucose, the enzyme amyloglucosidase is used in saccharification. Amyloglucosidase is an exo-α-amylase that catalyzes the hydrolysis of α-glucosidic bonds by successively recovering glucose from the nonreducing end of oligosaccharides in a stepwise manner. The enzyme is able to hydrolyze both α-1,4 and α-1,6 glucosidic bonds; however, the rate of hydrolysis of α-1,6 bonds is much slower than the rate of hydrolysis of α-1,4 bonds. It has been reported that amyloglucosidase hydrolyzes α-1,4 bonds between 20 and 30 times faster than α-1,6 bonds.

Table 23.5 Listing of the main saccharification enzyme required for various desired products

Product required	*Main saccharification enzyme*
Glucose	Amyloglucosidase
Maltose (50%)	Fungal α-amylase
Maltose ($\geq$ 50%)	Cereal β-amylase

Amyloglucosidase preparations are usually available at activities of either 200 or 300 U mL^{-1}. After liquefaction and dextrinization the solution is cooled to 60°C and the pHs adjusted to 4.2–4.6 with acid. After dextrinization, the dry substance should be between 25 and 37%. Amyloglucosidase (300 U mL^{-1}) is added, typically at dose rates of between 0.5 and 1.1 L ton^{-1} dry substance, depending on the required saccharification time. The saccharification reaction takes place in stirred reactors and is allowed to continue until peak dextrose is reached, typically 40–72 h. Given that the enzyme is able to hydrolyze both types of bonds present in starch, it would be reasonable to expect that amyloglucosidase is able to produce quantitative amounts of glucose from liquefied starch. At very low solid levels, glucose levels approaching 100% are possible, but at the higher solid levels used in industrial sacchari-fication this is not the case. Amyloglucosidase can polymerize glucose by the reverse reaction of hydrolysis. This reversion is a condensation reaction and is accompanied by the elimination of water.

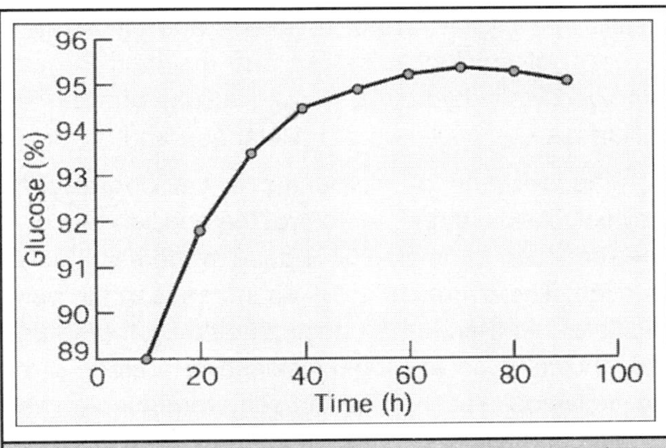

Figure 23.5 Typical saccharification profile with amyloglucosidase.

The products of this reversion are maltose and isomaltose. The reversion reaction takes place at a much slower rate than the hydrolysis reaction and is dependent on the dry substance level and the residual enzyme activity at the end of saccharification. At lower dry substance the rate of reversion is slower and higher levels of dextrose can be attained. Peak dextrose levels may also be reduced by the presence of a secondary enzyme activity within amyloglucosidase preparations, namely transglucosidase. This enzyme is a glucose transferase and catalyzes the formation of α-1,6 branched oligosaccharides such as isomaltose and panose from glucose and maltose. The buildup of isomaltose and panose during saccharification leads to an overall reduction in final dextrose levels. Transglucosidase activities have recently been essentially removed from amyloglucosidase preparations either after the fermentation process or by selection of an *Aspergillus niger* strain that does not produce the activity.

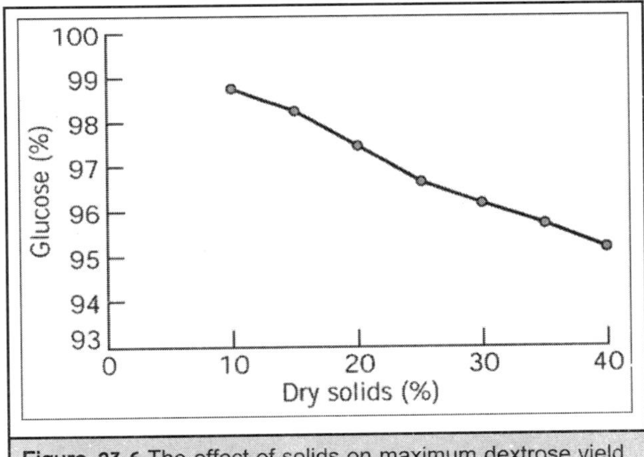

Figure 23.6 The effect of solids on maximum dextrose yield.

Amyloglucosidase itself is very inefficient at hydrolyzing the products of the reverse reaction, such as isomaltose, isomaltotriose, and panose. The formation of these products is the major factor in governing the dextrose level after saccharification. It is important, therefore, to limit the rate of the reversion reaction as much as possible. One of the ways to achieve this is to lower the level of amyloglucosidase in saccharification. The relative inefficiency of amyloglucosidase in hydrolyzing α-1,6 glucosidic linkages in the branched oligosaccharides is a major obstacle to dextrose production. An excess of enzyme is therefore required to achieve the desired saccharification rates. This excess results in a higher rate of reversion and ultimately lower dextrose levels. It is possible to reduce the level of amyloglucosidase and maintain the reaction rate by the addition of an enzyme that can specifically hydrolyze the α-1,6 links of the branched oligosaccharides. This enzyme is called pullulanase, and blends of pullulanase and amyloglucosidase are now available.

The addition of a specific α-1,6 debranching enzyme augments the α-1,6 activity of amyloglucosidase and, therefore, the addition rate of the glucose-producing enzyme may be lowered. The reduction in amyloglucosidase in the saccharification tank decreases the rate of the reversion reaction. The overall effect is an increase in the peak dextrose level. The dose rate and conditions of use of the blended products should be the same as the dose rate for amyloglucosidase alone. The choice of an amyloglucosidase or a blend of pullulanase and amyloglucosidase depends on the individual saccharification requirements. A fast saccharification can be achieved by a high addition of amyloglucosidase, but this in turn will mean quicker reversion rates; therefore, if dextrose levels are important, the enzyme blend is the preferred route. Often the residual oligosaccharides referred to as DP_4+ are considered undesirable, and again the enzyme blend has been shown to lower the levels of this oligosaccharide fraction.

The use of an amyloglucosidase–pullulanase blend gives greater flexibility and control over the saccharification process. Further alternatives are available to achieve specific saccharification requirements. For example, all amyloglucosidase preparations contain acidic α-amylase activity. The presence of this activity enhances the saccharification rate by degrading the higher molecular weight oligosaccharides into lower molecular weight fractions, which also results in a reduction of DP_4+ oligosaccharides in the final product. Higher dextrose levels may also be obtained by addition of an α-1,4-glucosyl transferase to amyloglucosidase. Unlike transglucosidase, which catalyzes the formation of α-1,6 glucosyl linkages, α-1,4 glucosyl transferase produces α-1,4 glucosyl linkages. Such an enzyme is produced by the fermentation of *Bacillus megaterium*. Substrates for the enzyme include isomaltose and isomaltotriose; these are combined with glucose to form molecules that can then be hydrolyzed by amyloglucosidase. A combination of amyloglucosidase and α-1,4 glucosyl transferase results in higher dextrose levels by hydrolysis of the products of the reversion reaction. The enzyme may be added at the start of saccharification along with amyloglucosidase, however, the pH optimum of Megadex is above that of amyloglucosidase and the enzyme maybe added to the process after a pH adjustment.

Maltose syrups

Syrups containing high levels of maltose have different properties than those containing high dextrose levels. Maltose has a different sweetness characteristic, and its syrups have a resistance to crystallization. Unlike the production of high-dextrose syrups, it is difficult to achieve maltose levels in excess of 80–85% at high dry substance levels. There are two main maltose-producing enzymes. The first is a maltogenic α-amylase produced by *Aspergillus oryzae* and is often referred to simply as fungal α-amylase. The second is a β-amylase, which is found in cereals such as barley, wheat, rye, oats, and sorghum, and also in soya beans and sweet potato. Recently a microbial β-amylase has been produced from a *Bacillus* species. The production of high-maltose-containing syrups using *A. oryzae* (fungal) maltogenic α-amylase follows a similar process as the production of high-dextrose syrups. After dextrinization using a thermostable α-amylase the solution is cooled to 55–60°C and the pH adjusted to 4.7–5.0. Fungal α-amylase is added to the stirred reactor at a level of between 0.3 and 0.7 L ton^{-1} dry solids. The reaction is allowed to continue for 24 to 40 h. After saccharification the crude syrup may be filtered and ion exchanged to remove ash. The syrup can then be decolorized by activated charcoal and finally evaporated to the desired solids level.

Syrups produced by fungal α-amylase can contain up to 60% maltose based on dry substance. Dextrose levels may be controlled by limiting dextrinization with the thermostable α-amylase. One disadvantage of maltose syrup production using a fungal α-amylase is the high level of maltotriose obtained. Maltose syrups produced by cereal β-amylases have lower levels of maltotriose than those made with fungal α-amylases. After dextrinization, the starch solution is cooled to 54–58°C and the pH lowered to 5.2–5.6. Cereal β-amylase is added between 0.25 and 0.75 L ton^{-1} dry solids and the reaction is allowed to continue for 36–48 h, depending on enzyme dose rate. The level of maltose produced by cereal β-amylases is limited to between 50 and 55% on dry substance. β-Amylases produce maltose sequentially from the non-reducing ends of the oligosaccharides formed by the action of the thermostable α-amylase. Once the α-1,6 branch points are reached, the enzyme is unable to continue. The major products of the reaction are limit dextrins and maltose itself.

Table 23.6 Carbohydrate spectrum of maltose syrups produced using fungal α-amylase and Cereal β-amylase

	Fungal α-amylase	*Cereal β-amylase*
Glucose	3	1
Maltose	50	60
Maltotriose	26	8
Higher dextrins	21	31

Syrups containing between 50 and 60% of maltose on a dry solids basis are referred to as high-maltose syrups. Higher levels of maltose may be achieved by addition of a debranching enzyme (pullulanase). The pullulanase hydrolyzes the α-1,6 glucosidic linkages in branched dextrins and amylopectin and allows the β-amylase to continue producing maltose. With careful control of the process, up to 75% maltose may be achieved. Such syrups are called very-high-maltose syrups (VHMS). Careful control of dextrinization is imperative to produce maltose in excess of 70%. After liquefaction, dextrinization is controlled to give a DE of between 7 and 10, the solution is cooled to 54–58°C, and the pH adjusted to 5.2–5.6. Cereal β-amylase is added at 0.5 and 0.75 L ton^{-1} dry solids and the pullulanase is added in a similar amount. The reaction is allowed to continue for 36–58 h and the product refined and evaporated. Maltose-producing enzymes do not catalyze the reverse reaction, so there is no reduction in maltose level if the reaction is allowed to proceed.

Maltotriose is a barrier to higher levels of maltose being achieved. β-Amylases are unable to hydrolyze maltotriose, so it is important to keep the level of this oligosaccharide as low as possible. One of the ways this can be achieved is by controlling the DE to a low level during dextrinization. An alternative is to use an enzyme that has some action toward maltotriose, unfortunately, these enzymes, in addition to producing maltose, also produce glucose. It is often a requirement of maltose syrups that the glucose level be kept low (<3.0% on dry substance).

Filtration Enzymes

Maize starch hydrolyzates generally exhibit good filtration characteristics. High filtration rates make it easy to remove undesirable, insoluble components from the hydrolysates in a relatively short time. However, wheat starch hydrolysates have very poor filtration characteristics due to low filtration rates and obstruction of the filter cake. To overcome this problem so-called filtration enzymes are commonly used in the industry. The major cause of the poor filtration has been identified as a monoacyl lipid compound that is thought to occur in inclusion complexes with the amylose fraction. In wheat the native starch contains about 0.8% of this internal lipid, the principal component of which is lysophosphatidylcholine (lysolecithin). Amylose molecules are helical, with a hydrophobic interior to the helix, which can accommodate certain lipophilic molecules, such as lysophosphatidylcholine. On hydrolysis of the amylose fraction of wheat starch, the lysophosphatidylcholine is released into the solution.

Lysophospholipids are water soluble and excellent emulsifiers. It has been reported that the lipid forms micelles above a concentration of 0.025 g kg^{-1}. When the critical micelle concentration is exceeded, the filtration rate of the hydrolysate is decreased. The critical micelle concentration is 40 times lower than the amount of lysophosphalipids in wheat starch hydrolysates. Therefore, the concentration of lysophosphalipids needs to be reduced to less than 2.5% of the original amount in wheat starch hydrolysates to achieve acceptable filtration rates. The reduction in lysophosphalipids can be achieved by the addition to the hydrolysate of the enzyme lysophospholipase, which cleaves the remaining fatty acid to give a free fatty acid and a glycerophosphatide. This cleavage destroys the emulsifying ability of the phospholipid.

The molecule no longer contains a hydrophobic part and a hydrophilic part, but instead a water-insoluble molecule (the fatty acid) and a water-soluble molecule (the glycerophosphatide) are formed. Lysophospholipase preparations are produced by the fermentation of *A. niger*. The enzymes are very low in amyloglucosidase activity and are therefore suitable for use in the production of maltose syrups. Manufacturers of glucose from wheat starch should choose an amyloglucosidase enzyme preparation that has an appreciable amount of lysophospholipase as a side activity. There should be no need to add further quantities of lysophospholipase to aid filtration of glucose syrups produced using amyloglucosidase. A second cause of poor filtration in wheat starch hydrolysates has been identified as the gummy substances from wheat starch, the arabinoxylans, often referred to as pentosans.

Pentosans are important macromolecules of wheat flour because of their ability to bind water. Because of the high water-binding capacity, the soluble pentosans form highly viscous solutions with intrinsic viscosities that are about 15–20 times higher than those of soluble proteins. The addition of an enzyme with the ability to break down the water-soluble pentosans will reduce the viscosity of a wheat starch hydrolysate, improving filtration performance. A second class of pentosans can be separated by means of centrifugation of a flour–water suspension in the so-called tailings, which form a layer above the primary starch. These are the water insoluble pentosans of wheat flour.

The structure of the water-insoluble pentosans is similar to that of the soluble pentosans, but unlike the soluble pentosans, on hydration the water-insoluble pentosans swell without going into solution. On contact with the filter bed these water-insoluble pentosans impede flow through the filter. A second type of pentosan-degrading enzyme can be added to degrade the insoluble pentosan fraction and improve the flow through the filter. The preferred pentosan degrading enzyme for wheat starch hydrolysate combines action toward both soluble and insoluble pentosans.

Isomerization

The worldwide demand for fructose syrups has increased in recent years due to the inclusion of enriched fructose corn syrup (EFCS) containing 55% fructose in popular soft drinks. Fructose is a naturally occurring sugar that is approximately 1.5 times as sweet as sucrose on an equal weight basis. Although fructose can be produced by the process of inversion of sucrose followed by separation of the fructose and glucose by chromatography, it is also produced by the enzymatic hydrolysis of starch. The production of fructose from starch initially involves the production of glucose, as already outlined. Glucose is converted into fructose by a process called isomerization. The enzyme used to catalyze this conversion is glucose isomerase and is usually used in an immobilized form suitable for a fixed-bed continuous column process.

The enzyme-catalyzed reaction is reversible, and under the process conditions the equilibrium constant is approximately 1. Glucose solution from the saccharification is refined using filtration, carbon treatment, and ion exchange and is evaporated to 40–50% dry matter. Calcium ions inhibit the activity of the enzyme, and their levels should be kept as low as possible. The effect of residual calcium ions can be counteracted by the presence of magnesium. Magnesium sulfate is added to give a magnesium ion concentration above 50 ppm and 20 times greater than the calcium ion concentration. The pH is adjusted to 8.1–8.3 and the temperature to between 55 and 58°C. The solution is then pumped through a fixed-bed column containing immobilized glucose isomerase. The flow rate through the column should be adjusted in proportion to the enzyme activity in such a way that the fructose content of the product remains constant. The enzyme activity in the column decreases during use and the flow rate must be increased over the lifetime of the column to maintain the fructose level.

Table 23.7 Carbohydrate profile for products of isomerization

Type of carbohydrate	*% w/w on dry solids*
Fructose	42
Glucose	50–55
Others	3–8

The product from the isomerization columns may be refined and concentrated to approximately 70% solids to give a high-fructose syrup at 42% fructose based on dry solids. The 55% fructose syrup can be achieved by concentrating this 42% high-fructose syrup to 90% fructose by a separation process, such as chromatographic adsorption. This concentrated 90% fructose syrup can then be blended with 42% high-fructose syrup to produce the required 55% enriched fructose syrup. The stability of the enzyme and its ability to retain its activity under process conditions are essential factors in the productivity of the immobilized enzyme. The operating conditions illustrate the feedstock parameters that have an influence on enzyme performance and therefore productivity. Table 23.8 is not an exhaustive list of parameters that effect enzyme performance, but it gives an indication of the conditions that will give optimal enzyme performance.

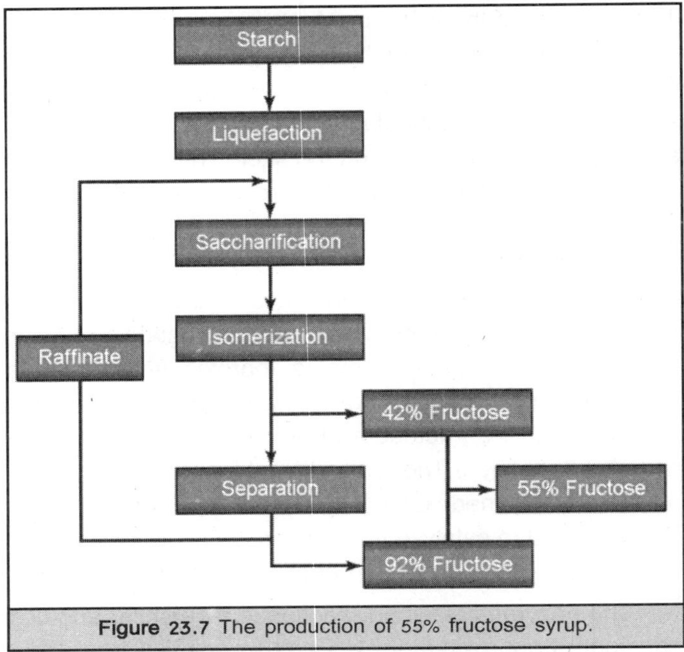

Figure 23.7 The production of 55% fructose syrup.

Secondary Product Processing

The processing of the secondary products of the starch industry can be improved with the addition of enzymes. The major use of enzymes in this area is in wheat starch processing. The yield of quality starch from wheat that is suitable for processing into maltodextrins and sweeteners is lower than the total starch content of the wheat itself. Between 10 and 15% of the total starch of wheat is not separated along with the large particle starch (A starch) during the separation process. This starch, depending on the separation process, is called the B starch or C starch fraction and is usually sold as a low-value animal feed ingredient. The B or C starch fraction contains high levels of protein and gums (pentosans) that make this fraction unsuitable for processing into sweeteners. These high levels of protein and gums lead to problems with color formation and evaporation of the syrup.

Table 23.8 Conditions required by certain parameters to promote optimal enzyme performance

Parameter	*Range*
Dry Substance (%)	40–50
Glucose (% dsb)	≥ 94
Fructose (% dsb)	≤ 3
Temperature (°C)	55–58
pH	7.8–8.3
Magnesium (ppm)	≥ 50
Sulfite (ppm)	≥ 125

As with other starches, B starch can be liquefied and dextrinized using a heat-stable α-amylase. A saccharification enzyme, such as amyloglucosidase, can be used to produce sugars. The value of the B starch in animal feed is often related to the DE of the syrup. The conditions of liquefaction, dextrinization, and saccharification are no different from those used in the production of nutritive sweeteners. To aid evaporation, a pentosanase enzyme is also added during saccharification to reduce the viscosity of the syrup. The level of protein can be reduced by introducing a heating step that allows the protein to coagulate. This protein can be dried and sold as a further product in animal feedstuffs. The enzyme used in the reduction of viscosity can be heat stable, such as that produced by *Penicillium emersonii*, which allows a continued action in the evaporators. Alternatively, there may be no need for a heat-stable enzyme if the saccharification time is long enough to allow the viscosity to be reduced sufficiently.

Using enzymes to reduce the viscosity of B starch syrups has the following benefits.

1. Higher solids content of the final evaporated syrup
2. Improved heat transfer during evaporation and less energy consumption
3. Less fouling of the evaporators, leading to more efficient cleaning
4. Higher evaporation capacity

Hemicellulases (xylanases) may also be used to aid the filtration of maize gluten. The enzyme reduces the viscosity of the gluten suspension and thereby improves the rate of filtration. The development of the enzyme-catalyzed conversion of starch has led to great advancements in the variety and quality of starch-derived sweeteners. The production and use of enzymes is still a relatively new technology, and with the help of modern biotechnology further improvements are not far in the future.

24

BIOTECHNOLOGY OF ANTIBIOTICS

The field of natural products owes its origins to the discovery of penicillin and its ability to kill bacteria. Since Fleming's original discovery in 1928 of the antibacterial effects of *Penicillium chrysogenum* and Flory and Chaim's efforts more than a decade later to bring that discovery to production, the field of natural products has undergone enormous changes. Vuillemin first introduced the concept of antibiotic activity in 1889, using the term *influences antibiotiques* (antibiotic influences) to describe negative interactions between plants and animals. Waksman later coined the term *antibiotic* to describe "a chemical substance derived from microorganisms which has the capacity of inhibiting growth, and even destroying, other microorganisms in dilute solutions". Natural product antibiotics are almost universally secondary metabolites, compounds generally produced at low growth rates or after growth has ceased, nonessential for growth of the producing organisms in pure culture, and typically possessing unusual structures not found in primary metabolites of central metabolism. Although natural product antibiotic biosynthesis is not essential to the producing organism in pure culture, these compounds are critical to the producing organisms in their natural environment, both for survival and competitive advantage. In more recent years, the term *antibiotics* sometimes has been used more broadly to define anti-infective agents, essentially in line with Waksman's definition, as well as *bioactive natural products*. This latter group includes immunomodulatory agents, statins, antitumor drugs, antiparasitic compounds, insecticides, and other bioactive natural products, many of which also possess weak antibacterial or antifungal activities.

Indeed, the immunosuppressive drug cyclosporin was originally discovered in an antifungal screening program. Additionally, the biosynthesis of bioactive antibiotics is often very similar to that of other bioactive natural products. As a result, cyclosporin, a weak antifungal antibiotic but an excellent immunosuppressive agent, is usually included in discussions of other peptide antibiotics. Similarly, when reviews are written about macrolide antibiotics, other bioactive natural products such as tacrolimus (formerly called FK-506) and rapamycin, both of which are immunosuppressive agents, are typically included in discussions with erythromycin, a bacteriostatic antibiotic; avermectin, a potent antiparasitic compound; and tylosin, an agricultural feed additive. This certainly makes sense because all these bioactive natural products possess similar modes of biosynthesis. An extensive table including all marketed antibiotics (both natural product-derived and chemosynthetic) as well as all other marketed bioactive natural products, as of December 1996, was recently compiled for the book entitled *Biotechnology of Antibiotics*. Additionally, a recent review by Shu highlighted most of the recent important discoveries in the field of natural products, ranging from new antifungal activities to natural products active as neuromodulators. Thus, this review will focus on natural product anti-infective and antitumor agents.

WORK DONE

Cragg et al. published in 1997 a well-cited review of the natural products field in which they analyzed the percentage of natural products in various fields. Of 520 new approved drugs between 1983 and 1994, 55.6% were chemically synthesized drugs, 24.4% were semisynthetic compounds, 5.8% were unmodified natural products, 8.8% were chemically synthesized drugs patterned after natural products, and 5.4% were "biologics". Thus, approximately 40% of all newly approved drugs over that period were derived from natural product leads. In breaking down the categories, however, natural products dominated the antibacterial markets (63% were derived from natural product leads) and the anticancer market, in which 61% of the new drugs were derived from natural product leads. Natural product leads also strongly contributed to the antihyperintensive market (48% were derived from natural product leads) and the anti-inflammatory market, in which one-third of the new drugs approved between 1983 and 1994 were derived from natural product leads.

According to Cragg et al., however, natural products did not contribute in the period of 1983–1994 to newly approved drugs in the antifungal, analgesic, anxiolytic, antihistamine, antidepressant, cardiotonic, or hypnotic fields. Strohl conducted a similar analysis of bioactive natural products with antibiotic activities, identifying a list of 252 natural and synthetic compounds used clinically and agriculturally in the United States with selected compounds included from other countries. Thus, although this analysis was not all-inclusive, it represented the majority of commercially important antibiotics and antimetabolic drugs. Of those compounds identified, 75 (30%) were natural products, another 75 (30%) were semisynthetic compounds derived from natural products, and the remaining 40% were chemically synthesized compounds. The discovery of new, novel natural products continues to rise dramatically. Based on an analysis of papers published in *Journal of Antibiotics*, Berdy indicated that about 11,900 antibiotics and bioactive natural products had been discovered through 1994, with increasing rates of discovery every year through 1990. He projected that the total of new, novel natural product structures discovered by the year 2000 would top 16,000.

The annual rate of new natural product descriptions published is now more than 500 per year, up from 200 to 300 per year 20 years ago. Nevertheless, there is the widespread perception in both the scientific literature and the popular press that the discovery of new natural products and natural product antibiotics is on the wane. This anomaly, however, can easily be explained. Because most new natural product drugs still are concentrated in the fields of anti-infectives and anticancer drugs, and natural product leads have been around long enough in both the anti-infectives and anticancer areas, the law of diminishing returns is most prevalent in these fields. For example, from 1962, when nalidixic acid was developed as the first quinolone antibiotic, to 1996, no new classes of clinical natural product-derived antibacterials have been introduced to the market. In 1993, only one new natural product antibiotic was approved by the Food and Drug Administration for use in the United States, none in 1994, and only a few in 1995 and 1996. Overall, from 1980 to 1990, the worldwide discovery of more than 4,600 new bioactive natural products was published, yet only 20 new natural-product-derived drugs (3 natural products and 17 semisynthetic products) were commercialized in that period. Finally, in the period spanning from World War II to 1975, approximately 1.6% of all natural product antibiotics reported (either by publication or patent) were eventually commercialized. Since 1975, this number has dropped to about 0.11%. Thus, fewer antibiotics are being commercialized from an increasingly larger pool of isolated candidate compounds. These data suggest that the Golden Era of antibiotics, generally recognized as having begun shortly after World War II ended, lasted nearly 30 years before the law of diminishing returns caught up to it.

Another major contributing factor for the dearth of commercialized natural product antibiotics from 1975 to the present is that the U.S. government, academic researchers, and pharmaceutical companies alike turned their efforts toward the defeat of cancer and heart disease, believing that infectious diseases were under control. On the other hand, in pharmacological areas in which natural products are making new inroads, many discoveries of bioactive natural products have resulted in compounds that are still in preclinical or clinical trials. Since the late 1960s, Umezawa and colleagues in Japan showed interest in discovering non-anti-infective bioactive natural products, particularly in the area of protease inhibitors. Although researchers in U.S. pharmaceutical companies showed interest in developing natural products as bioactive molecules in areas other than the anti-infective and anticancer markets (i.e., whole cell assays), many viewed the Japanese mechanism-based assay approach more as a curiosity than a serious discovery effort. It was not until the publication of Merck's patent and paper in *Proceedings of the National Academy of Science* in 1980 on lovastatin as a remarkable inhibitor of hydroxymethylglutaryl-CoA reductase, and ultimately an outstanding cholesterol-lowering drug, that the inclusion of natural products in mechanism-based screens was considered seriously. Since that time much has changed, and although some pharmaceutical companies have minimized natural product screening efforts because of the perceived cost-to-benefit ratio, others have sought to expand the scope of natural products in their screening programs.

As evidenced by publications in journals such as *Journal of Antibiotics*, patent applications, and clinical trial candidates, it has now become apparent that natural products can significantly augment synthetic chemical approaches for the discovery of new drug entities in a wide range of pharmacologically important areas. The two most significant recognized drawbacks of natural products in a broad-based screening program are the inclusion of crude extracts (which usually include salts, proteins, fatty acids and other potentially interfering materials) in in vitro assays (e.g., receptor-based, enzyme inhibition, or protein–protein interaction assays) and the time it takes to turn a crude-extract hit into a chemical structure. Despite these drawbacks, natural products now have been found to contribute to significant leads in most therapeutic areas. The sources of natural products are also likely to change. Berdy showed that of the 11,900 antibiotics that had been discovered through 1994, approximately 6,600 (55%) were produced by *Streptomyces* spp. The remainder of producing organisms included the filamentous fungi, which produced 2,600 (22%); nonactinomycete bacteria, which produced 1,400 (12%); and non-*Streptomyces* strains of actinomycetes, which produced 1,300 (11%). In the broader pharmacological field, it is widely recognized that a great percentage of pharmacologically derived products in use today are derived from or patterned after plant products. It can be expected that in the future a wider range of microorganisms, including well-known but under used groups such as the filamentous fungi, myxobacteria, bacilli, and pseudomonads, will contribute to the pool of natural products in larger ways.

Pharmaceutical Values

The total world sales for prescription pharmaceuticals was estimated in a September 1998 news release to be $308 billion. In a recent report of the 200 most-prescribed medicines in the United States, natural products accounted for 10 of the top 35 prescribed drugs. All the top-prescribed natural products or semisynthetic drugs derived from natural products are anti-infective agents except for three statins: simvastatin, lovastatin and pravastatin. Although it is difficult to identify the entire contribution of natural products to the world pharmaceutical market economy, it should easily total more than $30 billion, including the natural-product-derived anti-infective agents (>$20 billion), the natural-product-derived anticancer agents (in which the anthracyclines and taxanes alone represent more than $3 billion in market size), the statins (with a market size estimated at more than $6 billion and rising at a rate of more than 20% annually), and the large pool of plant-derived natural products or compounds patterned after plant products. The total world market for anti-infective agents in 1996 was reportedly $23 billion. The 1995 U.S. antiinfectives market alone was greater than $8 billion, with cephalosporins (45%), penicillins (15%), fluoroquinolones (11%), tetracyclines (6%), and macrolides (5%) comprising the majority of sales.

Of these top-selling anti-infective agents, only the fluoroquinolones lack the distinction of being natural products, and even they are patterned after the original quinolone, nalidixic acid. Thus, in deference to their origins, antibacterial drugs are still natural products, semisynthetic compounds derived from fermentation products, or patterned after natural products. The world antifungal market in 1995 was approximately $3 billion. Although this represents a much smaller market than the antibacterial market, the antifungal market is projected to grow at 20% annually. This rapid growth in antifungals unfortunately is driven by the dramatic increase in life-threatening fungal diseases, mostly as a result of opportunistic fungal pathogens infecting immunocompromised patients. The antiviral market was $1.8 billion in 1995, and the projected 1998 market is $3 billion. Although the top product in 1995 was acyclovir (50% of the market), more than 17 million people were infected with HIV, a number that is expected to reach 30 to 40 million by year 2000. Because of the development of successful new anti-HIV drugs such as the HIV-proteinase inhibitors, it is expected that the antiviral market should increase substantially within the next few years.

Biological Production of Antibiotics

Since the discovery and elucidation of the structure of penicillin, there has been an enormous effort placed into finding new natural product antibiotics, particularly those of the antibacterial, antifungal, and anticancer therapeutic classes. Natural product antibiotics, for the most part, fall into several major classes of structures, based on the precursors from which they are assembled and the mechanisms of assembly. Precursors common for the synthesis of natural products include short chain acylCoAs such as acetyl-CoA, malonyl-CoA, and propionyl-CoA, which are used to synthesize polyketides; activated amino acids, which are used to build peptides and β-lactams; various pyrimidines, which are the building blocks for nucleoside natural products; isoprene units, which form the basis of terpene, alkaloid, and substituted steroid natural products; and sugars and amino sugars, which are the building blocks for aminoglycosides as well as for the sugar moieties that decorate a wide variety of natural products.

Many natural products result from the fusion of moieties arising from two or more different biogenic classes. For example, 6-deoxyerythronolide B (DEB), the macrolide backbone of erythromycin, is synthesized by a type I modular polyketide synthase, but the molecule is not biologically active until the sugars desosamine, attached at C-5, and cladinose, attached at C-3, are added. Several other macrolides, including spiromycin and tylosin, have this same characteristic. Similarly, anthracyclines are comprised of the aglycone portions, derived from the activity of a type II iterative polyketide synthase system, and daunosamine, a 2,3,6-trideoxy-3-aminohexose synthesized from TDP-glucose.

Glycopeptides such as vancomycin also contain core units derived from the activity of a nonribosomal peptide synthetase that are decorated by sugar units to make bioactive molecules. An exaggerated example of this is the antifungal compound pradimicin, which is synthesized by a type II iterative polyketide synthase and then decorated both with sugar moieties and D-amino acid moieties. Other molecules, such as the rifamycins, contain a unique building block such as 3-amino-5-hydroxybenzoic acid (AHBA), which is used as the primer with which to initiate the synthesis of the macrocycle. Thus, many natural products are derived from mixed biogenic origins. For more detailed information on specific antibiotics, recent books edited by Vining and Stuttard and by Strohl cover the biogenesis and biosynthesis of virtually all the major classes of antibiotics and bioactive natural products.

ANTIBACTERIAL COMPOUNDS

The most widely prescribed, best known, best characterized, and perhaps also the most widely disparaged natural products are the antibacterials. After the discovery of penicillin by Flemming in 1928 and its subsequent development as a drug during World War II, the pharmaceutical industry and well-known academics such as Waksman of Rutgers University worked feverishly to discover new antibacterials that would eliminate bacterial infections of all kinds. The natural product antibiotics *gramicidin* (1939), *streptomycin* (1944), *chlortetracycline* (1948), *chloramphenicol* (1948), *cephalosporin C* (1948), *neomycin* (1949), *oxytetracycline* (1950), *erythromycin* (1952), *tetracycline* (1953), *vancomycin* (1955), *rifamycins* (1959), *lincomycin* (1962), *pristinamycin* (1962), and *nalidixic acid* (1962), the predecessor to *fluoroquinolones,* were all discovered by 1962. No new, truly novel classes of natural product antibacterial drugs have been marketed since then.

Even in the cases of modern synthetic antibacterials, all successful synthetic antibacterials have been biomimetically derived from natural product structures (the best example being the fluoroquinolones, which are based on nalidixic acid). There are new classes of candidate antibacterials, such as the oxazolidinones, which are not derived from natural products, but these have yet to break into the market, in part because of associated toxicities exhibited by this class of compounds. The targets for these drugs, including cell wall biosynthesis, protein biosynthesis, and nucleic acid biosynthesis, all have been targeted many times over by analogues of them, as well as by other compounds, both synthetic and naturally derived. As we enter the 21st century, there is a critical need for new, potent antibacterials to which resistance is not easily gained.

Commercial Value

With the emergence of new infectious diseases as well as the emergence of bacterial strains resistant to existing antibiotics, there is an enormous challenge to pharmaceutical companies, researchers, and governments to develop new methods for treating existing infectious diseases and those emerging as new health threats. As recently reviewed by Strohl, there are several reasons for the need to develop new antibiotics, preferably with novel structures and activities. First and foremost, pathogenic bacteria are acquiring or developing resistance to existing antibiotics and classes of antibiotics in direct correlation with the use of those antibiotics to treat them. As Julian Davies stated, "it is frightening to realize that one single base change in a gene encoding a bacterial *b*-lactamase can render useless $100 million worth of pharmaceutical research effort". Although this is an oversimplification of the problem, it succinctly sums up the problem with the development of antibiotic resistance by pathogenic bacteria. The most critical reason for the development of new antibiotics is the race to remain a step ahead of antibiotic-resistance development, which has proceeded at an alarming rate.

Even since the earliest use of antibacterial agents, the emergence of resistance to the drugs used has been observed. For example, within 4 years after the introduction of penicillin during World War II, several strains of penicillin-resistant bacteria were isolated from infected patients. After the war, outbreaks of dysentery in Japan, caused by sulfonamide-resistant *Shigella* spp., presented a serious health hazard to the recovering population. Within just a few years after introduction in 1950 of the newly developed antibiotics, streptomycin, tetracycline, and chloramphenicol to help quell the problem, resistance to each of these antibiotics was observed. By 1969, 69% of shigellae were resistant to sulfonamides, streptomycin, tetracycline, and chloramphenicol. Thus, in fewer than 20 years, nearly 70% of all *Shigella* spp. in Japan became resistant to four antibiotics to which there was minimal resistance before their use. Widespread antibiotic use has resulted in the rapid spread of multidrug resistant pathogens, and it has become both a local and global health hazard of epidemic proportions.

The incidence of methicillin-resistant *Staphylococcus aureus* (MRSA), vancomycin-resistant enterococci (VRE; particularly by *Enterococcus faecium* and *Enterococcus faecalis*), and β-lactam-resistant *Streptococcus pneumoniae* are among the most serious problems, but are by no means the only ones. The incidence of nosocomial infections (i.e., hospital-acquired) is on the rise, and of those, an increasingly greater percentage of the infectious agents (typically MRSA and VRE) are drug-resistant. The U.S. Centers for Disease Control and Prevention (CDC) estimates that of the 40 million hospitalizations in the United States each year, 2 million patients (5%) contract nosocomial infections, and that as many as 58,000 of them (0.15%) die of them. Unfortunately, VRE are nearly untreatable with current antibiotics, with the possible exceptions of the experimental drugs Synercid or teichoplanin, which have been used successfully to treat some cases of VRE. The mortality rate of patients infected with multiple-drug resistant enterococci is 70%, making this a serious health care problem. Second, new pathogens are evolving at an alarming rate. For example, today there are more than 30 new infectious diseases (e.g., AIDS, Ebola, Legionnaires disease, hantavirus, Lyme disease, and food-borne *E. coli* O157:H7 infections) that were unknown 20 years ago.

Third, even with the development of remarkable anti-infective drugs with outstanding biological activities, there are still certain pathogens that naturally defy even the best efforts to defeat them. Perhaps the best example of this is our current inability to treat successfully *Pseudomonas aeruginosa* infections resulting from cystic fibrosis. Finally, many potentially important antibiotics have associated toxicities that limit their use. The best example of this is gentamycin and other aminoglycosides, which are limited in effectiveness because of their associated nephrotoxicity and ototoxicity. Thus, critical challenges lay ahead for the entire antiinfectives field, many of which may be met by the discovery and development of new natural product leads. There are three critical needs in the antibacterial market. First, new parenteral drugs that can successfully treat MRSAs and VREs are absolutely required. This need will grow dramatically in the coming years as the once-rare incidence of vancomycin-resistant *S. aureus* strains, the single greatest worry of infectious disease experts, expands.

Second, even with a seemingly crowded pediatric market, there is still a critical need (and market place) for an orally absorbed, pediatric-safe (which eliminates fluoroquinolones, which are not considered safe for children) antibacterial agent that can successfully treat upper respiratory tract infections (primarily otitis media caused by β-lactam resistant streptococci, moraxellae, and *Haemophilus influenzae*) and that is not compromised by β-lactam, aminoglycoside, or MLS (macrolide-lincosamine-streptogramin) resistance or degraded by extended spectrum β-lactamases. The pediatric antibacterial market, which has grown sharply since the early 1980s, has become one of the most important anti-infective markets today.

Pediatric antibacterials typically are prescribed for middle-ear infections (otitis media) caused by *H. influenzae, Moraxella* spp., and *Streptococcus pneumoniae*. By the year 2000, it is expected that the U.S. pediatric market alone will be $2.2 billion, with annual increases of approximately 15%. The current leading pediatric antibacterials are amoxicillin, Augmentin (amoxicillin plus clavulanic acid), azithromycin, and clarithromycin. It is expected that the ketolides may make a significant impact on this market. The final critical need is for a new tuberculocidal drug. Antibiotic resistance in *Mycobacterium tuberculosis* is considered to be a major growing problem. Currently, a combination treatment with isoniazid and rifamycin (both tuberculostats) is the method of choice, but many strains of *M. tuberculosis* have recently been found that are resistant to one or both of these antibiotics, and approximately 14% of all tuberculosis cases reported today involve *M. tuberculosis* strains that are resistant to one or more antibiotics. The problem facing doctors today is that there is no proved, effective alternative yet on the market to treat isoniazid- and rifamycin-resistant *M. tuberculosis*.

Resistant Against Antibiotic

Antibiotic resistance is a major clinical problem that will only get worse with time. In the case of bacteria, resistance can be mediated by several different types of mechanisms, including modification of the target to make it unsusceptable to the antibiotic, enzymatic modification of the antibiotic to render it inactive, exclusion of the antibiotic by modification of membrane permeability, active export of the antibiotic, and bypass of the critical physiological mechanisms or structures inhibited by the antibiotic. Resistance to some these mechanisms, encoded by chromosomal mechanisms, is attained at a rate of 1×10^{-8}, and others are obtained at much higher frequencies through horizontal plasmid transfer. Plasmid-encoded TEM- and SHV-type β-lactamases are the most common mechanisms responsible for β-lactam resistance in Gram-negative bacteria.

These β-lactamases have generally broad substrate specificities for penicillins as well as some first- and second-generation cephalosporins and are responsible for much of the resistance to β-lactams observed in the clinics. It is well documented that as new β-lactams have been developed to resist plasmid-borne β-lactamases, the β-lactamases themselves have evolved. In an extraordinary example of natural directed evolution, TEM- and SHV-type β-lactamases recently have been found containing specific point mutations, all falling in close proximity to the active site cavity, which confer resistance to third-generation cephalosporins and monobactams. Vancomycin has long been the last line of defense against antibiotic-resistant bacteria, particularly MRSAs and β-lactam- and aminoglycoside-resistant enterococci. Just a little over a decade ago, it was thought that microorganisms would never become vancomycin resistant because of its unusual mechanism of action. This illusion was shattered, however, when VREs were reported in England and France in 1987. VREs now are being discovered at an alarming rate, with no new last line of defense substitute firmly entrenched on the market. Not all clinically significant antibiotic resistance is plasmid- or transposon-mediated. *S. pneumoniae,* the primary causative agent in pediatric otitis media and pneumonia as well as bacterial meningitis and bacteremia, remained sensitive to penicillin for many years.

Recently, however, an alarming number of penicillin-resistant *S. pneumoniae* strains have recently been isolated from infected patients, correlating with the use of penicillins to treat them. These penicillin-resistant *S.pneumoniae* were not found to produce β-lactamases, the common plasmid-borne mechanism for resistance. Instead, they had incurred chromosomal mutations giving rise to altered penicillin-binding proteins (PBPs), the target for penicillin activity. This discovery has a touch of irony, because *S. pneumoniae* is the bacterium with which plasmid-mediated transformation was first observed. Similarly, the Oxford strain of *S. aureus,* isolated 50 years ago, is sensitive to nearly every antibiotic tested against it. The staphylococci, however, acquired resistance to aminoglycosides very quickly, and by 1992, 32% of *S. aureus* strains tested were found to be resistant to methicillin, the drug of choice to treat staphylococcal infections, up from 2% in 1975.

Like pneumococci, MRSAs have altered PBPs, giving them a resistance mechanism that cannot be countered easily. Thus, vancomycin has become the drug of last resort to treat MRSAs. Ever since *Enterococcus faecalis* was shown to transfer vancomycin resistance to *S. aureus* in the laboratory in 1992, one of the greatest fears of health professionals universally has been the emergence of vancomycin-resistant, methicillin-resistant staphylococci. Recently, such doubly resistant, nearly untreatable strains have appeared in sporadic cases, confirming the greatest fears of infectious disease professionals. A new mechanism of antibiotic resistance recently was discovered in meropenem-resistant *Pseudomonas aeruginosa* strains. *P. aeruginosa* is an opportunistic pathogen that contributes to nosocomial infections, eye infections, and opportunistic infections of burn patients and those with cystic fibrosis. The mechanism was found to be based on the alteration of outer membrane proteins, causing a reduction in permeability of several antibiotics, including meropenem, through the outer membrane. This is a significant finding, because carbapenems (e.g., imipenem) have been considered the antibiotics of choice for the treatment of opportunistic pseudomonads.

β-Lactams

As is well known, the discovery of penicillin was the genesis of modern pharmaceutical efforts to produce antibiotics, the magic bullets that at one time were thought to cure the world of disease. β-Lactams function by binding to PBPs, especially transpeptidases, that normally function in peptidoglycan biosynthesis. Because peptidoglycan is novel to bacteria, these drugs specifically act on a bacterial target. Since bacteria cannot grow in normal environments without their cell wall, inhibition of cell wall biosynthesis results in killing of the bacteria; thus, all β-lactams by definition are bacteriocidal drugs. The total antibacterial market tops $20 billion annually, with nearly all the top-selling antibacterial drugs being natural products, semisynthetics derived from natural products, or compounds patterned after natural products. Even more than a half century after the first widespread use of penicillin during World War II to treat bacterial infections, β-lactams as a whole still make up 60%, or more than $12 billion worth, of the total antibacterial market. The commercial β-lactam antibiotics are comprised of many types of several different classes, ranging from penicillin G to the monobactam biomimetic drug, aztreonam. The natural product β-lactams can be separated into five separate biogenic subclasses: penams (e.g., penicillins), cephems (cephamycin C, cephalosporins), clavams (clavulanic acid, clavamycins), carbapenems (thienamycin), and monobactams (aztreonam).

The ring structures of the various β-lactams differ from the classical penicillin core as follows: cephems have an expanded 6-membered ring containing a double-bond but retaining the sulfur atom; clavams have an oxygen atom replacing the sulfur of penicillins, and they tend to differ at C-2 and C-6; in carbapenems, the sulfur is replaced by a carbon and a double bond is present; and the monobactams are monocyclic. β-Lactams of the penicillin, cephamycin, and cephalosporin subclasses share a common biogenic origin, i.e., the condensation of three activated amino acids, L-α-aminoadipate, L-cysteine, and L-valine by aminoadipyl-cysteinylvaline (ACV) synthase (*pcbAB* gene products) to form the tripeptide δ-L-(α-am inoadipyl)-L-cysteinyl-D-valine (ACV), which is then cyclized by the enzyme isopencillin N (IPN) synthase (*pcbC* gene product) to form isopenicillin N. ACV synthases comprise a subset of a larger class of non-ribosomal peptide synthases, which have been characterized in detail, both biochemically and molecularly.

Both ACV synthases and IPN synthases of all β-lactam producers are highly conserved, indicating common ancestral origins. IPN is either directly converted to penicillin G via IPN acyltransferase (*penDE* gene product), or via IPN epimerase (*cefD* gene product) to penicillin N, which is converted via deacetoxycephalosporin C (DAOC) synthase (*cefE* gene product; also called expandase because it expands the 5-membered ring of penicillins to the 6-membered ring of cephalosporins) to deacetoxycephalosporin C (DAOC).

Table 24.1 Important clinical and experimental β-lactams

Subclass	*Drugs on market*
Penicillins	Ampicillin, *amoxicillin*, bacampicillin, cloxacillin, floxacillin, mezlocillin, nafcillin, oxacillin, penicillin G, penicillin V
Penicillinase-resistant penicillins	*Methicillin*, dicloxacillin
Antipseudomonal penicillins	Carbenicillin indanyl, piperacillin, *ticarcillin*
First-generation cephalosporins	Cefadroxil, cefazolin, *cephalexin*, cephalothin, cephapirin, cephradine
Second-generation cephalosporins	*Cefaclor*, cefmetazole, cefonicid, cefotetan, *cefuroxime*
Third-generation cephalosporins	Cefamandole, *cefixime*, cefoperazone, cefotaxime, cefpodoxime proxetil, *cefprozil*, ceftazidime, *ceftibuten*, ceftizoxime, ceftriaxone
Oxycephams (third generation)	Flomoxef, latamoxef
Fourth-generation cephalosporins	*Cefepime*
Cefam	*Cefoxitin*
Carbacefem	*Loracarbef*
Carbapenems	*Imipemen/cilistatin*, meropenem, MK-826, panipenem (Japan only)
Monobactams	Aztreonam
Clavams (β-lactamase inhibitors)	*Clavulanate*, sulbactam, tazobactam
Penicillins/β-lactamase inhibitors	*Amoxicillin/clavulanate (Augmentin)*, ampicillin/sulbactam (Unasyn), pipericillin/tazobactam (Zosyn), ticarcillin/clavulante (Timentin)

The clavam β-lactamase inhibitor, clavulanic acid, is produced by *Streptomyces clavuligerus*, which also produces cephamycins. Interestingly, the biogenesis of the bicyclic nucleus of clavulanic acid occurs via a completely different mechanism than does the biosynthesis of other β-lactams. In *S. clavuligerus*, the clavulanic acid biosynthesis genes are just downstream of the cephamycin C biosynthesis genes. Not all β-lactam-producing taxonomic groups of organisms synthesize all subclasses of β-lactams, as depicted in Table 24.2. Filamentous fungi, for example, produce β-lactams only of the penam and cephem classes, whereas various actinomycetes produce β-lactams of all subclasses except the penams and cephalosporins of the cephem subclass. The cephems in general are produced by the taxonomically widest range of microorganisms, including filamentous fungi, Gram-negative bacteria (e.g., lysobacters), and actinomycetes.

The cephem products accumulated by each group differs, however, likely indicating evolutionary branches in terms of enzymes present. The Gram-negative bacteria, *flavobacterium* sp. SC 12.154

and *Lysobacter lactamgenes* accumulate a 7-formylamino analogue of cephalosporin called cephabacin, an unusual natural β-lactam structure that is highly resistant to β-lactamases. The actinomycetes *Streptomyces clavuligerus* and *Nocardia lactamdurans* produce cephamycin C, which is formed from DAOC by the enzymes (gene names in parentheses) DAOC hydroxylase (*cefF*), deacetylcephalosporin C (DAC) *O*-carbamoyltransferase (*cmcH*), cephalosporin 7-α-hydroxylase (*cmcI*), and 7-α-hydroxy-*O*-carbamoyl-deacetoxycephalosporin C methyltransferase (*cmcJ*). The filamentous fungus, *Cephalosporium acremonium*, on the other hand, converts DAOC to cephalosporin C in two steps: DAOC hydroxylase (*cefF*), and DAC acetyltransferase (*cefG*). As is well known, penicillin was the first widely used clinical antibiotic. The first oral penicillin, used primarily to treat Gram-positive bacteria, was the fermentation product phenoxymethylpenicillin (penicillin V), still sold today under the name Veetids.

Table 24.2 Biogenic classes of β-lactams

		β-Lactams produced by		
Subclass	*Example compounds*	*Fungi*	*G$^+$ bacteria*	*G$^-$ bacteria*
Penams	Penicillin G	*Penicillium, Aspergillus*	—	—
Cephems	Cephalosporin C	*Cephalosporium*	—	—
	DAOC	—	—	Various strains
	Cephabacins	—	—	*Lysobacter*
	Cephamycin C	—	*Streptomyces, Nocardia*	—
Carbapenems	Thienamycin	—	*Streptomyces cattleya*	*Serratia, Erwinia*
Monobactams	Aztreonam	—	*Nocardia*	*Pseudomonas*
Clavams	Clavulanic acid	—	*Streptomyces*	—
	Clavamycins	—	*Streptomyces*	—

Second-generation penicillins include the aminopenicillin group of ampicillin, amoxicillin, and bacampicillin, which exhibited much better gastrointestinal tract absorption and possess longer half lives. These aminopenicillins also exhibit greater activity against Gram-negative bacteria at the expense of activity against Gram-positives. Unfortunately, these penicillins are highly susceptible to β-lactamases. Development of β-lactamase inhibitors such as clavulanic acid, sulbactam, and tazobactam, however, have greatly extended the use of these broad-spectrum penicillins. In fact, Augmentin (oral amoxicillin/clavulanic acid), Unasyn (injectable ampicillin/sulbactam), Timentin (injectable ticarcillin/clavulanic acid), and Zosyn (injectable pipericillin /tazobactam) are widely used antibiotics. Carbenicillin indanyl, the first oral antipseudomonal β-lactam, and ticarcillin, a potent parenterally administered drug, are third-generation penicillins. The development of antipseudomonal fluoroquinolones, however, has rendered carbenicillin indanyl obsolete. Pipericillin, azlocillin, and mezlocillin are parenteral fourth-generation penicillins that have not found wide usage clinically.

First-generation cephalosporins typically inhibited Gram-positive bacteria other than the enterococci. Second- and third-generation cephalosporins typically inhibit Gram-negative bacteria at the expense of activity against Gram-positives. A new fourth-generation parenteral cephalosporin, cefepime (Maxipime), possesses activity against both Gram-positives and Gram-negatives.

Additionally, cefipime penetrates bacteria well, is not susceptible to β-lactamases, and has a lower propensity to induce β-lactamases. Unfortunately, this broad-spectrum cephalosporin does not effectively treat either VREs or MRSAs. Ceftriaxone was very recently approved as a one-time-only injectable for treatment of pediatric otitis media. The concept behind the development of this drug is that the greatest problem, and reason for failure, with treatment of otitis media is patient compliance; the option of a one-time-only injectable removes this as an obstacle. Carbapenems are very broad spectrum, β-lactamase-resistant β-lactams that contain a carbon atom instead of a sulfur atom in the 5-membered ring.

Imipenem/cilistatin (Primaxin) was the first clinical carbapenem antibiotic, developed in the mid-1970s and made available commercially since 1986, as a synthetic biomimetic analogue of the natural product thienemycin. Imipenem is administered with cilistatin, an inhibitor of human renal dehydropeptidase, to protect the antibiotic from enzymatic degradation. Imipenem has broad-spectrum activity against Gram-negatives and anaerobic bacteria and moderate activity against Gram-positive bacteria. Unfortunately, imipenem also is limited by potential renal toxicity. Meropenem (Merrem) is a new carbapenem antibiotic approved in 1996 for use as a broad-spectrum injectable antibiotic for treatment of intra-abdominal infections and for pediatric bacterial meningitis. The advantage of meropenem over imipenem is that meropenem is stable in the presence of renal dehydropeptidase, thereby bypassing the need for cotherapy with cilistatin. A third carbapenem called panipenem is available in Japan but not in the United States. Merck currently has a new, very potent, broad-spectrum, long-acting carbapenem, MK-826, in phase-II clinical trials. This new carbapenem drug candidate has been demonstrated to be highly resistant to β-lactamases, including plasmid-encoded extended spectrum β-lactamases (ESBLs).

Nevertheless, certain rarely pathogenic bacteria (particularly *Acinetobacter baumannii*) contain specific carbapenem-degrading enzymes known as carbapenemases, which may, through horizontal gene transfer, enter the broader pathogen field in the future. An interesting new parenteral β-lactam drug candidate has recently been described. The new Glaxo-Wellcome experimental compound, GV129606, called a trinem, is a highly potent antibacterial with efficacy against Gram-positives, Gram-negatives, aerobes and anaerobes alike, plus marked resistance to extended spectrum, clinically relevant β-lactamases. This compound may mark a new path to be followed in the β-lactam field.

Antibiotics Produced by Actinomycetes

The macrolide antibiotics are a group of compounds produced by actinomycetes via type I modular polyketide synthases. The core aglycone of the prototype macrolide, erythromycin, is synthesized by the huge multifunctional enzyme, 6-deoxyerythronolide B synthase (DEBS), which consists of three, multidomain subunits, each subunit with a molecular weight around 300,000 Da. The macrolide chain is synthesized by a set of enzymatic functional domains within the proteins in a linear, stepwise fashion. Interestingly, the enzymatic active sites are arranged in the polypeptides in the same order as their apparent function. Although macrolides theoretically can have ring sizes ranging from as few as 6-membered to over 30-membered, most of the clinically relevant macrolides possess 14-membered or 16-membered rings. Current clinical antibacterial macrolides are mostly derived from erythromycin, although a wide variety of biologically active macrolides are used for several purposes across several therapeutic areas. The macrolide antibiotic, erythromycin, was first discovered in 1952, the year the double-helical structure of DNA was first announced.

Macrolides, which are bacteriostatic agents, inhibit bacterial protein synthesis by structure-based mechanisms: 14-membered macrolides such as erythromycin block translocation of peptidyl-tRNA, whereas 16-membered macrolides inhibit peptidyltransfer reactions. There are apparently additional interactions of macrolides of both groups with ribosomes, but even after 30 years of research, these are not fully understood. Resistance to macrolides is most often via the MLS resistance mechanism, which is based on mechanism of action. MLS resistance, conferred by several different genes, results in methylation of adenosine-2058 of 23S rRNA. MLS-I resistance, exemplified by the *tlrD* gene, is result of monomethylation, whereas MLS-II (*ermE*-type) is the result of dimethylation. Erythromycin itself is a poorly soluble, acid-labile molecule that nevertheless has enjoyed a long history of successful use.

Erythromycin has been an important orally active antibiotic for many years. The efforts for years in the macrolide field have been aimed at developing new drugs that are acid-stable, possess a longer half-life, have broader ranges of activity, and are not subject to MLS resistance mechanisms. Recently, several semisynthetic macrolides arising from modifications of erythromycin, including clarithromycin (Biaxin) the 15-membered azalide, azithromycin (Zithromax), and most recently, dirithromycin (Dynabac), have made significant contributions to the antibacterials market. Among them, these macrolides represent more than $2 billion in market share. The most exciting research area in antibacterials today may be the discovery of ketolides, erythromycin-analogue macrolides in which the L-cladinose at the 3-position of erythromycin has been replaced by a keto moiety. The ketolides are particularly interesting because they do not induce MLS-resistance mechanisms in pathogenic organisms.

This is critical because MLS resistance is typically not constitutive. The two leading ketolide drug candidates, RU64004 (also known as HMR 3004) and RU66647 (also known as HMR 3647), have been shown to be potent inhibitors of Gram-positive bacteria, particularly the pneumococci, including those that show marked resistance to other macrolides such as erythromycin and azithromycin. The ketolides, however, show poor activity against Gram-negative bacteria. Recently, a series of new, 14-membered macrolide antifungal compounds known collectively as galbonolides was discovered, of which one, rustmicin, has been investigated in more detail. The data indicate that rustmicin and, by extension, possibly the other galbonolides appear to interact with the α-mannan of the cell wall by a potentially novel mode of action.

Several years ago, interest was sparked when a macrolide known as FK506 (Tacrolimus) was found to possess significant immunosuppressive action. Additional compounds of interest in this area included rapamycin, which is expected to be approved soon, and immunomycin, which was dropped because of a narrow efficacy-to-toxicity window. Both the milbemycins and avermectins have been used for several years as antiparasitic drugs. Ivermectin has been used and approved to treat human diseases such as river blindness and intestinal threadworm. Additionally, macrolides such as tylosin have long been important as animal-feed additive growth promotants in the agricultural area. During the past few years, interest in macrolides as bioactive natural products has grown significantly in areas other than the anti-infective area. There currently is active experimental interest in macrolides in multiple therapeutic areas, including antibacterial, antifungal, antiparasitic, anticancer, immuno-suppressive, neuroimmunophilin, gastrointestinal, respiratory, and agricultural areas. There are five human therapeutic areas in which macrolides have not been used clinically, but in which there is currently considerable interest. An acid-catalyzed breakdown product of erythromycin, called motilide, has been shown to be a potent agonist of the motilin receptor, but it is unstable.

Abbott has the lead drug candidate in this area, ABT-229, a 12-deoxyerythromycin homologue possessing modification in the sugars, which is stable and has reasonably high potency. It has recently been demonstrated that tacrolimus (formerly FK506) and rapamycin possess additional activities other than those related to immunosuppression. Both tacrolimus and rapamycin have been demonstrated to promote neuron cell growth under conditions in which they can still bind the FK-binding protein in the absence of the second ligand (calcineurin or RAFT, respectively). The macrolide, roxithramycin, currently an investigational drug for treatment of asthma, potentially has multiple modes of action for this indication, including inhibition of glycoconjugate secretion to increase mucous clearance; clearance of *Chlamydia pneumoniae*, a possible causative or implicated agent in asthma; suppression of T-cell cytokines interleukins 2, 3, and 4; and inhibition of cholinergic neuroeffector transmission in airway smooth muscle cells. Finally, a macrolide produced by the myxobacterium, *Sorangium cellulosum*, was found to possess tubulin-stabilizing activity similar to Taxol, a $1 billion per year anticancer drug. Several companies and research laboratories have shown strong interest in trying to develop epothilone analogues as potential antitumor drugs.

New Combination

For more than 25 years, pristinamycin, a combination of two antibiotics, pristinamycin I_A (a peptidic macrolactone belonging to the streptogramin B group) and pristinamycin II_A (a polyunsaturated macrolactone belonging to the streptogramin group A), has been used in Europe as a drug to combat severe infections caused by both Gram-positive and Gram-negative bacteria, without development of resistance against it. Each component of pristinamycin is bacteriostatic with poor efficacy, but the combination is bacteriocidal with good potency against a broad range of bacteria. The usefulness of pristinamycin, however, has been limited because of its poor solubility. In February 1996, however, Rhone-Poulenc Rhorer introduced Synercid, a new variation on pristinamycin. A mixture of the two water-soluble, semisynthetic streptogramins, quinupristin (RP57669) and dalfopristin (RP54476), Synercid (RP59500) is currently in phase III clinical trials worldwide and is currently available in the United States and Europe in an emergency-use program. More than 700 patients have received Synercid in this program.

In one study, 70% of 115 patients with VRE infections were successfully treated (i.e., infection cleared) with Synercid. Synercid has been found to be bactericidal for most Gram-positive bacteria, but bacteriostatic for *Listeria monocytogenes*. In one study, 96% of all staphylococci tested were sensitive to Synercid in vitro, suggesting that this drug may become the "vancomycin of the future," a drug of last resort for use against severe bacterial infections. One hopeful aspect concerning Synercid is that it is unaffected by classical rRNA-methylase-conferred MLS resistance mechanisms, the most common resistance mechanisms found in staphylococci. Other MLS resistance mechanisms such as acetylation or hydrolysis of streptogramins still affect these compounds, however, so they are not completely impervious to bacterial resistance mechanisms. A recent study showed that Synercid at 2 μg/mL inhibited nearly 99% of vancomycin-resistant *E. faecium* clinical isolates tested in the United States, giving rise to hopes that Synercid will help to contain the growing VRE problem.

Earliest Antibacterial Antibiotics

The *tetracyclines* are among the earliest of antibacterial antibiotics to be discovered, with chlortetracycline being discovered in 1948. The natural tetracyclines are produced by various species of actinomycetes; *Streptomyces aureofaciens* produces both chlortetracycline and tetracycline, the production of each which can be optimized by specific fermentation conditions, *Streptomyces rimosus* produces oxytetracyclines, and dactylocyclines are produced by *Dactylosporangium* spp. and *Actinomadura brunnea*. Tetracyclines function as antibiotics by inhibiting protein translation, specifically by disrupting codon-anticodon interactions of tRNA and mRNA at the acceptor site of the ribosome. Because they inhibit protein synthesis, tetracyclines, like macrolides, are bacteriostatic. Tetracyclines are widely used not only in human health, but also in animal health programs; perhaps the widest current use of chlortetracycline and oxytetracycline is as prophylactic drugs in the fish husbandry industry. Bacteria typically develop resistance to tetracyclines by acquiring a plasmid- and transposon-borne resistance gene (*tetA*) encoding an efflux pump. Thus, tetracycline resistance is one of the primary resistances observed as being transmitted horizontally among different groups of bacteria. Additionally, a second mechanism, mediated by TetM, gives resistance by inhibition of the interaction of the tetracyclines with the ribosome.

This, unfortunately, has led to the diminution of use of the classical tetracyclines for most human health indications. Some of the more modern tetracyclines, such as minocycline, are important for treating MRSAs. The most exciting advance in the tetracycline area has been the development of the glycylglycine tetracyclines. Two new semisynthetic glycylglycines, *N,N*-dimethylglycylamido-9-aminominocycline (DMG-MINNO) and 9-amino-6-demethyl-6-deoxytetracycline (DMG-DMDOT), have been developed by Lederle for their very broad spectrum of activity against Gram-positive, Gram-negative, aerobic, and anaerobic bacteria. Their spectrum includes, perhaps surprisingly, bacteria as widely disparate as mycoplasmas, MRSAs, VREs, *Neisseria* spp., *Haemophilus* spp., and *Moraxella* spp. Most importantly, glycylglycine tetracyclines bind the ribosome more tightly than traditional tetracyclines so that TetM, which mediates ribosomal resistance mechanism, cannot interfere with their activity, and they are not substrates for the TetA-mediated efflux mechanisms. Thus, there currently is no effective resistance mechanism yet known for these modified tetracyclines.

Another Antibacterial Drug

Vancomycin was first isolated in the early 1950s, making it another of the group of long-standing, well-known *antibacterial drugs*. For a long time, vancomycin was not used to a significant degree, but as the incidence of infections with drug-resistant organisms increased, the use for vancomycin became significantly greater. Because vancomycin is absorbed poorly by the gastrointestinal tract, it is rarely used as an oral drug. Instead, its primary use today comes as a parenteral drug as the last line of defense against many potentially lethal pathogenic bacteria, including MRSAs and antibiotic-resistant *Enterococcus faecium*. In 1987, however, after 30 years of vancomycin being on the market, resistant strains of *E. faecium* (VREs) were finally observed. The increased incidence of vancomycin resistance, especially in *S. aureus*, is of enormous concern. Little more than a decade ago, resistance to vancomycin was unheard of and totally unsuspected, because its mechanism of action was so different from that of any other antibiotic class. Nevertheless, resistance to vancomycin and other glycopeptides was discovered and is on the rise.

Vancomycin and other glycopeptides exert their activity by inhibiting peptidoglycan biosynthesis, making them potent bacteriocidal drugs. Peptidoglycan consists of a sugar backbone to which pentapeptide chains are attached. The terminal residues of the pentapeptide chain are D-alanine residues, which are critical to processing the peptidoglycan matrix as it is built. Vancomycin and other glycopeptides bind, through multiple hydrogen bonding, to the terminal D-alanine residue of the pentapeptide chain and prevent both transglycosylation and transpeptidation. Resistance occurs when a bacterium contains special genes (*vanAHXY*) that allow for incorporation of a D-lactic acid residue in place of the terminal D-alanine residue of the peptidoglycan pentapeptide chains, rendering the peptidoglycan unrecognizable to vancomycin binding. Two genetically distinct forms of resistance have emerged, although both operate by the same mechanism. VanA-mediated resistance gives high-level resistance to both vancomycin and teichoplanan, whereas VanB-mediated resistance is restricted to vancomycin.

A recent study of clinical VRE isolates in the United States showed that 72% possessed the VanA phenotype, whereas only 28% possessed the VanB phenotype. There are several relatively new analogues of vancomycin on the market to combat vancomycin resistance. Teichoplanan has been around for several years, but its use is restricted to Europe so far. Ramoplanan, another glycopeptide that has been around for a few years, is currently in phase II clinical trials at BioSearch Italia. Eli Lilly has developed a new chlorobiphenyl analogue of vancomycin, LY333328, that is considerably more potent than vancomycin. Moreover, LY333328 appears not only to inhibit bacterial growth, but to kill existing bacteria, which may give it a distinct clinical advantage over other existing antibiotics. LY333328 is very potent against VRE in vitro, giving hope for treatment of vancomycin-resistant pathogens.

Aminoglycosides

Aminoglycosides are not widely used in clinical therapy in the United States because of their relative toxicity profile as compared to other available antibacterials and to the rapid development of plasmid-borne resistance mechanisms through pathogen populations. *Streptomycin* and *neomycin* were first described in 1944 and 1949, respectively, so this group of antibacterials has a rich history. An incredible amount of knowledge about the biosynthesis, the genes, and the molecular mechanisms for regulation of aminoglycoside production is now known, and Piepersberg, the architect of much of this effort, has coalesced this enormous wealth of information in a recent review. One new aminoglycoside worth mentioning is the development in Japan of arbekacin, a new aminoglycoside that has potent activity against methicillin-, gentamycin- and tobramycin-resistant

S. aureus strains. The incidence of methicillin-resistance in Japan has risen to nearly 60%, and only vancomycin and the newly developed arbekacin are approved for treatment of these strains. The new aminoglycoside is apparently effective because the widely distributed resistance mechanism, a bifunctional aminoglycoside phosphotransferase (APH [2″])-6′-acetyltransferase (AAC[6′]) which confers resistance to other aminoglycosides such as gentamycin and tobramycin, is relatively ineffective at modifying the new drug. It is probable that within a few years, APH(2″)-AAC(6′) will evolve so that it will be able to modify arbekacin.

Other Antibacterial Compounds and New Activities of Interest

Several new antibacterial drugs recently have been introduced to the market or in clinical trials that hold hope for containing the problems of newly emergent pathogens and increased resistance to existing drugs. A comparison of the minimal inhibitory concentrations (MICs) to kill 90% of *S. pneumoniae* treated in vivo (MIC_{90}s, in μg/mL) was made of several of the new antibacterials with the following results (in order of efficacy): ketolides (new macrolides containing keto groups at C-3), 0.0 16 to 0.25; everninomicins (novel oligoglycoside from marine organism), 0.06 to 0.25; glycylglycines (modified tetracyclines), 0.12 to 0.5; new fluoroquinolones (e.g., sparfloxacin and trovafloxacin), 0.12 to 0.5; oxazolidinones (new class of chemosynthesized antibacterials), 0.5 to 1.0; and new streptogramins (e.g., Synercid), 0.25 to 2.0. These can be compared to MIC_{90}s for the existing drugs on various (and variably resistant) strains of *S. pneumoniae*: amoxicillin-clavulanate (Augmentin), 0.06 to 1.0; cefuroxime (semisynthetic cephalosporin), 0.12 to 2.0; cefpodoxime (semisynthetic cephalosporin), 0.25 to 2.0; and cefixime (semisynthetic cephalosporin), 0.5 to 16. As important as MICs are a low MIC does not alone make a good drug.

The other critical factor is the development of antibacterials that can be taken orally in once-per-day doses to increase patient compliance. Thus, the time for a particular drug dose to maintain in vivo efficacy above MIC (T > MIC) value is also critical to dial into the structure of newly developed drugs. Most of the new antibacterials on the market are modified analogues of existing, well-known classes of antibacterials: azithromycin and clarithromycin, as well as the new experimental ketolides, are based on modifications to the erythromycin core structure; trovafloxacin and sparfloxacin are new fluoroquinolones; cefepime is a new cephalosporin; meropenem and MK-826 are new carbapenems; trinem is a novel β-lactam; LY33328 is a new glycopeptide; and Synercid is comprised of two new streptogramins. Only the oxazolidinones, a chemical synthesis class, represents a new, novel class of antibacterials. Pharmacia-Upjohn is testing two new oxazolidinones for their ability to treat multiple-drug-resistant strains of *Mycobacterium tuberculosis*. These new chemically synthesized antibiotics inhibit protein synthesis in a manner different from natural product protein synthesis-inhibitory antibiotics, have novel structures unrelated to structures of other antibiotics, and have a wide spectrum of activity.

Experiments have shown that, at least in in vitro tests, pathogens do not form resistance against these antibiotics at significant frequencies. Mupirocin (Bactroban) was approved in 1997 by the U.S. FDA for use as a topical agent, especially against nasal MRSA. Mupirocin has been used in the U.K. since 1985 and is currently approved in 90 countries. Produced as pseudomonic acid by *Pseudomonas fluorescens*, mupirocin is interesting because of its novel mode of action; it inhibits bacterial protein synthesis specifically through its binding to isoleucyl tRNA synthetase. Unfortunately, mupirocin-resistant MRSA have emerged, which is not surprising since the target is a single enzyme. Ziracin is an interesting, if still-evolving story.

The everninomycins have been known since the early 1960s as bacterial protein synthesis inhibitors, but they have attracted little attention for many years. Ziracin, an aromatic moiety-containing oligosaccharide, was first isolated in 1979 from a marine actinomycete, *Micromonospora cabonacea*. At that time, the everninomycin was not pursued because of its relatively narrow spectrum of being potent against only Gram-positive bacteria. With the increased resistance of Gram-positive bacteria to existing drugs, interest in the everninomycin, Sch27899, was revived. Now, Sch27899 is in clinical development as a novel antibacterial. The problem is that a similar everninomycin, avilamycin (MaxusG), has been used for several years in the European Union as a feed additive in the agricultural market. Recently, avilamycin-resistant isolates of *E. faecium* have been found in broiler chickens that are cross-resistant to Ziracin. This discovery is of great concern, potentially cutting short the effectiveness, and certainly the life-span, of Ziracin if it is ever marketed.

ANTIFUNGAL AGENTS

The need for new antifungal drugs has never been greater for a variety of reasons. Fungal infections fall into two major categories, those that are localized, dermal, and mostly annoyances, such as ringworm, athlete's foot, and nail infections (onychomycoses), and those that are disseminated and life threatening. Life-threatening, disseminated fungal diseases have risen dramatically during the past few decades, making this an important area for new drug development. Typically, life-threatening fungal infections occur in patients with impaired defenses, such as patients with AIDS; with other debilitating diseases; who are under treatment with immunosuppressive drugs (usually after organ transplants, to prevent tissue rejection); who are receiving chemotherapy; or with severe burns. Additionally, with the potentially broader use of immunosuppressive drugs for other indications such as arthritis, the potential for fungal infections may be expanded. The most important pathogenic fungi are *Candida albicans* (disseminated candidiasis), *Candida* spp., *Aspergillus fumigatus* (systemic aspergillosis), *Cryptococcus neoformans* (disseminated cryptococcosis leading to meningoencephalitis), *Pneumocystis carinii* (*P. carinii* pneumonia [PCP]), *Histoplasma capsulatum* (histoplasmosis), and *Coccidioides immitis*, the final example being a fungal pathogen very rich in chitin.

The most important fungal pathogens associated with immunocompromised patients, such as those with AIDS or receiving immunosuppressive therapy, are *P. carinii, C. neoformans*, and *Candida* spp. Current potent antifungal drugs have problems that make them less than optimal for clinical use. For example, amphotericin B, the antifungal drug of choice in life-threatening cases, causes acute toxicity, resulting in it not being well tolerated in a large percentage of patients, although it is efficacious against PCP. Azoles (e.g., clotrimazole, fluconazole, ketoconazole), which target membrane ergosterol synthesis, are static in vitro; they stop fungi from growing but do not kill them. Such agents, which depend on the immune system to clear the existing fungal infection, are used in immunocompromised patients but, because they do not clear the infections, require lifelong therapy to prevent recurrence. Additionally, *Aspergillus* spp. are naturally resistant to azoles, preventing their use in typically aggressive systemic aspergillosis infections that result in high mortality rates. Finally, the incidence of azole-resistant fungi is increasing, particularly among *Candida* spp. such as *C. glabrata* and *C. krusei*. Because most of the current antifungal drugs suffer from either toxicity (amphotericin B) or the limitations of being static in vitro and increasing resisted by target populations (azoles), there is a critical need for new antifungal agents that act against new targets. Azoles target lanosterol demethylase, a P450 enzyme involved in the synthesis of ergosterol, a sterol found in fungal cell membranes but not mammalian cell membranes.

Antifungal polyenes, including amphotericin B, nystatin, candicidin, and faerifungin, also target the cell membrane of fungi, but in this case by the formation of transmembrane pores and other membrane-associated effects. Although polyene antifungal agents are typically fungicidal, they also are not generally well tolerated by a large majority of patients, making them drugs used as last resorts. Drug resistance in fungi, which more closely resembles drug resistance in tumor cells, is inherently different from drug resistance in bacteria. First, there is no evidence for plasmid-borne drug resistance mechanisms in fungi, so resistance is limited to chromosomal and nontransferable resistance genotypes. Most fungal drug resistance is effected by changes in membrane permeases, leading to exclusion of the drugs, induction of membrane-bound efflux mechanisms, modifications of activases required for the activation of some antifungal drugs into their active forms, and site-specific mutations of target enzymes.

Degradation of antifungal drugs by target organisms (e.g., similar to the action of β-lactamases) or modification of antifungal drugs to inert forms (e.g., aminoglycoside phosphotransferases) has not been observed thus far. With the sequencing of the entire *Saccharomyces cerevisiae* genome, a study was conducted to search for novel targets, unique to fungi or different enough in fungi, which might be used to develop new antifungal agents with novel mechanisms of action (95). Of 6,000 genes analyzed, only a dozen or so potential targets were identified, including chitin synthetase, β-1,4-glucan synthetase, tubulin, elongation factor 2,*N*-myristoyl transferase, acetyl-CoA carboxylase, inositol phosphoryl ceramide synthase, membrane ATPase, mannosyl transferase, tRNA synthetases, lanosterol dehydrogenase, lanosterol synthase, and squalene epoxidase.

The most widely considered new targets for fungicidal activity are all involved with cell wall biosynthesis or structure. Yeast cell walls, which make up 15 to 25% of the cell dry weight, are typically comprised of β-1,3- and β-1,6 glucan polymers (about 50%), α-mannan (highly glycosylated mannoproteins; about 50%), and chitin (about 2%). Although different species of fungi contain different percentages of these components in their cell walls, the major constituents are typically present. Thus, the biosynthesis or structural integrity of β-1,3-glucans, β-1,6-glucans, and α-mannans represent potential excellent targets. The pneumocandins and echinocandins, novel lipopeptides produced by *Glarea lozoyensis* and similar fungi, inhibit the synthesis of β-1,3-glucans. Two new drug candidates, Merck's MK-0991 and Eli Lilly's LY303366, are semisynthetic lipopeptide echinocandin derivatives that specifically target the Rholp dissociable subunit of the heterodimeric transmembrane 1,3-β-(D)-glucan synthase. Because there is apparently only a single gene that is responsible for the synthesis of this subunit, this is considered an excellent target.

The pneumocandin experimental drug, MK-0991, has been shown to possess efficacy aganst *Aspergillus fumigatus, C. albicans, H. capsulatum* and *P. carinii* (but not against *Cryptococcus neoformans*), suggesting that this would make an outstanding potential broad-spectrum candidate for disseminated, life-threatening fungal diseases, particularly in immunocompromised patients. Other targets of interest that have been pursued recently include chitin and sphingolipid biosynthesis. Fungi contain three chitin synthetases (CSI, CSIi, CSIII) that catalyze the synthesis of the polymer from individual units of UDP-*N*-acetylglucosamine. The leading antifungal chitin biosynthesis-inhibitor drug candidate is nikkomycin Z, a nucleoside dipeptide substrate analog of UDP-*N*-acetylglucosamine which competitively inhibits chitin synthases. Although nikkomycin Z is effective in vitro against *C. albicans, C. immitis,* and other fungi with relatively high chitin contents, it does not kill *Saccharomyces cerevisiae*.

The polyoxins, also peptidyl-nucleoside natural products, also inhibit chitin synthase, but none of these has been developed yet as a clinical candidate. Rustmicin (also called galbonolide A), a 14-membered macrolide produced by *Micromonospora chalcea,* inhibits inositol phosphoceramide synthase at pM levels, resulting in the loss of all complex sphingolipids and the accumulation of ceramide. Although rustmicin was reasonably efficacious in a murine cryptococcosis model, it suffers from severe in vivo instability, brought about by serum protein binding, and by being an excellent substrate for *S. cerevisiae* efflux pump Pdr5. The pradamycins are D-amino-acid-substituted benzonaphthacene quinone glycosides that appear to act by binding to the saccharide moieties of the mannoproteins in fungal cell walls, leading to a loss of intracellular potassium that results in gross morphological changes. Another unusual antifungal compound is sordarin, which exerts antifungal activity via inhibition of the elongation step in protein synthesis. Glaxo-Wellcome has developed a semisynthetic sordarin, GM237354, that possesses a broad spectrum of activity and high in vivo potency.

NATURAL ANTICANCER COMPOUNDS

The search for natural product anticancer agents began in the late 1950s and became a reality with the discovery in 1964 of daunorubicin, a cytotoxic, antitumor drug of the anthracycline class with efficacy against a wide variety of solid tumors and leukemias. Five years later, doxorubicin, the 14-hydroxy analogue of daunorubicin, was discovered at Farmitalia. Doxorubicin, which showed a wider range of efficacy for treatment of various cancers, has been a mainstay in chemotherapy for more than 25 years. It was shown recently that doxorubicin as well as many other cytotoxic antitumor drugs (e.g., etoposide) act by trapping topoisomerase II with DNA in what has been referred to as the cleavable complex, resulting in multiple DNA strand breaks. Another class of anthracyclines, exemplified by aclarubicin (aclacinomycin A), were recently shown to inhibit by a different mechanism of action, by inhibition of topoisomerase II–DNA interaction. Other natural product antitumor chemotherapy agents used for several years, but more restrictively than doxorubicin, include mithramycin, discovered in the early 1950s; the mitomycins, first described in 1956; bleomycin, described by Umezawa in 1965; and the vinca alkaloids, vinblastine and vincristine, which were first described in the period of 1958–1962. In all these cases, the mode of action is cytotoxicity through drug-DNA interactions.

With the ultimate desire to get away from cytotoxic drugs that are detrimental to all cells and not just cancerous cells, these cytotoxic natural product antitumor drugs are considered to be drugs of the past and not of the future. Interest in natural product antitumor drugs was renewed, however, with the discovery of taxanes from the pacific yew tree. Paclitaxel (Taxol), a complex diterpenoid alkaloid, was discovered as a potential antitumor agent in 1971, and it has taken nearly three decades to bring it to the market.

One of the unique features that helped move the taxanes from experimental drugs to clinical candidates was the finding that they acted as tubulin stabilizers, a novel mode of action. Docetaxel (Taxotere) and paclitaxel (Taxol) have excellent activity against ovarian and breast cancers. Additionally, Taxol and semisynthetic taxanes may be efficacious against a wide range of tumors. Thus far, the greatest impediment to the development of these drugs on a large scale was that they are only found at levels of about 0.01%(w/w) in the bark of several related yew trees, *Taxus brevifolia* (Pacific yew, an endangered species), *Taxus baccata, Taxus cuspidata* (Japanese yew), and others. Moreover, the chemical synthesis of paclitaxel is very difficult, requiring more than 30 steps. Thus, paclitaxel is currently made semisynthetically from 10-deacetylbaccatin III, produced by the needles of the Himalayan yew. Other novel antitumor drugs developed in recent years include irinotecan (Camptosa) and topotecan (Hycamptin), both which are inhibitors of topoisomerase I.

CONCLUDING REMARKS

Several approaches are being taken to address the issues of MRSA, VRE, other antibiotic-resistant bacteria, newly emerging infectious diseases and related problems. One approach is the use of genomics to help to identify new targets amongst bacteria and fungi. As mentioned previously, sequencing of the yeast genome helped to identify about a dozen unique targets for new antifungal drugs. It appears that the incidence of fungal resistance to new drugs targeted against single, critical enzymes, such as glucan synthase, is very low. Similarly, it is hoped that genomic analysis of bacteria is as useful in identifying new targets for novel antibacterials. Unfortunately, there is a significant difference between bacteria and fungi and between bacterial infections and fungal infections that must be considered. First, any single-site bacterial target for which single mutations can render resistance cannot be seriously considered because of the high frequency of mutations (10^{-8}) that occur in bacteria. Second, many resistance mechanisms are passed horizontally among bacterial populations by plasmid- and transposon-mediated transfer, which renders many antibiotics useless.

Even the fluoroquinolones, which for years were thought to target only DNA gyrase, are now known to target at least two separate enzymes. This is critical, too, because single mutations in bacterial gyrases have been shown recently to yield resistance to several fluoroquinolones. What is the best answer? Although there is probably no single best answer for all resistance problems, the clear-cut best approach is still to inhibit cell wall biosynthesis, preferably with a novel mechanism of action or at several sites. Alternatively, inhibition of systems or processes, such as protein or nucleic acid synthesis, again at multiple sites, appears to be back in vogue. The conundrum, however, is that protein synthesis inhibitors are bacteriostatic agents, which in recent years have been thought not to be desired. Nevertheless, the newer macrolides, dirithromycin, clarithromycin, and azithromycin, all protein synthesis-inhibiting bacteriostats, are effective drugs.

25

MUTATION AND GENETIC STABILITY

The description *genetic instability* is used in slightly different ways by different authors. Some would consider it synonymous with a high mutation rate. This can affect all genes in the organism or only particular genes and may occur only in particular growth conditions. Ideally the mutation rate should be measured as the number of new mutants arising per genome per generation. In practice it is often difficult or impossible to measure the mutation rate per generation. For example the presence of multiple genomes in fragments of a mycelial organism often prevents quantitation. In many cases the proportion of mutants in a population is measured, and this quantity depends on both the mutation rate and the selective forces acting on the population. For most organisms the frequency of spontaneous mutants for a particular gene in the population is under 10^{-4}. An operational definition of genetic instability is that the frequency of mutants is considerably higher than this. In *Streptomyces,* genes are considered unstable if spontaneous mutants are present at 0.1% per spore or higher. Defining genetic instability in terms of the number of mutants in the population is useful for industrial microorganisms because the mutations that can cause problems during fermentation are those that accumulate to significant levels in the population. Genetic instability occurs in brewer's yeast (*Saccharomyces cerevisiae*).

In this case, *petite* mutants, which are defective in mitochondrial function, occur at high frequency (about 1% of colonies). However, the mutants grow much slower than wild type and do not cause any problems in fermentations. Such genetic instability is not very important for industrial applications. In contrast, many industrial strains undergo a rather ill-defined process of "strain degeneration" in which undesirable properties such as reduced production level accumulate. This is normally countered by careful strain maintenance and testing. Strain degeneration likely reflects the fact that desirable properties for production (e.g., high levels of secondary metabolites) probably have a negative influence on the ability of the strain to compete, so mutants outgrow it. This will happen even at low mutation rates because of the presence of strong selective forces. Little literature is available about genetic instability in industrial strains. This reflects that such data can be valuable proprietary knowledge, and the fact that genetic instability is often ignored as a phenomenon during fermentation development. One important industrial organism in which some studies have been carried out is *Streptomyces rimosus,* the oxytetracycline producer. Some high producers are genetically unstable, and the instability can be recognized by the frequent appearance of morphological mutants.

Two classes of mutants with undesirable properties have been recognized: those in class I retain resistance to the antibiotic but have much lower production, whereas those in class II have also lost resistance. Because high levels of oxytetracycline are produced during industrial fermentation, class II mutants are irrelevant to the production process; they are killed by the antibiotic. However, class I mutants survive, and competition experiments show that they outgrow the parent production strain. If care is taken with inoculum preparation it is possible to reduce problems from genetic instability to a manageable level.

Experience shows that the way in which the strain is grown can have large effects on genetic instability. Thus, if the strain is propagated via spores rather than as mycelial fragments, then a much higher frequency of mutants occurs. This is shown in figure in which plates are shown from cultures propagated as mycelial fragments and compared with those from spores. The colonies are the high producers; it can be seen that a variety of lighter variants occur—most of these are class I variants. Another important factor is the medium. More instability is seen on malt agar than on soya-mannitol agar. A third factor is growth temperature: higher temperatures (e.g., 37°C) lead to much higher levels of genetic instability, compared with lower temperatures (e.g., 28°C). Thus, careful choice of growth conditions can be used to control instability.

SELECTIVE MUTATIONS

The proportion of mutants in the population is governed both by the frequency of generation of mutants and by selection effects acting on the mutants. In general in batch processes both of these parameters will vary throughout the process, making simple models such as have been used for chemostat populations invalid. If the population is small compared with the mutation rate, then the frequency of mutations will be governed by the time needed to generate the first mutant and there will be a lot of variation between parallel cultures. In most industrial fermentations the cultures should be large enough to make stochastic effects unimportant. However, large fluctuations in yield in small cultures (e.g., at the shake flask stage) might point to genetic instability being present. Experimental studies with chemostats show that mutants usually displace the initial population within 100 generations. The biological basis for this behavior is that strains are not well adapted to growth in the chemostat, so very strong selection pressures result. This tends to limit the usefulness of continuous culture for many industrial applications despite the many desirable features for plant utilization.

Unless the selected properties in the continuous culture conditions are the same as those desired, genetic instability will cause problems. One area in which this may be true is in biomass production. In some cases continuous culture can be used with recycling of biomass. In this case there is little growth, so the number of generations remains relatively small. Usually during scale-up of batch processes one hopes to alter the physiological conditions between the large and small fermenters as little as possible. Thus, scale-up can be considered, at a first approximation, as just growing the culture for more generations. For example, scaling up from 1 L to 200,000 L corresponds to about 18 more generations. This increase can lead to genetic instability becoming a problem. An example of genetic instability occurring during scale-up was provided by the bicyclomycin fermentation using *Streptomyces sapporonensis*. There was a drastic reduction in yield compared with shake flasks on scaling up to 4,000 L in a pilot plant, and this was shown to be due to the accumulation of low-producing mutants. The mutants could be recognized on the basis of colony morphology because they were deficient in aerial mycelium production.

There is very little information in the literature on the frequency of genetic instability causing problems during scale-up. This is partly because it is not generally recognized as a problem and it is unusual to do genetic studies on the population of cells in a fermenter. We suspect that genetic instability is not uncommon in industrial fermentations, but that the empirical optimization of yield will result in growth conditions that reduce the instability to acceptable levels. However, this would have the consequence that the conditions will probably be suboptimal for product yield by the nonmutated cells. This suggests that it would be valuable to monitor the proportion of mutants so as to provide an extra parameter for the fermentation development.

As most large scale fermentations are carried out in a multistage process with the inoculum being introduced to a series of three or more fermenters, on increasing scale it would be possible, for instance, to try to optimize the genetic stability in the smaller fermenters and then optimize the yield in the final production fermenter. This sort of strategy depends on methods for detecting mutants, which will be discussed later. A genetic approach to increased stability can also be taken. This will to some extent occur automatically in a strain selection program because obviously unstable strains will be rejected. A second approach that has been used in some cases is to carry out genetic crosses between unstable high producers and stable low producers to obtain high producers with improved stability.

In principle it should be possible to approach the problem in a directed rational way. If the mechanism for generation of the mutants is known, it ought to be possible to obtain derivatives with a lower mutation rate. However, this will usually involve some disruption of the DNA processing in the cell and is likely to lead to lower viability and production. A second approach is to introduce a system that selects against mutants. Thus, if mutation is caused by deletions it ought to be possible to move a selectable gene into the deletogenic region and thus ensure that the mutants will have a growth disadvantage. In many cases the strategies to combat instability involve manipulation of the selection effects rather than the mutation rate. The use of immobilized cells or biomass recycling should greatly reduce the potential for genetic instability because little growth will occur. However, even in these cases there may be high mutagenesis rates in a subpopulation of stationary-phase cells and differential growth of a subpopulation. This has been demonstrated in colonies on agar, but the role of genetic instability in industrial applications using nongrowing cells is not yet known.

BACTERIA

Many different types of mutations can occur, and the different classes often involve different molecular mechanisms. Mutations can be classified into point mutations or DNA rearrangements.

Point mutations were originally defined as mutations that map as a single point in genetic crosses. In molecular terms they are usually taken as mutations that affect a single base or a small number of bases. In contrast, DNA rearrangements affect a larger number of bases. One type of point mutation is a base change in which a single base in the DNA is changed to another base. In a *transition*, a purine is substituted for another purine (i.e., A $\rightarrow$ G or G $\rightarrow$ A), whereas a *transversion* substitutes a purine by a pyrimidine (i.e., A $\rightarrow$ C, A $\rightarrow$ T, G $\rightarrow$ C, G $\rightarrow$ G). The other important type of point mutation is the addition or deletion of a small number of bases in the DNA. These are often referred to as frameshift mutations because the addition or deletion of one or two bases in a protein coding area causes a shift of the reading frame such that all distal codons will be read wrongly. Important DNA rearrangements in bacteria are deletions (loss of a DNA segment), insertions (insertion of a segment of DNA into another sequence), and inversions (a segment of DNA is inverted with respect to the neighboring sequences).

The effect of any mutation depends on its location. In bacteria the genes are usually closely packed with little space in between. This means that most point mutations will occur in genes or their control regions and therefore have a genetic effect. If deletions or insertions affect genes, they will usually result in inactivation. In some cases insertions can carry promoters and activate neighboring genes after insertion. Inversions will have a clear cut phenotypic effect only if the inversion end points lie in genes or operons. However, as the transcription direction of genes with respect to the replication origin is highly nonrandom it is to be expected that many inversions will have subtle effects on the growth of strains. A major problem in understanding genetic changes that might occur in genetic instability is that the functions of about two-thirds of the genes in a microorganism are unknown. Thus even if a particular mutation in a specific gene can be associated with genetic instability, it may be difficult to understand why. The mutation frequency in a strain may be altered by genetical or environmental factors. Mutator mutations can lead to a greatly enhanced rate of mutagenesis.

Many mutator mutations lie in genes involved in DNA repair and may increase mutation rates up to 1,000-fold. Most mutator strains grow more slowly than their nonmutator parent strains. However, despite this, they win against nonmutator strains in chemostat competition experiments. This demonstrates the importance of mutation in conditions with high selection and indicates that inappropriate selection strategies could lead to enrichment of mutator strains. In *E. coli* and *Salmonella* a subpopulation of cells in stationary phase can enter a hypermutable state. This leads to the creation of new mutants in the absence of growth and can produce mutants that can grow and be enriched in the absence of growth of other cells in the population. It is likely that such mutagenic states exist in most microorganisms.

It is thus important to culture and store industrial microorganisms in conditions that minimize the chance to enter such a state. In some cases there are programmed mutation events that switch between expression of different sets of genes. One well known example is that of phase variation in *Salmonella* strains, in which a site-specific recombination system causes inversion of a segment of DNA.

The invertible segment carries a promoter that reads out of it and can activate a neighboring operon; inversion leads to switching off this operon. In *Salmonella* this leads to a switch in the expression between two flagellar antigens that probably helps to defeat the host's immune system, but similar systems might control other functions in other bacteria. In some cases it is known that certain regions of the chromosome are particularly subject to instability and may not be suitable sites for introducing foreign DNA during strain construction. The replication terminator region of the *E. coli* chromosome is subject to a high frequency of DNA rearrangements. The end regions of the linear chromosomes of *Streptomyces* are also subject to genetic instability. This can lead to large deletions. (Deletions of over 1 Mb of the 8-Mb chromosome can occur without very dramatic phenotypic effects!) There can also be tandem DNA amplification of chromosomal sequences: the sequences can vary from 2 to 250 kb in size, and the copy number can be over 100.

The amplifications may be accompanied by deletion of neighboring chromosomal sequences up to the chromosome end. These unstable regions may affect secondary metabolite production. The mutations are usually deleterious for growth, compared with wild-type strains, but may give an advantage in comparison with high-production parents. The mutants can often be recognized by changes in colony morphology. It may be difficult to detect mutations in bacteria because they may not show very obvious morphological or auxotrophy properties, and it is likely that the phenomenon of genetic instability is often not recognized.

If the changes involve large-scale genome rearrangements, then pulsed-field gel electrophoresis (PFGE) may prove useful. PFGE can separate linear DNA fragments in size ranges up to about 10 Mb (bacterial genomes range in size from about 0.6 to 10 Mb). It is often possible to find rarely cutting enzymes that restrict the whole genome into 5–30 fragments that can be conveniently separated by PFGE. The gels can also be used for Southern blotting to detect specific DNA sequences. PFGE can be used to screen strains for DNA rearrangements (deletions, duplications, and sometimes inversions).

Phage infection can be a major problem in some industrial fermentations. In addition to lyticphages that kill all the cells, there are many lysogenic phages that maintain themselves within the host cells either by integrating into the host chromosome or by establishing a plasmid state. Lysogenic phages exist within some industrial strains (e.g., the phages RP2 and RP3 of *S. rimosus*) and may cause changes in the behavior of a strain. Some phages (e.g., phage Mu in *E. coli*) can integrate into random sites in the chromosome and inactivate genes, in other words, they are transposable elements).

It is possible that some changes observed in bacteria are of an epigenetic nature, that is, they are stable changes that do not depend on changes in the DNA sequence. One candidate for such an effect is the mycelial fragment size in *Streptomyces* fermentations. In many species fast growth is associated with larger fragment sizes. If a continuous culture in a chemostat is started from fast-growing mycelium, then large fragments are stably maintained at a variety of dilution rates. However, if the chemostat is started from a slow-growing culture entering stationary phase, then a small fragment size is maintained over a range of dilution rates. It is possible to switch from small fragments to large fragments by a short period of growth at high dilution rate, and from small fragments to large fragments by a short period at a low dilution rate. At dilution rates between these critical rates the form depends on the history of the culture and remains stable for many generations.

EXTRANUCLEAR DNA

Plasmids are common in most groups of bacteria. Most plasmids are circular DNA molecules, but linear plasmids are present in some species. Natural plasmids may be of significance in some industrial strains, but the most important role of plasmids is for constructing recombinant strains. Natural plasmids come in a large range of sizes (2–1,000 kb) and copy numbers (1–500/chromosome) and have been used to construct vectors with different copy numbers for different applications. Two important classes of genetic instability that can affect plasmids are segregational instability (plasmid loss), and structural instability (DNA rearrangements in the plasmid). Low copy number plasmids have systems to ensure that plasmid copies are distributed to both daughter cells on division. Many natural plasmids carry conjugation systems that would reintroduce the plasmid into any member of the population that had lost it. However, biological containment considerations and the desire to obtain small vectors make this undesirable for recombinant plasmids. It is likely that many high copy number plasmids do not need an active system to segregate them to daughter cells but instead rely on the low statistical chance of producing plasmid-free progeny. (If there are N copies of a plasmid at cell division then the chance of producing a plasmid-free cell is 2^{-N}; for $N = 20$ this is 10^{-6}.) However, the situation is complicated if plasmid multimers are produced (i.e., circles containing two or more copies of the plasmid as a direct repeat).

The copy number control determines the number of replication origins, so a plasmid dimer would have only half the number of molecules per cell as a monomer. Because multimers are generated by homologous recombination systems, plasmids often have site-specific recombination systems that break down multimers into monomers and result in a higher segregational stability (e.g., the *xer* system of ColE1). It is thus important to ensure that the stability-enhancing systems of natural plasmids are retained during vector construction; many common laboratory cloning vectors have lost such systems because they were constructed with minimal-size replicons. Choice of fermentation conditions can make a large difference in the segregational stability of plasmids. The choice of limiting nutrient in chemostat cultures can radically affect plasmid loss. Choice of host may also have a significant effect; at present this can be tested only empirically because the basis for differences is unknown.

The stability of plasmids is often affected by the growth rate of cultures. There are cases in which plasmids are more stable at high growth rates. Inocula for fermenters are typically prepared in several stages involving increasing volumes. An improvement in stability can be obtained by avoiding slow growth during inoculum preparation (i.e., transferring to a larger volume before the culture enters stationary state). The replication systems of plasmids may differ significantly from that of the chromosome, and this may lead to a significantly higher frequency of mutation including DNA rearrangements. Some plasmids have a rolling-circle replication mode with single-stranded DNA intermediates. This system is more prone to DNA rearrangements than other replication systems. DNA rearrangements can include deletions, insertions, duplications, and inversions.

The chromosomal replication control system is a positive-control system in which every replication origin in the cell is activated once in the cell cycle, and all origins are activated almost simultaneously. In contrast most plasmids have negative-control systems, which initiate replication at an origin at times spread through the cell cycle. A random pool mechanism functions, which means that replication may be initiated on a particular copy of the origin several times in a cell cycle, while another copy is not replicated. This means that any mutated copy of a plasmid with a small advantage for selection for replication can rapidly replace a parent plasmid. DNA rearrangements in plasmids are usually easily detected using restriction digests and agarose gel electrophoresis.

The selection of segregational or structural mutants can lead to rapid overgrowth of the original population. A plasmid-carrying strain may grow more slowly than one that has lost the plasmid, but this is by no means universally true for natural plasmids. However, it is likely that most constructs for industrial use will negatively affect the growth of the host strain. Although larger derivatives of a plasmid are often less stable than small derivatives, it is not clear that this is mainly due to the size per se. It seems much more the case that the presence of promoters on the plasmid can lead to selection against the insert. In some plasmids, transcription from insert promoters into the vector can lead to stability problems, possibly by interfering with vector replication. Therefore it has become common practice to insert transcriptional terminators into vectors to prevent read-through from the insert into the vector; this was one key step in constructing successful vectors for *Bacillus*.

It is also desirable only to transcribe the cloned gene when product is required because this often reduces selection against the insert. Various inducible promoter systems have been developed, such as the recombinant *tac* promoter, which combines the desirable strength of the *trp* operon promoter with the control of the *lac* promoter (induction with IPTG). Other promoters are the lambda PR and PL promoters, which can be controlled by temperature by using a temperature-sensitive repressor. Another possibility is to use a promoter from phage T7 that can be transcribed only when the special T7 RNA polymerase is expressed. By using a combination of plasmid stability systems and control of the transcription of cloned genes it is possible to achieve good results in *E. coli*. However, even in *E. coli* there is a need for considerable empirical work, and many surprises will occur. In other bacterial systems the state of knowledge about plasmid stability is considerably worse, and a good portion of luck is necessary to achieve success.

YEAST AND FUNGI

Many of the same considerations apply to yeast and fungi as to bacteria. However, the more complex eukaryotic nature of these organisms leads to some further mechanisms for generating mutants. Eukaryotes have a number of linear chromosomes. Many yeast and fungi are haploid, but there are also diploid and dikaryotic (i.e., two separate haploid nuclei in the same cytoplasm) strains. There can be changes in the number of chromosomes (either in the number of copies of the genome present, "ploidy"; or in the loss or gain of copies of single chromosomes, "aneuploidy"). There can also be transfer of parts of chromosomes to other chromosomes (translocation). Most yeast and fungi also have sexual systems in which meiosis takes place. Such sexual cycles are usually associated with stress conditions (e.g., nutrient stress) and may be avoidable by suitable strain maintenance conditions.

Meiosis can lead to changes in the genetic make up of cultures when diploid or dikaryotic organisms are used. Although many of the present important industrial strains are haploid, this may become more of a problem in the future as a wider range of organisms is used in biotechnology.

Sexual systems usually involve the interaction of two different mating types; some organisms have only two sexes, but many fungi have several mating types. The mating types may be able to change: the best-characterized example of this is the mating-type switch in *S. cerevisiae* in which it is possible for a strain to switch backward and forward between the two mating types, a and α. There are strains that switch at high frequencies (homothallic strains) due to the presence of a gene, *HOM*, encoding an endonuclease, whereas other strains switch only at low frequency (heterothallic).

The ability to switch between mating types can thus be seen as a sort of genetic instability and may have consequences for fermentation if one mating type is selected under fermenter conditions. Transposable elements are also present in yeast and fungi. One class that has not been found in bacteria the retrotransposons, which resemble retroviruses. They transpose via an RNA intermediate and transcribe into neighboring genes. The example of *Ty1* in yeast shows that this can activate genes by providing a new promoter. This ability to activate as well as inactivate genes could be relevant to genetic stability. Such elements are present in many copies in the cell (>30 copies), which makes it unlikely that they can be removed by genetic manipulation. Apart from the ability to change gene expression patterns by transposition, it is possible that mutations in control proteins will result in new patterns.

The *areA* gene in *Aspergillus nidulans* controls ammonium repression for a large number of genes involved in the catabolism of various nonfavored nitrogen sources. Many mutations in *areA* are known that differentially change the pattern of control (i.e., some genes are silenced and some are overexpressed). In one case a duplication within the *areA* gene resulted in another gene becoming subject to *areA* control. This mutational flexibility would allow easy evolution of new expression patterns and could contribute to genetic instability. Another feature of eukaryotes that makes them genetically more complex than bacteria is the possession of mitochondria, which also have their own genome (mtDNA). In *S. cerevisiae* there are very frequent deletions in the mtDNA that cause a loss of mitochondrial function.

The organisms can still grow, but the slower growth results in small colonies (which is why they are called *petite* mutants). These mutants account for 1% of colonies or more, depending on the strain. Another genetic phenomenon caused by mtDNA is senescence in the fungus *Podospora anserina*. This species cannot grow indefinitely vegetatively and, after a certain amount of mycelial growth, death occurs. The senescence is caused by the accumulation of a high number of copies of a plasmid formed by excision from the mtDNA. Such senescence phenomena may cause problems in the maintenance of industrial strains. Problems of genetic instability can also arise in culture collections. For example, it is by no means uncommon for the secondary metabolite spectrum of a filamentous fungus to change radically if it is examined with the same strain from a culture collection after a gap of several years. Although this problem can arise with any organism, many bacteria can be kept in nongrowing states (e.g., lyophilization, storage at –70°C in glycerol), which limits the scope for selection to act. However, most filamentous fungi cannot be maintained in this way. In view of the number of different strains that must be maintained, there is probably no real answer to the problem.

It is likely that adaptive mutation occurs in most microorganisms as a response to certain stresses and might be one cause for such strain degeneration. Experience can guide the choice of media and storage conditions to try to minimize these effects. Epigenetic effects can also be important in yeast and fungi. The DNA of most eukaryotes contains some 5-methyl cytosine, which is formed by methylation after DNA replication; this usually occurs at CG dinucleotides, and so the complementary strand can also be methylated. If a methylated sequence is replicated, two hemimethylated daughter molecules (i.e., each methylated in only one strand) are produced. Because DNA methylase enzymes usually prefer such hemimethylated substrates this results in an inheritance of the methylation patterns.

Methylation of genes is associated with their nonexpression, but it is not clear how the de novo methylation arises. In some cases the introduction of foreign genes into fungi can result in their inactivation by methylation. The reason why this occurs is not known. In some cases methylation seems to be targeted at duplicate copies of a gene in an organism, that is, introduction of an extra copy of a gene will result in inactivation of either the introduced copy or the original copy. The methylation may also lead to mutation. 5-Methyl cytosine can be spontaneously deaminated to produce thymine and, in the absence of repair, this gives a C $\rightarrow$ T transition. This can result in a "RIP effect" in some fungi in which a very high mutation rate is observed after introduction of an extra copy of a gene such that one copy is inactivated by multiple C $\rightarrow$ T transitions. A further epigenetic effect that has been observed in *S. cerevisiae* is the existence of prions. These are proteins of altered structure that can catalyze the change of copies of normal structure to the altered structure. This can function as a sort of "infective" process and is a potential source of instability.

ANIMAL AND PLANT CELLS

Animal cells have been used for the growth of viruses for the production of vaccines and in more recent years for the production of recombinant proteins. Plant cells are not yet used on an industrial scale, but there has been interest in this possibility. Smaller-scale cultivation of animal cells has been used extensively in laboratories. Most primary cell cultures have a limited life span in the sense that they can divide only a limited number of times. However, immortalized cell lines, which do not have this limitation, are of most interest for large-scale culture. These are tumour cell lines that have often changed considerably in their properties as compared with the original cells from which they derive. Cultured animal cells are often genetically unstable and undergo frequent karyotype changes. These include polyploidy (changes in the number of chromosome sets from diploidy to triploidy and tetraploidy), aneuploidy (loss or gain of copies of particular chromosomes), and frequent chromosomal rearrangements (deletions, inversions, and translocations). Chinese hamster ovary (CHO) cells are an immortalized cell line that has been used to produce recombinant proteins at a 10,000-L scale. They possess good growth properties for large-scale culture and pose less safety problems with respect to viruses than cell lines more closely related to humans.

An important strategy to obtain increased production of recombinant products is to work with high copy numbers of the introduced genes. This can be done by using a cell line that lacks dihydrofolate reductase (DHFR) and transforming with the required gene coupled to the DHFR gene. The antibiotic methotrexate inhibits DHFR, and selection for growth on increasing levels of DHFR results in the amplification of the DHFR gene and coupled sequences. The amplifications can have a complex structure and be either extrachromosomal or integrated in the chromosome. Continued selection usually results in integrated copies, but the positions of integration vary more or less randomly between independent clones.

Control of stability requires the careful construction of master cell banks and working cell banks and the avoidance of too many generations of growth. Various methods can be used to check genetic stability. One method is to use Southern blots. However, this does not show heterogeneity in the population. A better method is to use karyotype-based methods. In particular, fluorescent in situ hybridization (FISH) can locate the amplified sequences and show heterogeneity. If chromosome-specific probes are used it is possible to monitor the copy number of a chromosome in interphase nuclei. If metaphase cells are used, details of chromosome rearrangements can be monitored with appropriate probes. It would normally be necessary to carry out FISH and measurements of productivity at several times during growing cultures.

Flow cytometry can be used to show variations in DNA content in a population of cells that can indicate the extent of polyploidy and, to some extent, aneuploidy. The fact that the amplifications are selected with methotrexate and that lower producers might be expected to overgrow the best producers would suggest that stability would be enhanced if methotrexate were present in the medium. However, there are data showing an increased instability with methotrexate. One explanation is that methotrexate leads to a reduction of thymidine synthesis, which could allow more dUTP incorporation into the DNA. The resulting repair and strand cutting might stimulate instability. In fact an experiment with a heterogeneous population of clones in the absence of methotrexate suggested that such cultures may be surprisingly stable in the absence of selection.

VIRUSES

Viruses are propagated for vaccine manufacture. This includes the use of live attenuated viruses and killed viruses. Various culture systems are used, such as cell cultures and chicken eggs. Viruses show a bewildering range of different genome structures (single- and double-stranded DNA or RNA) with specialized replication methods. Virus replication may be much less accurate than host replication (especially for RNA viruses) and thus show a high mutation rate. Genetic instability has vital biological safety considerations because a reversion of attenuated viruses to higher pathogenicity or changes that reduce the efficiency of killing in killed-virus vaccines could be disastrous. Any release of viruses from the cells used to culture the virus could also have serious implications. The main weapon against these problems is to reduce the number of growth cycles as much as possible. It is also important to monitor the cultures and product.

CONCLUDING REMARKS

Good manufacturing practice aims to guarantee reproducibility of production runs and avoid product contamination. The details are dependent on the organism and product, but there are some general considerations:

1. A master cell bank (MCB) is derived from a single colony or single cell. It consists of many replicates kept in the most stable form possible (e.g., lyophilized or frozen at –70°C or in liquid

nitrogen). Each ampoule is opened only once. It is important to make sufficient replicates because generation of a new MCB will require an application to the regulatory authority. It is also usual to store the material at more than one location in case of destruction (e.g., freezer problems, fire).

2. The working cell bank (WCB) is derived from the MCB and is used to inoculate a production batch.
3. Detailed records of both MCBs and WCBs must be kept, and all possible efforts must be made to limit the number of growth cycles (number of passages) to minimize the effects of genetic instability.
4. Both MCBs and WCBs must be rigorously tested. The tests for MCBs will be particularly exhaustive and depend on the process. A genotypic characterization using various DNA tests is usual (e.g., DNA fingerprinting to confirm the strain, and testing the structure of recombinant DNA constructs). There will also be phenotypic tests and tests that normal production levels occur. There may be tests for contamination (e.g., viruses in cell culture).

These procedures aim to confirm the reproducibility of the starter cultures and the production process. As long as the build up of any variants is fairly reproducible, genetic instability during the fermentation will not conflict with these aims but will lead to a reduction in production compared with the ideal case. However, any evidence for genetic instability and the occurrence of heterogeneity during production fermentations is likely to lead to problems with the regulatory authorities. It is likely that this also contributes to the reluctance to consider genetic instability in many fermentation sections of companies.

INDEX

T

X